BIBLIOGRAPHIC RECORD TARGET

Graduate Library
University of Michigan

Preservation Office

Storage Number: ______________________

ACM2715
UL FMT B RT a BL m T/C DT 07/18/88 R/DT 07/18/88 CC STAT mm E/L 1
035/1: : |a (RLIN)MIUG86-B26911
035/2: : |a (CaOTULAS)160432294
040: : |a RPB |c RPB |d MiU
100:1 : |a Jellett, John Hewitt, |d 1817-1888.
245:03: |a An elementary treatise on the calculus of variations. |c By the Rev. John Hewitt Jellett.
260: : |a Dublin, |b J. McGlashan; |a London, |b W. S. Orr & Co., |c 1850.
300/1: : |a xx, [1]-377 p. |b I fold. diagr. |c 23 cm.
650/1: 0: |a Calculus of variations
998: : |c DPJ |s 9124

CALCULUS OF VARIATIONS.

AN

ELEMENTARY TREATISE

ON

THE CALCULUS OF VARIATIONS.

BY THE

REV. JOHN HEWITT JELLETT, A. M.,

FELLOW OF TRINITY COLLEGE, AND PROFESSOR OF NATURAL PHILOSOPHY IN THE UNIVERSITY OF DUBLIN.

DUBLIN

PRINTED AT THE UNIVERSITY PRESS.

JAMES McGLASHAN, 21 D'OLIER-STREET.

WILLIAM S. ORR AND COMPANY, LONDON.

MDCCCL.

DUBLIN:
M. H. GILL, PRINTER TO THE UNIVERSITY.

PREFACE.

The want of a systematic treatise upon the Calculus of Variations has long been felt. The researches of Poisson, Jacobi, Ostrogradsky, and Delaunay, which have added so much to the completeness of the science as it came from the hands of Lagrange, are only known through the medium of scientific journals and Transactions of learned Societies, and are thus inaccessible to many readers, and inconvenient of access to all. The student has also to complain of the want of a sufficient number of examples to illustrate the principles of the science, a defect which renders these principles, from their very abstract nature, exceedingly difficult to be understood by a beginner. These deficiencies it is the object of the present work to supply.

The plan which has been adopted may be briefly stated as follows:

After a short introductory sketch of the origin and history of the science, the author has, in the first Chapter, endeavoured to give a clear statement of its principles, considered as a branch of pure Analysis. In the following Chapter these principles have been applied to the investigation of the variation of those expressions with which, in the present state of mathematical and physical science, we are most generally concerned, namely, differential coefficients and definite integrals, the attention of the reader being directed solely to functions of one independent variable.

In the third Chapter the author has considered, under the same limitation, the important problem of maxima and minima. Of this problem some examples have been given in immediate connexion with the general methods of solution. But as the most interesting examples of the Calculus of Variations are to be found in its applications to particular sciences, it has been thought most expedient to place them in separate chapters under the head of the science to which they respectively belong. These examples will be found in Chapters IV., VIII., and IX., containing respectively the applications of the Calculus of Variations to the Theory of Curves, Theory of Surfaces, and Mechanics.

In Chapters V., VI., and VII., is discussed the case of functions of more than one independent variable, and the extension of the methods of the Calculus of Variations to such quantities.

Finally, in Chapter X. the author has given the application of the Calculus of Variations to the integration of functions of one or more independent variables, a branch of the science which has not met with much attention, but which appears to be of considerable importance.

Besides the ordinary treatises upon this subject, the author has been much indebted to the two memoirs of M. Delaunay, published in the Journal de l'Ecole Polytechnique,* and in Liouville's Journal,† respectively, as also to a memoir published in the Transactions of the Academy of St. Petersburgh,‡ by M. Ostrogradsky, all of

---

* Journal de l'Ecole Polytechnique, tom. xvii.

† Journal de Math., tom. vi.

‡ Mem. de l'Acad. de St. Petersb. A. D. 1838.

which he would strongly recommend to the attention of his readers.

The author desires to take this opportunity of returning his sincere thanks to the Board of Trinity College, for the liberality with which they have contributed to defray the expense of the present work.

6, Trinity College,
March, 1850.

ERRATA.

Page 22, line 1, after "equation" read (A).

—— 46, line 15, before the second = read v.

—— 83, line 9, for $\left(\frac{d^n \delta y}{dx^n}\right)^2$ read $\frac{d^{2n}\delta y}{dx^{2n}}$.

—— 92, line 15, for $\sqrt{(1+C_1^2)}$ read $(1+C_1^2)^{\frac{3}{2}}$.

—— 105, line 11, for u_1, read u'; and line 14, in the second member of the equation, for δ'' read δ'''.

—— 161, line 2, for a^2, read $-a^2$.

Pages 179, 180, 184, 190, 191, for ρ, ρ', read $-\rho$, $-\rho'$.

Page 240, line 13, after $-\frac{dV_{y^2}}{dy}$ read $-\frac{dV_{xy}}{dx}$.

CALCULUS OF VARIATIONS.

CONTENTS.

INTRODUCTION.

Although the Calculus of Variations, properly so called, is justly due to the genius of Lagrange, many of the principles, as well as the results of the science, are of a more ancient date. Like the Cartesian Geometry, and many of the other analytic methods in use among mathematicians of the present day, its origin is to be sought, not in any systematic treatise, but in the investigation and solution of some particular problem. This, indeed, is the ordinary history of the great improvements in mathematical science. Some problem, physical or mathematical, is proposed, which is found to be insoluble by known methods; and in the solution of such a problem a new principle is necessarily introduced. It is soon observed that this principle is not limited in its application to the particular question which occasioned its discovery, and it is then stated in an abstract form, and applied to problems of gradually increasing generality. Other principles, similar in their nature, are added; and the original principle itself receives such modifications and extensions as are from time to time suggested by the various problems to which it is applied. Finally, these several parts are grouped together, a uniform system of notation is adopted, and the principles of the new method become entitled to rank as a distinct science. The mathematical historian cannot, of course, expect to be able in all cases to trace this process. The several steps may be the work of a single mind, and the author, in giving his finished discovery to the public, may not think it necessary to detail the method by which his own mind was conducted to it. But, from the experience of those who *have* left us such a detail, as well as from the history of sciences which cannot be traced to an individual discoverer, we are warranted in concluding that the history of mathematical discovery is, generally at least, such as has been described above.

In estimating the truth or falsehood of such a conclusion, and, more generally, in examining the laws which regulate the progress of the human mind in the discovery of truth, the most important evidence is derived from sciences which have been at their first promulgation most incomplete, and have owed their subsequent advancement rather to the successive labours of several, than to the efforts of a single mind. An individual discoverer seldom gives us the results of his labours in the same form in which they first presented themselves to his own mind. Still more rarely are the steps, by which he desires to conduct the mind of his reader to the perception of a truth, identical with those by which he himself arrived at it. These last are commonly tedious and inelegant, and when the conclusion has been once reached, it is generally possible to discover some more compendious mode of arriving at it. And if *both* these discoveries be the work of a *single* mind, the first is seldom given to the reader. Thus while much is gained in the brevity and elegance of the published demonstration, future inquirers are deprived of a most important aid to discovery, by the suppression of the process through which the mind of the author actually passed. It is unnecessary to say that such a suppression is impossible, where the original discovery and the finished demonstration have emanated from different persons. Hence the historical importance of those sciences whose principles have been given to the public, not in a complete and systematic form, but gradually, and by methods more or less tedious or imperfect. As there is, perhaps, no science which furnishes a better example of this than the Calculus of Variations, it may not be unprofitable to trace briefly the several steps of its progress.

In the month of June, 1696,* John Bernouilli proposed to the mathematicians of his day the following problem:

"PROBLEMA NOVUM,

Ad cujus solutionem Mathematici invitantur.

Datis in plano verticali duobus punctis A et B, assignare mobili M viam AMB per quam gravitate sua descendens, et moveri incipiens a puncto A, brevissimo tempore perveniat ad alterum punctum B."

The novelty of this problem, which appeared to differ essentially from any previously solved question of maxima and minima, attracted immediate attention, and we find three of the most illustrious mathematicians of the day, Leibnitz, James Bernouilli, and De l'Hôpital, engaged in the attempt to solve it. The first of these appears to have succeeded in obtaining a solution within the allotted time.† This, however, he did not publish, being, as he states, desirous that other mathematicians should be encouraged to attempt the solution. He, therefore, merely transmitted it to John Bernouilli, receiving in return the solution which that mathematician had previously obtained, to be published at the proper time. Subsequently, Leibnitz, influenced by the same

* Previously to this, Newton had solved a problem of a similar nature, namely, the determination of the solid of least resistance. But he did not publish the method by which his result had been obtained, and no impulse seems to have been given by it to the researches of other mathematicians. The history of the Calculus of Variations, therefore, properly begins with the problem of the brachystochrone.

† James Bernouilli states that he also had solved the problem within the allotted period, but that, on learning that the proposer had extended it, he reserved his solution, with the intention of investigating and adding to it certain other problems of a similar nature. The fact that his Isoperimetrical Problem—a problem which is greatly in advance of the brachystochrone—was published along with his solution of the latter, is of itself almost sufficient to prove the truth of this statement.

motive, requested that the time originally prescribed by the author might be extended, a request with which John Bernouilli complied, and again proposed the problem in a Programma published at Groningen in January, 1697. Three solutions of the problem appeared in the Acta Eruditorum for May, 1697, emanating respectively from the proposer, James Bernouilli, and the Marquis de l'Hôpital. The solution given by the two Bernouillis (that of De l'Hôpital being without demonstration) rest substantially upon the same principle, namely, that "whatever maximum or minimum property is possessed by the entire curve must belong also to every element of it." But James Bernouilli is undoubtedly in advance of his brother, both in adopting a more direct process, and in stating, in abstract terms, the new principle upon which that process is founded. This, besides subjecting the assumed principle to a more rigid scrutiny, brings us a step nearer to a *general* analytical method.

In the problems which James Bernouilli proposes at the close of the same paper, he gives the germs of two other important branches of the new method. These problems are:—1. Among all curves drawn from a given point to a given vertical line to determine the curve of quickest descent. 2. Of all curves of the same length described upon a given base, to determine a curve such that the area of a second curve, each of whose ordinates is a given function of the corresponding ordinate or arc of the first, may be a maximum.

In the difference between the first of these and the original question of the brachystochrone, we recognise the germ of the general problem subsequently considered by Lagrange, namely, "If a given definite integral receive a maximum or minimum value, what are the conditions to be fulfilled at the limits of integration?"

In the second, or isoperimetrical problem, we see the first step to the general question of *relative* maxima and minima, in which the maximum or minimum curve is to be determined, not from among all possible curves, but from among those only which possess some given property. The unsuccessful attempts of John Bernouilli to solve the second case of this latter problem showed at once the inaccuracy of the principle upon which the original question had been discussed. It had been there assumed that the curve might be considered as a rectilinear polygon, having an infinite number of sides, and that whatever maximum or minimum property belongs to the entire curve belongs also to each consecutive *pair* of these sides. By the application of this principle to the second case of the isoperimetrical problem, John Bernouilli obtained continually erroneous results; nor does he appear to have been sensible of the cause of his mistake until the publication of James Bernouilli's "Analysis magni Problematis Isoperimetrici," in which the original principle is modified by the supposition that *three* elements of the curve (considered as a polygon) vary simultaneously.

This valuable memoir appeared in the Acta Eruditorum for May, 1701. It contains two important steps in the progress of the method, namely:—1. The modification of the original principle just alluded to, by which it was rendered (although not universally true) more general in its application than it had previously been. 2. The method of taking into account the *isoperimetrical* condition. Much was done afterwards by John Bernouilli in simplifying his brother's demonstration, as well as in establishing a more uniform system of equations for the solution of such problems.* But the actual limits of the power of the new method

* The following principle, which was afterwards extensively used, is due to John Bernouilli:

If an equation of the form

$$f\left(x, y, \frac{dy}{dx}, \text{\&c.}\right) = f\left(x', y', \frac{dy'}{dx'}, \text{\&c.}\right)$$

do not appear to have been materially extended until the subject was taken up by Euler.

It would be impossible, in a brief sketch like the present, to give an adequate idea of the labours of this illustrious mathematician, to whom the Calculus of Variations is perhaps even more indebted than to Lagrange. We must, therefore, content ourselves with a rapid view of the principal additions made by him to the researches of the Bernouillis and some of their contemporaries.

Euler commences by a classification of the problems which he proposes to consider, founded on the *number* of properties (exclusive of the maximum or minimum property) which the sought curve is required to have.

In the first class are to be reckoned those problems in which, as in the case of the brachystochrone, a curve is sought possessing a maximum or minimum property, but not restricted by the necessity of possessing any other given property. In the second class, including the isoperimetrical problems of James Bernouilli, he places those problems in which a *second* condition, such as a given length, a given area, &c., is attached to the curve. In problems of the third class, *two* such conditions are supposed to be added. By the introduction of problems of this last class, Euler, in his first memoir, extended the limits of the method considerably beyond the position in which the Bernouillis had left them. He also greatly facilitated the solution of all the ordinary problems, by the construction of a table of formulæ, which are of very extensive application. Lastly, we may discover in Euler's first

---

hold between any two consecutive points of a curve, the functions on each side being similar in form, we must have

$$f\left(x, y, \frac{dy}{dx}, \text{\&c.}\right) = \text{const.}$$

This principle, the truth of which is self-evident, was first used by John Bernouilli, and subsequently by Euler and others, in the solution of problems of maxima and minima.

memoir the germ of that general method of solving problems of relative maxima and minima, which he himself afterwards completed, and which is now in general use.

Euler's second memoir was published in 1736. In this memoir we can trace the rapid advance of the method to complete generality. The use of the table of formulæ, which had been itself a vast improvement upon the previous methods, was now superseded by the discovery of a *single* equation, of so comprehensive a nature, that no subsequent generalization of the science has removed it from the place it occupies, as the general solution of all cases in which the maximum or minimum property is capable of being expressed by a formula such as

$$\int f\left(x,\, y,\, \frac{dy}{dx},\, \frac{d^2y}{dx^2}, \ldots \frac{d^ny}{dx^n}\right) dx.$$

The principle upon which all previous solutions had been founded, namely, that what is true of the entire curve is also true of each of its elements, was rigorously examined, and shown to be, in an important class of problems, untrue. Thus the way was prepared for a more general method. Much was also done in facilitating the processes employed in the former memoir.

Still greater advances to a complete system are made in the "Methodus inveniendi Lineas curvas Proprietate maximi minimive gaudentes." In this great work, displaying an amount of mathematical genius almost unrivalled, Euler arranged, in a regular and systematic method, his former discoveries. The problems which he proposes to solve are divided into two great classes, *absolute* and *relative*; in the former of which it is required to determine the maximum or minimum curve from among all curves whatever; while in the second this curve is to be selected from those curves only which possess one or more given properties.

The principles of the method are stated in clear and logical order, and the conceptions of the student are aided by a great number of illustrations and examples. The defect, arising from the want of generality in the principle noticed in his previous memoir, is supplied, and equations of solution are given for the cases in which that principle fails. Finally, a general method of solution is given for problems of *relative* maxima and minima, which remains in use to the present day.

To estimate the importance of the labours of Euler, it is only necessary to compare with the "Methodus Inveniendi" the last memoir of John Bernouilli, which marks the extent to which the method had been carried before the author of the former commenced his investigations. From being little more than the solution of a particular problem, it had almost become a complete science. General methods had taken the place of the consideration of individual questions, the principles of the science had been clearly defined, and the whole had been arranged into a regular didactic treatise. Still, however, much remained to be done. The method of maxima and minima, as it came from the hands of Euler, wanted that which is essential to every analytic method, namely, an analytic foundation. In deriving its principles from geometry, Euler established a connexion which was not natural, and whose inevitable tendency was to limit the extent to which those principles were capable of being carried. The method, too, by which he arrives at his conclusions, is tedious and difficult, and in the hands of a less accomplished mathematician would probably be unsuccessful There was also wanting in the method of Euler a definite system of notation by which its distinctive character might be marked. The omission of the "definite equations," as they were termed by Lagrange, without which the problem would be in general indeterminate, is a serious defect in Euler's theory; and, lastly, there seems to have been no attempt made to

extend it to *surfaces* possessed of a maximum or minimum pr
perty, or, in other words, to cases in which the property i
question is expressed by an integral of a degree higher than th
first.

These defects were removed by the genius of Lagrange. I
separating the principles of the Calculus of Variations from th
geometrical considerations from which his predecessors had de
rived them, he not only placed the science upon its true and leg
timate foundation, but opened a new and extensive field for i
future applications. By the invention of a simple and definit
notation he gave distinctness and permanence to the new metho
securing it from being confounded with the other infinitesim
methods, to which it is in some degree similar.

In the investigation of the definite equations, the general me
thod of maxima and minima, where the variables are not inde
pendent of one another, and, above all, in his applications of i
principles to Mechanics, Lagrange increased so largely both th
completeness and the extent of the new science, that he is, for a
these reasons, justly reputed the inventor of the Calculus of Va
riations. It is, however, the less necessary to enter minutely int
the principles laid down by Lagrange, as it is the object of th
following treatise to develope them. For although much has bee
done by Poisson, Jacobi, Ostrogradsky, Sarrus,* Delaunay, an
others, to extend its methods and supply its deficiencies, the Cal
culus of Variations is now, in its essential principles, the same a
when it came from the hands of Lagrange.

In concluding this brief sketch of the history of the Calculu
of Variations, the author would refer any of his readers, who ar

* I regret much that, in consequence of the delay in the publication of M. Sarru
Memoir, which was crowned by the Academy of Sciences in 1843, I have been unable
consult it previously to the publication of the present Treatise.

desirous of more extensive information on this part of the subject, to Woodhouse's Isoperimetrical Problems, in which the various improvements made by successive mathematicians are detailed with great clearness, and from which the preceding account has been in a great measure compiled.

CALCULUS OF VARIATIONS.

CHAPTER I

DEFINITIONS AND PRINCIPLES.

1. One variable quantity is said to be a function of any number of others, when there exists between them a certain relation, rendering the value of the first dependent on the values of the others, which are termed *independent* variables. The nature of the relation subsisting between the first, or dependent variable, and the others, or independent variables, is termed the *form* of the function. This is ordinarily expressed by the notation $u = \phi\,(x_1, x_2, x_3 \ldots .)$, where x_1, x_2, &c., are the independent variables, u is the dependent variable, and ϕ is a general symbol, denoting the form of the function.

2. From this definition it is evident that the value of a dependent variable or function depends upon two different things, viz.: 1. the values of the independent variables; 2. the form of the function, or nature of the relation by which it is connected with them. A change in either of these will change the value of the function. Thus, for example, let the function be $u = \sin . x$, and it is evident that the value of u may be changed, either by a change in the value of x, or by a change of the functional symbol sin. into any other, such as cos., tan., log., &c. Functions may, therefore, be divided into—1. Determinate, or those whose form is not supposed to change; 2. Indeterminate, or those whose form is variable. This division is analogous to that of ordinary quantities into constant and variable. The changes in value of which

determinate functions are susceptible, arising solely from a change in the value of some one or more of the independent variables, have been already fully discussed in the Differential Calculus. But the changes in value of which indeterminate functions are susceptible, arising, as they do, from a different cause, require to be treated in a different way, the rules of the Differential Calculus not being (as will be seen hereafter) universally applicable to them. It is with such changes in value, those, namely, which arise from a change in form, that we shall be, in our present subject, principally* concerned, inasmuch as these changes only are peculiar to the Calculus of Variations.

3. It is evident that the form of one function may be so connected with the form or forms of one or more others, that if the form or forms of the latter be determined, that of the first is determined also. Thus, for example, the form of the differential coefficient of any function depends on, and may be deduced from, the form of the function itself. This species of relation between functions may be denoted by giving the name of *primitive* functions to the latter, whose forms are independent, and that of *derived* function to the former, whose form depends on those of the primitives. We shall denote it analytically by the symbols F, F', &c.; i. e. we shall use the symbol $F.\phi$ to represent a function whose form depends on that of the function ϕ. Now if the form of one or more of the primitive functions be supposed to change, it is evident that the form of the derived function will undergo a corresponding change; and if the relation between the forms of the primitive and derived functions be supposed invariable, the change in the form of the latter will not be arbitrary, but connected by a fixed relation with the change in the form of the former. To deduce this relation, or, in other words, "*to investigate the change in a derived function, in consequence of a change in the form of its primitive,*" is the object of the Calculus of Variations.

4. As it is essential no less to the problems of the Calculus of Variations, than to those of the Differential Calculus, that the

* I say *principally*, because, as will be seen, many problems in the Calculus of Variations involve the consideration of increments of both kinds.

increments assigned to the variable quantities should be susceptible of indefinite diminution, it becomes a question of importance to determine a method of giving such an increment to a function by means of a change in its form. It is, moreover, essential that this increment should continue indefinitely small for all values of the independent variables. Hence if $u = \phi(x_1, x_2 \ldots .)$ be the given function in its original state, and if $u' = \phi'(x_1, x_2 \ldots .)$ be the function after having undergone the required change of form, and if i be a constant quantity of whatever degree of smallness $u' - u$ is intended to have, it is evident that

$$\frac{\phi'(x_1, x_2 \ldots .) - \phi(x_1, x_2 \ldots .)}{i}$$

must be finite for all values of x_1, x_2, &c., which are consistent with the conditions of the problem. Assume

$$\psi(x_1, x_2 \ldots .) = \frac{\phi'(x_1, x_2 \ldots .) - \phi(x_1, x_2 \ldots .)}{i}$$

and we have $u' - u = i\psi(x_1, x_2 \ldots .)$, or $u' = u + i\psi(x_1, x_2 \ldots .)$, where $\psi(x_1\, x_2 \ldots .)$ is a function subject to no other restriction than that of not becoming infinite for any values of x_1, x_2, &c., which are consistent with the conditions of the problem.

Hence it appears that the required method of assigning to a given primitive function an increment susceptible of indefinite diminution for all values of the variable, is to add to it another function (subject to the foregoing restriction), multiplied by a constant quantity, which is to be assumed of whatever order of magnitude the increment is required to have. Such an increment is properly termed a *variation*, which may, therefore, be defined to be "*The indefinitely small change in value which a function receives in consequence of a change in its form.*" From this and the preceding number it immediately appears that the variation of a primitive function is perfectly arbitrary, and that the variation of a derived function depends on that of its primitive.

5. Let $u = \phi(x_1, x_2 \ldots .)$, an indeterminate function of any number of variables, and let $v = F.u$,* a function derived from u,

* It must be carefully kept in mind that v is not a function of u, as the symbol F denotes a relation between *forms*, not between *magnitudes*.

i. e. a function whose form depends on the form of u. Let it be required to find the variation of v, i. e. the change which v undergoes in consequence of an indefinitely small change in the form of u.

Substitute, according to the method in (4), for

$$\phi(x_1, x_2 \dots), \quad \phi(x_1, x_2 \dots) + i\psi(x_1, x_2 \dots),$$

and let the operation denoted by F be performed on the function so changed, sufficiently far to obtain the coefficient of the first power of i. If this coefficient be denoted by ω, $i\omega$ will be the required variation. This appears by precisely the same reasoning as that employed in the Differential Calculus in the investigation of a differential.

This is the most general problem of the Calculus of Variations. But as, in the present state of mathematical science, we are concerned with but two species of derived functions, *sc.*, those which are derived by the process of differentiation, denoted by the symbol d, and those which are derived by the process of integration, denoted by the symbol $\int$, the investigation of so general a problem is quite unnecessary. We shall, therefore, proceed to consider a particular case, which is, however, sufficiently general for all purposes to which the Calculus of Variations has been hitherto applied.

Let the symbol of derivation, F, be *distributive*, i. e. such as to satisfy the equation

$$F.\phi + F.\phi_1 = F.(\phi + \phi_1),$$

where ϕ and ϕ_1 are functions of any number of variables; and let it be required to find the variation of $v = F.\phi$. Substitute, according to the method given above, in $F.\phi$, $\phi + i\psi$ for ϕ. Now since, by the equation of condition,

$$F(\phi + i\psi) = F.\phi + F.i\psi,$$

it is evident that the total increment of $F.\phi$ is $F.i\psi$. But $F.i\psi = iF.\psi$ (*vid.* Note A). Hence, according to the principle above stated, the variation of $F.\phi$ is $iF.\psi$, or $F.i\psi$. It is easily seen that the equation of condition,

$$F.\phi + F.\phi_1 = F(\phi + \phi_1),$$

is satisfied by the two modes of derivation represented by the symbols d and $\int$.

6. Hitherto we have only considered that species of increment which is peculiar to the Calculus of Variations, viz., that which a function receives in consequence of a change in its form. But as the problems with which we are concerned in this science frequently involve the consideration of the increments which a function receives in consequence of change in one or more of its independent variables, the following principle is necessary to the determination of the complete change on the function.

If, from the operation of any cause whatever, a quantity receive an increment which is indefinitely small, in comparison with the quantity itself; and if, from the operation of another cause, the same quantity receive another such increment, and so on for any number of causes, the increment which it would receive from the combined action of all these causes is the sum of the increments which each produces when acting separately.

The truth of this principle appears at once from the supposition that the increment is indefinitely small when compared with the quantity increased.

It is, in fact, perfectly analogous to the principle of the superposition of small motions in Mechanics, and may be proved in a similar manner.

As the increments with which the Calculus of Variations is concerned are of two essentially different species, it may be well, before proceeding further, to establish a distinct notation to express them. This is the more necessary, as many writers on the Calculus of Variations have been led into considerable difficulties by an unsteady use of the symbol δ, a symbol which they employ sometimes to express the increment which a function receives in consequence of a change of form *only*, and sometimes to express the increment which it receives from the variation, not only of its form, but also of its independent variables.

We shall, then, use the symbol δ to denote that species of increment which is peculiar to the Calculus of, Variations, that, namely, which a function receives in consequence of a change in its form *only*. We shall, as in the Differential Calculus, denote by the symbol d that increment which a function receives in consequence of a change in the magnitude of its independent variables.

Finally, we shall use the symbol D to express the total increment of which any function is susceptible, arising from the variation of every quantity connected with it which can be varied. Hence, if u be a determinate function of variable quantities,

$$Du = du\,;$$

if u be an indeterminate function of constant quantities,

$$Du = \delta u\,;$$

and if u be an indeterminate function of variable quantities,

$$Du = \delta u + du.$$

As an independent variable is capable of but one species of increment, it is immaterial what symbol be employed to express it. We shall, however, in general, denote the increment of an independent variable by the symbol d; and whenever it may be necessary to vary this notation, we shall be careful to state it distinctly, so as to preclude the possibility of any mistake as to the meaning of the symbol employed.

7. We shall now proceed to apply the principles which have been established, to the investigation of the variations of the several species of quantities with which we are concerned in this science.

(1.) Let

$$u = f(x_1, x_2, \ldots . x_n),$$

a determinate function of any number of independent variables, and let it be required to find its complete increment. Here the form of the function not being supposed to vary,

$$Du = du = \frac{du}{dx_1} . dx_1 + \frac{du}{dx_2} . dx_2 + \&c. + \frac{du}{dx_n} . dx_n.$$

(2.) Let

$$u = \phi\,(x_1, x_2 \ldots . x_n),$$

a primitive indeterminate function. In this case

$$Du = \delta u + du.$$

But

$$du = \frac{du}{dx_1} \, . \, dx_1 + \&c. + \frac{du}{dx_n} \, . \, dx_n,$$

and

$$\delta u = i\psi \, (x_1, x_2, \ldots . x_n) \, ;$$

hence

$$Du = \frac{du}{dx_1} \, . \, dx_1 + \frac{du}{dx_2} \, . \, dx_2 + \&c. + \frac{du}{dx_n} \, . \, dx_n + i\psi \, (x_1, x_2 \ldots . x_n).$$

(3.) Let

$$u = F.\phi \, (x_1, x_2, \ldots . x_n),$$

and it is evident that

$$Du = \frac{du}{dx_1} \, . \, dx_1 + \frac{du}{dx_2} \, . \, dx_2 + \&c. + \frac{du}{dx_n} \, . \, dx_n + \delta u,$$

where δu is to be determined according to the method in Art. 5.

(4.) Let

$$u = F. \phi \, (x_1, x_2 \ldots . x_n),$$

a derived function, in which the symbol F satisfies the equation

$$F.\phi + F.\phi_1 = F.(\phi + \phi_1) \, ;$$

and since (Art. 5),

$$\delta u = F.i\psi = F.\delta\phi,$$

$$Du = \frac{du}{dx_1} \, . \, dx_1 + \frac{du}{dx_2} \, . \, dx_2 + \&c. + \frac{du}{dx_n} \, . \, dx_n + F.\delta\phi.$$

(5.) Let

$$V = f(x_1, x_2, \ldots . u_1, u_2 \ldots .),$$

a determinate function of the quantities within the brackets, of which x_1, x_2 .... are independent variables, and u_1, u_2 .... indeterminate functions of any or all of these variables; and let it be required to find the complete increment of V, i. e. $D.V$.

Here it is evident that V may vary either in consequence of a change in one or more of the independent variables, x_1, x_2, ... or in consequence of a change in the form of one or more of the functions u_1, u_2 ....

If, for example, x_1 vary, while everything else remains constant, the corresponding change in V will be

$$\left(\frac{dV}{dx_1} \, . + \frac{dV}{du_1} \, . \, \frac{du_1}{dx_1} \, . + \frac{dV}{du_2} \, . \, \frac{du_2}{dx_1} + \&c. \right) . \, dx_1 \, ;$$

similarly, for x_2,

$$\left(\frac{dV}{dx_2} + \frac{dV}{du_1}\cdot\frac{du_1}{dx_2} + \frac{dV}{du_2}\cdot\frac{du_2}{dx_2} + \&c.\right) dx_2 :$$

and so on for the other independent variables.

Now, let the form of the function u_1 vary, while everything else remains constant, and, since V is a function of u_1, the corresponding change in V will be

$$\frac{dV}{du_1}\cdot \delta u_1 ;$$

the truth of this theorem depending solely on the fact that V is a function of u_1, and being altogether independent of the *species* of increment assigned to u_1. Similar terms being introduced by the functions $u_2, u_3, \ldots$, the complete increment will be

$$\begin{aligned} D.V = &\left(\frac{dV}{dx_1} + \frac{dV}{du_1}\cdot\frac{du_1}{dx_1} + \frac{dV}{du_2}\cdot\frac{du_2}{dx_1} + \&c.\right) dx_1, \\ &+ \left(\frac{dV}{dx_2} + \&c. \qquad\qquad\qquad\qquad\qquad \right) dx_2, \\ &+ \&c. \\ &+ \frac{dV}{du_1}\cdot\delta u_1 + \frac{dV}{du_2}\cdot\delta u_2 + \&c. \end{aligned}$$

The expression for the *variation*, properly so called, will evidently be

$$\delta V = \frac{dV}{du_1}\delta u_1 + \frac{dV}{du_2}\delta u_2 + \&c.$$

(6.) Finally, let it be required to find the complete variation of $U = F.V$, where

$$V = f(x_1, x_2 \ldots . u_1, u_2 \ldots .),$$

and F is a symbol of derivation, which satisfies the condition

$$F.\phi + F.\phi_1 = F.(\phi + \phi_1).$$

We shall proceed, as before, to determine separately the variations arising from changes in the several variable elements which U contains. Let x_1 vary, while all the other elements, viz., the independent variables, $x_2, x_3 \ldots$, and the forms of the several functions, $u_1, u_2 \ldots$, remain constant, and let it be required to find the corresponding change in U. Now, since V is a determinate function of $x_1, x_2 \ldots u_1, u_2 \ldots$, it is evident that, as long as the forms of the functions $u_1, u_2 \ldots$ remain unchanged, V will be a determinate function of the independent variables $x_1, x_2 \ldots$; and since the symbol of derivation F is determinate, U will also be a determinate function of $x_1, x_2 \ldots$. Hence, if x_1 vary while everything else remains constant, the corresponding change in U will be

$$\frac{dU}{dx_1} \cdot dx_1,$$

$\frac{dU}{dx_1}$ denoting the complete differential coefficient of U, with regard to x_1. Similar terms will, of course, be introduced by the variation of $x_2, x_3 \ldots$. It remains then to find the *variation*, properly so called, *sc.*, that part of the complete increment which depends upon the change in the form of any of the functions $u_1, u_2 \ldots$.

Now it might be at first sight supposed that the variation arising from a change of any kind in u_1 should be, as before,

$$\frac{dU}{du_1} \cdot \delta u_1.$$

But it must be remembered that the truth of this theorem of the Differential Calculus depends entirely upon the supposition that U is a *function* of u_1, i. e. a quantity whose *magnitude* depends upon the magnitude of u_1. But U is not a function of u_1, inasmuch as the relation between them is a relation of *form*, not of *magnitude;* and it is therefore no longer true that the increment of U is

$$\frac{dU}{du_1} \cdot \delta u_1.$$

But although U is not a function of u_1, it is a function derived from u_1, for it is evident that the form of U depends upon the form of V, which itself depends upon the form of u_1. Since, then, $U = F.V$, we have (Art. 5) $\delta U = F.\delta V$; and since, from the preceding paragraph,

$$\delta V = \frac{dV}{du_1} \delta u_1 + \frac{dV}{du_2} \delta u_2 + \&c.,$$

it is evident that the part of δU which results from a change in the form of u_1 is

$$F.\frac{dV}{du_1} \delta u_1.$$

Hence the complete expression is

$$DU = \frac{dU}{dx_1} . dx_1 + \frac{dU}{dx_2} . dx_2 + \&c.$$

$$+ F . \left(\frac{dV}{du_1} \delta u_1 + \frac{dV}{du_2} \delta u_2 + \&c. \right)$$

We have in this proposition an instance of the caution requisite in applying the principles of the Differential Calculus to any cases in which the variation which we consider arises from a change in the form of the function.

8. Having thus stated the general principles of the Calculus of Variations, we shall next proceed to consider the several cases to which, in the actual state of mathematical and physical science, they may be applied; those cases, namely, in which the functions are derived from one another by the processes of differentiation and integration.

CHAPTER II.

FUNCTIONS OF ONE INDEPENDENT VARIABLE.

Prop. I.

9. To find the complete variation of the differential coefficient $\frac{d^n y}{dx^n}$, y being a function of the single independent variable x.

It is evident that there are but two modes in which such a differential coefficient can be varied, *sc.*, either by a change in the independent variable, or by a change in the form of the function. The complete variation will, therefore, according to the principle stated in Art. 6, Chap. I., be found by taking the sum of the increments produced by the separate action of these causes. Now if the independent variable x receive the increment dx, the form of the function y remaining unchanged, it is plain that the corresponding increment of $\frac{d^n y}{dx^n}$ will be $\frac{d^{n+1} y}{dx^{n+1}} \cdot dx$. Hence, according to the notation established in Art. 6, we shall have for the complete variation,

$$D.\frac{d^n y}{dx^n} = \frac{d^{n+1} y}{dx^{n+1}} \cdot dx + \delta . \frac{d^n y}{dx^n}.$$

But since the symbol of derivation, $\frac{d^n}{dx^n}$, satisfies the equation

$$F.\phi + F.\phi_1 = F.(\phi + \phi_1),$$

it is evident that

$$\delta . \frac{d^n y}{dx^n} = \frac{d^n . \delta y}{dx^n}.$$

Substituting this value in the expression for $D . \frac{d^n y}{dx^n}$, we have, finally,

$$D. \frac{d^n y}{dx^n} = \frac{d^{n+1} y}{dx^{n+1}} \cdot \delta x + \frac{d^n . \delta y}{dx^n}.$$

It is necessary to notice here a restriction upon the quantity δy. It will be remembered that in the preceding chapter, where we found $\delta u = i\psi(x_1, x_2 \ldots .)$, it was noticed, as a necessary restriction upon the function ψ, that it should not become infinite for any values of the independent variables consistent with the conditions of the question. This is sufficient as long as we are concerned only with the function itself; but when the question involves the consideration of functions derived from the original, it is necessary that the functions similarly derived from ψ should also be finite for all admissible values of the independent variables. In other words, when it is stated in Art. 7, No. (4), that

$$\delta . F . \phi = F . i\psi = iF . \psi,$$

it is, of course, supposed that $F . \psi$ does not become infinite for any values of x_1, x_2, &c. which are consistent with the conditions of the question. In the present case, where there is but one independent variable,

$$\delta y = i\psi x, \quad \delta \frac{d^n y}{dx^n} = i \frac{d^n . \psi x}{dx^n}.$$

It is, therefore, necessary to assume the function ψ of such a form as to render $\frac{d^n . \psi x}{dx^n}$ finite for all admissible values of x. This remark will be found of great importance in the application of the Calculus of Variations to the theory of maxima and minima.

Prop. II.

10. To find the complete variation of

$$V = f\left(x, y, \frac{dy}{dx}, \frac{d^2y}{dx^2}, \ldots \frac{d^n y}{dx^n}\right),$$

a determinate function of x, y, and its first n differential coefficients.

This is evidently a case of the general question discussed in Art. 7, No (5), of the preceding chapter, and its solution may be derived from the formula there given, by reducing the number of independent variables to one, and putting $u_1 = y$, $u_2 = \frac{dy}{dx}$, &c. If these substitutions be made, and if, for the sake of brevity, we assume

$$M=\frac{dV}{dx},\quad N=\frac{dV}{dy},\quad P_1=\frac{dV}{d\,.\,\frac{dy}{dx}},\quad P_2=\frac{dV}{d\,.\,\frac{d^2y}{dx^2}},\ \&c.,\quad P_n=\frac{dV}{d\,.\,\frac{d^ny}{dx^n}},$$

we shall have

$$D\,.\,V=\left(M+N\frac{dy}{dx}+P_1\frac{d^2y}{dx^2}+\&c.+P_n\frac{d^{n+1}\,.\,y}{dx^{n+1}}\right)dx$$

$$+N\delta y+P_1\delta\,.\frac{dy}{dx}+P_2\delta\,.\frac{d^2y}{dx^2}+\&c.+P_n\delta\,.\,\frac{d^ny}{dx^n};$$

or, if we substitute for

$$\delta\,.\frac{dy}{dx},\quad \delta\,.\frac{d^2y}{dx^2},\ \ldots.\ \delta\,.\frac{d^ny}{dx^n},$$

their values derived from the preceding proposition,

$$D\,.\,V=\left(M+N\frac{dy}{dx}+P_1\frac{d^2y}{dx^2}+\&c.+P_n\frac{d^{n+1}y}{dx^{n+1}}\right)dx$$

$$+N\delta y+P_1\frac{d\delta y}{dx}+P_2\frac{d^2.\,\delta y}{dx^2}+\&c.+P_n\frac{d^n\delta y}{dx^n}.$$

The variation δy being, as before, expressed by $i\psi x$, it appears from the concluding remark of the preceding proposition, that it is necessary to assume the function ψ of such a form, that neither itself, nor any of its first n differential coefficients, become infinite for any value of x consistent with the conditions of the question.

Prop. III.

11. To find the complete variation of

$$U=\int_{x_0}^{x_1}Vdx,\ \text{where}\ V=f\left(x,y,\frac{dy}{dx},\ldots\frac{d^ny}{dx^n}\right).$$

The value of a definite integral may be varied in one of three ways: *sc.* 1. By a change in the superior limit x_1. 2. By a change in the inferior limit x_0. 3. By a change in the form of the function to be integrated.* The complete variation will be, as before,

* It is usual, in investigating the variation of a definite integral, to assign an increment to the independent variable. But as this quantity does not enter into the final result,

the sum of the partial variations so obtained. Let the value of V at the superior limit be denoted by V_1, and let that limit be increased by dx_1, the form of the function V remaining unchanged. The corresponding increment of U is $V_1 dx_1$. Similarly, if the inferior limit x_0 receive the increment dx_0, the corresponding increment of U will be $-V_0 dx_0$. Hence

$$D \,.\, U = V_1 dx_1 - V_0 dx_0 + \delta \,.\int_{x_0}^{x_1} V dx.$$

It remains, therefore, to find $\delta \,.\int_{x_0}^{x_1} V dx$, i. e., the change in the value of the definite integral produced by a change in the form of the function V. Now, since the operation denoted by the symbol $\int_{x_0}^{x_1} (\quad) \,.\, dx$ satisfies the equation

$$F \,.\, \phi + F \,.\, \phi_1 = F \,.\, (\phi + \phi_1),$$

we have (Art. 5, Chap. I.)

$$\delta \,.\int_{x_0}^{x_1} V dx = \int_{x_0}^{x_1} \delta V \,.\, dx.$$

As V is a determinate function of x, y, $\frac{dy}{dx}$, $\frac{d^2y}{dx^2}$, &c., its form, considered as a function of the independent variable x, can be changed only by a change in the form of the function y. The value of δV will, therefore, be, as found in the preceding proposition,

$$\delta V = N\delta y + P_1 \frac{d\delta y}{dx} + P_2 \frac{d^2\delta y}{dx^2} + \text{\&c.} + P_n \frac{d^n\delta y}{dx^n},$$

and, therefore,

$$\delta \,.\int_{x_0}^{x_1} V dx = \int_{x_0}^{x_1} \left(N\delta y + P_1 \frac{d\delta y}{dx} + P_2 \frac{d^2\delta y}{dx^2} + \text{\&c.} + P_n \frac{d^n\delta y}{dx^n}\right) dx.$$

This may be reduced by the method of integration by parts, as follows:

which depends entirely upon the values of the limits and the form of the function to be integrated, it may naturally be expected that the complete variation will contain no term depending upon the increment of the independent variable, as distinguished from the increments of its limiting values. Accordingly, if an increment be assigned to the independent variable, it will be found that the coefficient of that increment, under the sign of integration, will vanish of itself.

$$\int_{x_0}^{x_1} P_1 \frac{d\delta y}{dx} = (P_1\delta y)_1 - (P_1\delta y)_0 \, . - \int_{x_0}^{x_1} \frac{dP_1}{dx} \, . \, \delta y \, . \, dx,$$

where $(P_1\delta y)_1$, $(P_1\delta y)_0$ denote the values of $P_1\delta y$ at the superior and inferior limits respectively. Similarly,

$$\int_{x_0}^{x_1} P_2 \frac{d^2\delta y}{dx^2} = \left(P_2 \frac{d\delta y}{dx}\right)_1 - \left(P_2 \frac{d\delta y}{dx}\right)_0 - \int_{x_0}^{x_1} \frac{dP_2}{dx} \, . \, \frac{d\delta y}{dx} \, . \, dx \, ;$$

or, if we integrate the last term again,

$$= \left(P_2 \frac{d\delta y}{dx} - \frac{dP_2}{dx} \delta y\right)_1 - \left(P_2 \frac{d\delta y}{dx} - \frac{dP_2}{dx} \delta y\right)_0 + \int_{x_0}^{x_1} \frac{d^2P_2}{dx^2} . \, \delta y . dx.$$

Similarly, if we integrate n times successively the term

$$\int_{x_0}^{x_1} P_n \frac{d^n\delta y}{dx^n} \, . \, dx,$$

we shall find

$$\int_{x_0}^{x_1} P_n \frac{d^n\delta y}{dx^n} = \left(P_n \frac{d^{n-1}\delta y}{dx^{n-1}} - \frac{dP_n}{dx} \, . \, \frac{d^{n-2}\delta y}{dx^{n-2}} + \&c. + (-1)^{n-1} \frac{d^{n-1}P_n}{dx^{n-1}} \, . \, \delta y\right)_1$$

$$- \left(P_n \frac{d^{n-1}\delta y}{dx^{n-1}} - \&c. \qquad \right)_0$$

$$+ (-1)^n \, . \int_{x_0}^{x_1} \frac{d^nP_n}{dx^n} \, . \, \delta y.$$

Collecting the coefficients of the several quantities,

$$\delta y_1, \left(\frac{d\delta y}{dx}\right)_1, \; \&c.,$$

we find

$$\delta \, . \int_{x_0}^{x_1} V dx = \left(P_1 - \frac{dP_2}{dx} + \frac{d^2P_3}{dx^2} - \&c. + (-1)^{n-1} \frac{d^{n-1}P_n}{dx^{n-1}}\right)_1 \delta y_1$$

$$- \left(P_1 - \frac{dP_2}{dx} + \&c.\right)_0 \delta y_0$$

$$+ \left(P_2 - \frac{dP_3}{dx} + \&c.\right)_1 \left(\frac{d\delta y}{dx}\right)_1 - (P_2 - \&c.)_0 \left(\frac{d\delta y}{dx}\right)_0$$

$$+ \&c.$$

$$+ \left(P_n \frac{d^{n-1}\delta y}{dx^{n-1}}\right)_1 - \left(P_n \frac{d^{n-1}\delta y}{dx^{n-1}}\right)_0$$

$$+ \int_{x_0}^{x_1} \left(N - \frac{dP_1}{dx} + \frac{d^2P_2}{dx^2} - \&c. + (-1)^n \frac{d^nP_n}{dx^n}\right) \delta y \, . \, dx \, ;$$

and, therefore,

$$D\,.\int_{x_0}^{x_1} V dx = V_1\, dx_1 - V_0\, dx_0$$

$$+\left(P_1 - \frac{dP_2}{dx} + \&c.\right)_1 \delta y_1 - \left(P_1 - \frac{dP_2}{dx} + \&c.\right)_0 \delta y_0$$

$$+ (P_2 - \&c.)_1 \left(\frac{d\delta y}{dx}\right)_1 - (P_2 - \&c.)_0 \left(\frac{d\delta y}{dx}\right)_0$$

$$+ \&c.$$

$$+\int_{x_0}^{x_1}\left(N - \frac{dP_1}{dx} + \frac{d^2P_2}{dx^2} - \&c. + (-1)^n \frac{d^nP_n}{dx^n}\right)\delta y\,.\,dx.$$

12. This expression, as will be seen, consists of three parts, essentially distinct, viz.: 1. The terms

$$V_1\, dx_1 - V_0\, dx_0,$$

which are independent of the change in the form of the function, and depend solely upon the variation of the limits. 2. The terms

$$(P_1 - \&c.)_1\, \delta y_1 - \&c.,$$

which depend on the change in the form of the function, not for every value of x, but for the limiting values only of that variable. 3. The terms under the sign of integration, *sc.*,

$$\int_{x_0}^{x_1}\left(N - \frac{dP_1}{dx} + \&c.\right)\delta y\,.\,dx,$$

depending on the general change in the form of the function. The nature of this difference will appear most clearly, if we recollect that $\delta y = i\psi x$. For it appears immediately—1. That the value of the first class of terms is wholly independent of the form of the function ψ. 2. That for the determination of the value of the terms of the second class it is not necessary to determine the *form* of the function ψ, but only the *values* which that function and its first $n-1$ differential coefficients have at the limits. 3. That the value of the term

$$\int_{x_0}^{x_1}\left(N - \frac{dP_1}{dx} + \&c.\right)\delta y\,.\,dx$$

depends on the form of the function ψ, and cannot be determined as long as that form remains arbitrary.

Prop. IV.

13. If V contain explicitly the limiting values of any number of the quantities $x, y, \frac{dy}{dx}$, &c., i. e. if

$$V = f\left\{x, y, \frac{dy}{dx} \dots \frac{d^n y}{dx^n}, x_1, y_1, \left(\frac{dy}{dx}\right)_1 \dots x_0, y_0, \left(\frac{dy}{dx}\right)_0 \dots\right\},$$

find the complete variation of $\int_{x_0}^{x_1} V dx$.

As y_1, x_1, and y_0, x_0, are connected by the same general relation as y, x, it is plain that the integral $\int_{x_0}^{x_1} V dx$ can be varied only by one of the three methods given in the foregoing proposition. Now if x_1 receive the increment dx_1, the form of the function y remaining unchanged, the corresponding increase on the integral will be

$$\left\{V_1 + \int_{x_0}^{x_1}\left(\frac{dV}{dx_1} + \frac{dV}{dy_1}\cdot\left(\frac{dy}{dx}\right)_1 + \frac{dV}{d\cdot\left(\frac{dy}{dx}\right)_1}\cdot\left(\frac{d^2y}{dx^2}\right)_1 + \text{\&c.}\right)dx\right\} dx_1.$$

Similarly, if x_0 receive the increment dx_0, the corresponding change in $\int_{x_0}^{x_1} V dx$ will be

$$\left\{-V_0 + \int_{x_0}^{x_1}\left(\frac{dV}{dx_0} + \frac{dV}{dy_0}\cdot\left(\frac{dy}{dx}\right)_0 + \frac{dV}{d\cdot\left(\frac{dy}{dx}\right)_0}\cdot\left(\frac{d^2y}{dx^2}\right)_0 + \text{\&c.}\right)dx\right\} dx_0.$$

Now let the form of the function change, while everything else remains the same, and it is evident that the change in $\int_{x_0}^{x_1} V dx$ will be

$$\delta U = \int_{x_0}^{x_1}\left(N\delta y + P_1\frac{d\delta y}{dx} + \text{\&c.}\right)dx + \delta y_1 . \int_{x_0}^{x_1}\frac{dV}{dy_1}dx + \delta y_0\int_{x_0}^{x_1}\frac{dV}{dy_0}dx$$

$$+ \left(\frac{d\delta y}{dx}\right)_1 . \int_{x_0}^{x_1}\frac{dV}{d\cdot\left(\frac{dy}{dx}\right)_1}dx + \left(\frac{d\delta y}{dx}\right)_0 . \int_{x_0}^{x_1}\frac{dV}{d\cdot\left(\frac{dy}{dx}\right)_0}dx$$

$$+ \text{\&c.}$$

Assume

$$\mu_1 = \frac{dV}{dx_1} \quad \nu_1 = \frac{dV}{dy_1} \quad \pi_1 = \frac{dV}{d.\left(\frac{dy}{dx}\right)_1} \text{ \&c.}$$

$$\mu_0 = \frac{dV}{dx_0} \quad \nu_0 = \frac{dV}{dy_0} \quad \pi_0 = \frac{dV}{d.\left(\frac{dy}{dx}\right)_0}.$$

Integrating by parts, as before, and adding the three expressions just given, we find the complete variation.

$$DU = \left\{ V_1 + \int_{x_0}^{x_1}\left(\mu_1 + \nu_1\left(\frac{dy}{dx}\right)_1 + \pi_1\left(\frac{d^2y}{dx^2}\right)_1 + \text{\&c.}\right)dx\right\} dx_1$$

$$+\left\{ - V_0 + \int_{x_0}^{x_1}\left(\mu_0 + \nu_0\left(\frac{dy}{dx}\right)_0 + \pi_0\left(\frac{d^2y}{dx^2}\right)_0 + \text{\&c.}\right)dx\right\} dx_0$$

$$+\left\{\left(P_1 - \frac{dP_2}{dx} + \text{\&c.}\right)_1 + \int_{x_0}^{x_1}\nu_1 dx\right\}\delta y_1$$

$$-\left\{\left(P_1 - \frac{dP_2}{dx} + \text{\&c.}\right)_0 - \int_{x_0}^{x_1}\nu_0 dx\right\}\delta y_0$$

$$+\left\{(P_2 - \text{\&c.})_1 + \int_{x_0}^{x_1}\pi_1 dx\right\}\left(\frac{d\delta y}{dx}\right)_1 - \left\{(P_2 - \text{\&c.})_0 - \int_{x_0}^{x_1}\pi_0 dx\right\}\left(\frac{d\delta y}{dx}\right)_0$$

$$+ \text{\&c.}$$

$$+\int_{x_0}^{x_1}\left(N - \frac{dP_1}{dx} + \frac{d^2P_2}{dx^2} - \text{\&c.}\right)\delta y.$$

Prop. V.

14. To find the complete variation of

$$U = \int_{x_0}^{x_1} V dx, \text{ where } V = f\left(x, y, \frac{dy}{dx} \ldots \frac{d^ny}{dx^n}, z, \frac{dz}{dx} \ldots \frac{d^mz}{dx^m}\right),$$

y and z being indeterminate functions of x.

The value of this integral may be changed either—1. by a change in x_1; 2. by a change in x_0; 3. by a change in the form of the function y; or, 4. by a change in the form of the function z. The complete variation will, therefore, be found by taking the sum of the partial variations arising from each of these causes.

Let, as before,

$$M = \frac{dV}{dx}, \quad N = \frac{dV}{dy}, \quad P_1 = \frac{dV}{d \cdot \frac{dy}{dx}}, \ldots . P_n = \frac{dV}{d \cdot \frac{d^n y}{dx^n}},$$

and

$$N' = \frac{dV}{dz}, \quad P'_1 = \frac{dV}{d \cdot \frac{dz}{dx}}, \ldots . P'_m = \frac{dV}{d \cdot \frac{d^m z}{dx^m}};$$

and it will appear, by the application of similar principles to those employed in Prop. III., that

$$DU = V_1 dx_1 - V_0 dx_0$$

$$+ \left(P_1 - \frac{dP_2}{dx} + \&c.\right)_1 \delta y_1 - \left(P_1 - \frac{dP_2}{dx} + \&c.\right)_0 \delta y_0$$

$$+ (P_2 - \&c.)_1 \left(\frac{d\delta y}{dx}\right)_1 - (P_2 - \&c.)_0 \left(\frac{d\delta y}{dx}\right)_0$$

$$+ \&c.$$

$$+ \left(P_n \frac{d^{n-1}\delta y}{dx^{n-1}}\right)_1 - \left(P_n \frac{d^{n-1}\delta y}{dx^{n-1}}\right)_0$$

$$+ \int_{x_0}^{x_1} \left(N - \frac{dP_1}{dx} + \frac{d^2 P_2}{dx^2} - \&c. + (-1)^n \frac{d^n P_n}{dx^n}\right) \delta y$$

$$+ \left(P'_1 - \frac{dP'_2}{dx} + \&c.\right)_1 \delta z_1 - \left(P'_1 - \frac{dP'_2}{dx} + \&c.\right)_0 \delta z_0$$

$$+ (P'_2 - \&c.)_1 \left(\frac{d\delta z}{dx}\right)_1 - (P'_2 - \&c.)_0 \left(\frac{d\delta z}{dx}\right)_0$$

$$+ \&c.$$

$$+ \left(P'_m \frac{d^{m-1}\delta z}{dx^{m-1}}\right)_1 - \left(P'_m \frac{d^{m-1}\delta z}{dx^{m-1}}\right)_0$$

$$+ \int_{x_0}^{x_1} \left(N' - \frac{dP'_1}{dx} + \frac{d^2 P'_2}{dx^2} - \&c. + (-1)^m \frac{d^m P'_m}{dx^m}\right) \delta z.$$

Similarly, if V contained any number of indeterminate functions, each of these functions would introduce into the expression for δU or DU a series of terms depending on the variation of that function, and precisely similar to those found in Prop. III. These expressions are equally true, whether it be supposed that the functions y, z, &c., are really independent, or that these functions are connected by any equation or equations.

Prop. VI.

15. To find the variation of

$$U=\int_{x_0}^{x_1} V dx, \text{ where } V=f\left(x, y, \frac{dy}{dx} \dots \frac{d^n y}{dx^n}, z, \frac{dz}{dx} \dots \frac{d^m z}{dx^m}\right),$$

the functions y and z being connected by the equation (differential or other) $L=0$.

The expression arrived at in Prop. V. is, as appears from the concluding remark, applicable to the present case, the truth of the principle on which it was obtained being unconnected with the dependence or independence of the quantities δy, δz. But as it is essential, in the application of the Calculus of Variations, that these quantities should be considered as being really independent of each other, the expression of Prop. V. requires certain modifications before it can be made use of. Now, if the equation $L=0$ can be solved for either of the functions y or z, so as to give a result of the form $z=f\left(x, y, \frac{dy}{dx}, \text{\&c.}\right)$, it is evident that the values of the several quantities,

$$\frac{dz}{dx}, \frac{d^2 z}{dx^2} \dots \frac{d^m z}{dx^m},$$

may be obtained by simple differentiation; and if the values so found be substituted in V, that quantity will become simply a function of x, y, and its differential coefficients. In this case, then, the variation of V is found as in Prop. III. But as the equation $L=0$ is, in general, a differential equation, not susceptible of integration, this method can rarely be applied, and it is, therefore, necessary to seek some other mode of removing the difficulty. This has been furnished by the illustrious Lagrange, the principle of whose method we shall partly state here, reserving a fuller explanation of it till we come to the application of the Calculus of Variations to the theory of maxima and minima. At present it will be sufficient to show that this method may be made to furnish an expression for $\delta . \int V dx$, in which but one of the variations, δy or δz, enters under the sign of integration.

16. Assume

$$a = \frac{dL}{dy}, \quad \beta = \frac{dL}{d . \frac{dy}{dx}}, \quad \gamma = \frac{dL}{d . \frac{d^2 y}{dx^2}}, \text{ \&c.}$$

$$a' = \frac{dL}{dz}, \quad \beta' = \frac{dL}{d . \frac{dz}{dx}}, \quad \gamma' = \text{\&c.}$$

Now, since the equation

$$L = 0$$

must, according to the hypothesis, be satisfied by all forms of the functions y and z, which are admissible in the question under consideration, it is plain that we must have $\delta L = 0$, or

$$a \delta y + \beta \frac{d \delta y}{dx} + \gamma \frac{d^2 \delta y}{dx^2} + \text{\&c.} + a' \delta z + \beta' \frac{d \delta z}{dx} + \text{\&c.} = 0. \qquad \text{(A)}$$

If this equation can be integrated so as to furnish a value for either of the quantities δy or δz in terms of the other,—if, for example, δz can be found in terms of δy,—the quantities $\frac{d \delta z}{dx}$, $\frac{d^2 \delta z}{dx^2}$, &c., may be deduced by simple differentiation in terms of δy, $\frac{d \delta y}{dx}$, &c.; and if these values be substituted in the expression for δU found in Prop. V., that expression will be found to contain but one arbitrary variation, δy, and its differential coefficients, $\frac{d \delta y}{dx}$, $\frac{d^2 \delta y}{dx^2}$, &c. In this case the expression for δU would have received the necessary modification, and the solution of the problem would be so far complete. But as it is rarely possible to integrate equation (A), this method is generally inapplicable.

The value of δV being

$$\delta V = N \delta y + P_1 \frac{d \delta y}{dx} + P_2 \frac{d^2 \delta y}{dx^2} + \text{\&c.} + N' \delta z + P'_1 \frac{d \delta z}{dx} + P'_2 \frac{d^2 \delta z}{dx^2} + \text{\&c.},$$

it is evident that, without altering the truth of this expression, we may add to it the left hand member of equation (A) multiplied by an indeterminate quantity λ. The expression for δV will then become

$$\delta V = (N + \lambda a)\,\delta y + (P_1 + \lambda\beta)\,\frac{d\delta y}{dx} + \&c.$$

$$+ (N' + \lambda a')\,\delta z + (P_1 + \lambda\beta')\,\frac{d\delta z}{dx}$$

$$+ \&c.$$

and hence it is easy to see that the expression for δU will become

$$\delta U = \left(P_1 + \lambda\beta - \frac{d(P_2 + \lambda\gamma)}{dx} + \&c.\right)_1 \delta y_1$$

$$- \left(P_1 + \lambda\beta - \frac{d(P_2 + \lambda\gamma)}{dx} + \&c.\right)_0 \delta y_0$$

$$+ (P_2 + \lambda\gamma - \&c.)_1 \left(\frac{d\delta y}{dx}\right)_1 - (P_2 + \lambda\gamma - \&c.)_0 \left(\frac{d\delta y}{dx}\right)_0$$

$$+ \&c.$$

$$+ \int_{x_0}^{x_1} \left(N + \lambda a - \frac{d(P_1 + \lambda\beta)}{dx} + \&c.\right) \delta y$$

$$+ \left(P'_1 + \lambda\beta' - \frac{d(P'_2 + \lambda\gamma')}{dx} + \&c.\right)_1 \delta z_1$$

$$- \left(P'_1 + \lambda\beta' - \frac{d(P'_2 + \lambda\gamma')}{dx} + \&c.\right)_0 \delta z_0$$

$$+ (P'_2 + \lambda\gamma' - \&c.)_1 \left(\frac{d\delta z}{dx}\right)_1 - (P'_2 + \lambda\gamma'_2 - \&c.)_0 \left(\frac{d\delta z}{dx}\right)_0$$

$$+ \&c.$$

$$+ \int_{x_0}^{x_1} \left(N' + \lambda a' - \frac{d(P'_1 + \lambda\beta')}{dx} + \&c.\right) \delta z.$$

Now if it be required to find an expression for δU, in which but one of the indeterminate variations, δy or δz, shall appear under the sign of integration, this may be effected by means of the indeterminate quantity λ. For if that quantity be assumed such as to satisfy the equation

$$N' + \lambda\alpha' - \frac{d(P'_1 + \lambda\beta')}{dx} + \&c. = 0,$$

the expression will be independent of the variation δz; and if it be assumed such as to satisfy the equation

$$N + \lambda\alpha - \frac{d(P_1 + \lambda\beta)}{dx} + \&c. = 0,$$

it will be independent of δy.

We shall now proceed to give some examples of the application of this method.

Prop. VII.

17. To find the complete variation of

$$U = \int_{x_0}^{x_1} V dx, \text{ where } V = f\left(x, y, \frac{dy}{dx}, \dots \frac{d^n y}{dx^n}, \int v dx\right),$$

and

$$v = f_1\left(x, y, \frac{dy}{dx}, \dots \frac{d^m y}{dx^m}\right).$$

Assume, as before,

$$M = \frac{dV}{dx}, \quad N = \frac{dV}{dy}, \quad P_1 = \frac{dV}{d \cdot \frac{dy}{dx}}, \quad P_2 = \&c.,$$

and

$$\mu = \frac{dv}{dx}, \quad \nu = \frac{dv}{dy}, \quad \pi_1 = \frac{dv}{d \cdot \frac{dy}{dx}}, \quad \pi_2 = \&c.$$

Assume also,

$$z = \int v dx, \text{ and } N' = \frac{dV}{dz}.$$

The equation

$$L = 0$$

becomes, in this case,

$$v - \frac{dz}{dx} = 0;$$

hence

$$\alpha = \nu, \quad \beta = \pi_1, \quad \gamma = \pi_2, \&c. \qquad \alpha' = 0, \quad \beta' = -1, \quad \gamma' = 0, \&c.$$

Making these substitutions in the general expression given in Prop. VI., we find

$$\delta U = \left(P_1 + \lambda_1\pi_1 - \frac{d(P_2 + \lambda\pi_2)}{dx} + \&c.\right)_1 \delta y_1$$

$$- \left(P_1 + \lambda_1\pi_1 - \frac{d(P_2 + \lambda\pi_2)}{dx} + \&c.\right)_0 \delta y_0$$

$$+ (P_2 + \lambda\pi_2 - \&c.)_1 \left(\frac{d\delta y}{dx}\right)_1 - (P_2 + \lambda\pi_2 - \&c.)_0 \left(\frac{d\delta y}{dx}\right)_0$$

$$+ \&c.$$

$$+ \int_{x_0}^{x_1} \left(N + \lambda\nu - \frac{d(P_1 + \lambda\pi_1)}{dx} + \frac{d^2(P_2 + \lambda\pi_2)}{dx^2} - \&c.\right) \delta y$$

$$- (\lambda_1\delta z_1 - \lambda_0\delta z_0) + \int_{x_0}^{x_1} \left(N' + \frac{d\lambda}{dx}\right) \delta z.$$

It is evident that, as $P'_1 = 0$, $P'_2 = 0$, &c., no terms will occur containing $\left(\frac{d\delta z}{dx}\right)_1$, $\left(\frac{d\delta z}{dx}\right)_0$, &c. It is evident also that the complete variation, DU, is found by adding $V_1 dx_1 - V_0 dx_0$ to the expression given above.

To reduce this expression to one in which but one of the arbitrary variations, δy, shall remain under the sign of integration, let λ be assumed such as to satisfy the equation

$$N' + \frac{d\lambda}{dx} = 0.$$

This will give

$$\lambda = -\int N'dx = i \text{ (suppose).}$$

Substituting this value, and adding the terms $V_1 dx_1 - V_0 dx_0$, we have ultimately

$$DU = V_1 dx_1 - V_0 dx_0 + \left(P_1 + i\pi_1 - \frac{d(P_2 + i\pi_2)}{dx} + \&c.\right)_1 \delta y_1$$

$$- \left(P_1 + i\pi_1 - \frac{d(P_2 + i\pi_2)}{dx} + \&c.\right)_0 \delta y_0$$

$$+ (P_2 + i\pi_2 - \&c.)_1 \left(\frac{d\delta y}{dx}\right)_1 - (P_2 + i\pi_2 - \&c.)_0 \left(\frac{d\delta y}{dx}\right)_0$$

$$+ \&c.$$

$$+ i_1\delta z_1 - i_0\delta z_0$$

$$+ \int_{x_0}^{x_1} \left(N + i\nu - \frac{d(P_1 + i\pi_1)}{dx} + \frac{d^2(P_2 + i\pi_2)}{dx^2} - \&c.\right) \delta y.dx.$$

Similarly if V were a function of

$$x,\ y,\ \frac{dy}{dx},\ \text{\&c.}\ \int^p v\,dx^p,$$

the equation

$$L = 0$$

would become

$$v - \frac{d^p z}{dx^p} = 0.$$

It is easy to see, then, that the coefficient of δz, under the sign of integration, is

$$N' + \frac{d^p\lambda}{dx^p}.$$

The expression will, therefore, be reduced to the same form as before, if we assume

$$\lambda = -\int^p N' dx^p = i_p.$$

The quantities P'_1, P'_2, &c. being, as before, each $= 0$, the preceding formula will become, by the substitution of i_p for i,

$$DU = V_1 dx_1 - V_0 dx_0 + \left(P_1 + i_p\pi_1 - \frac{d(P_2 + i_p\pi_2)}{dx} + \text{\&c.}\right)_1 \delta y_1$$

$$- \left(P_1 + i_p\pi_1 - \frac{d(P_2 + i_p\pi_2)}{dx} + \text{\&c.}\right)_0 \delta y_0$$

$$+ (P_2 + i_p\pi_2 - \text{\&c.})_1 \left(\frac{d\delta y}{dx}\right)_1 - (P_2 + i_p\pi_2 - \text{\&c.})_0 \left(\frac{d\delta y}{dx}\right)_0$$

$$+ \text{\&c.}$$

$$+ (-1)^{p-1} \left\{ \left(\frac{d^{p-1} i_p}{dx^{p-1}}\right)_1 \delta z_1 - \left(\frac{d^{p-1} i_p}{dx^{p-1}}\right)_0 \delta z_0 \right.$$

$$\left. - \left(\frac{d^{p-2} i_p}{dx^{p-2}}\right)_1 \left(\frac{d\delta z}{dx}\right)_1 + \left(\frac{d^{p-2} i_p}{dx^{p-2}}\right)_0 \left(\frac{d\delta z}{dx}\right)_0 \right\}$$

$$+ \text{\&c.}$$

$$+ (i_p)_1 \left(\frac{d^{p-1}\delta z}{dx^{p-1}}\right)_1 - (i_p)_0 \left(\frac{d^{p-1}\delta z}{dx^{p-1}}\right)_0$$

$$+ \int \left(N + i_p\nu - \frac{d(P_1 + i_p\pi_1)}{dx} + \text{\&c.}\right) \delta y\, dx.$$

Prop. VIII.

18. To find the variation of a function of

$$x, y, \frac{dy}{dx}, \&c.,$$

whose form is given by a differential equation of the first order.

Let

$$U = f\left(x, y, \frac{dy}{dx}, \frac{d^2y}{dx^2} \ldots\right),$$

the form of the function f being such as to satisfy the equation

$$L = f_1\left(x, y, \frac{dy}{dx}, \&c. \quad U, \frac{dU}{dx}\right) = 0.$$

The equation

$$L = 0$$

being supposed to hold for all forms of the function y, it is clear that we must have

$$\delta L = 0,$$

which, if we put as before,

$$\alpha = \frac{dL}{dy}, \quad \beta = \frac{dL}{d \cdot \frac{dy}{dx}}, \quad \&c.$$

and

$$N' = \frac{dL}{dU}, \quad P' = \frac{dL}{d \cdot \frac{dU}{dx}},$$

will become

$$\alpha \delta y + \beta \frac{d\delta y}{dx} + \&c. + N'\delta U + P' \frac{d\delta U}{dx} = 0;$$

or, assuming, for the sake of brevity,

$$\delta u = \alpha \delta y + \beta \frac{d\delta y}{dx} + \&c.$$

$$N'\delta U + P' \frac{d\delta U}{dx} + \delta u = 0.$$

Multiply this equation by an indeterminate factor λ, and integrate by parts. This will give

$$\lambda P'\delta U + \int\left(N'\lambda - \frac{d.P'\lambda}{dx}\right)\delta U + \int\lambda\delta u = 0.$$

As we may always suppose the equation $L = 0$ to have been solved for $\frac{dU}{dx}$, it is evident that we may suppose $P' = 1$; and as the factor λ is indeterminate, it may be assumed such as to satisfy the equation

$$N'\lambda - \frac{d\lambda}{dx} = 0,$$

giving

$$\lambda = e^{\int N'dx}.$$

This reduces the equation to

$$\delta U . e^{\int N'dx} = -\int e^{\int N'dx}\,\delta u\,;$$

or, finally,

$$\delta U = -\,e^{-\int N'dx}\int e^{\int N'dx}\,\delta u.$$

We have thus δU expressed in terms of δy, $\frac{d\delta y}{dx}$, &c., combined with x, y, $\frac{dy}{dx}$, &c., U. But the form in which this variation is here found renders it comparatively useless, and although we may derive from this expression the value of

$$\delta . \int U dx,$$

it will be always better to investigate this directly by the method of Prop. VI.

ON SUCCESSIVE VARIATION.

19. Hitherto we have made no particular hypothesis as to the *invariability* of the form of the function ψ or δy. And the conclusions at which we have arrived would hold equally whether we suppose that function to be of a constant or of a variable form.

Thus, for example, if the symbol of derivation, F, satisfy the condition

$$F.\phi + F.\phi_1 = F.(\phi + \phi_1),$$

it is equally true that

$$\delta F.\phi = F.\delta\phi = F.i\psi,$$

whether it be supposed that the form of the function ψ does or does not change.

But this circumstance is no longer indifferent when we come to consider the *second* variation, i. e. the variation of the variation. Thus, in the example just given, we should have, in general,

$$\delta^2 F.\phi = F.\delta^2\phi = F.i\delta\psi.$$

If, therefore, the form of the function ψ were invariable, this would become

$$\delta^2 F.\phi = F.0 = 0.^*$$

This we shall, in general, suppose to be the case, and shall, therefore, define a primitive function to be one whose variation is of arbitrary, but *invariable* form. In other words, if u be a primitive indeterminate function of any number of variables, we shall suppose that the variation δu is such as to satisfy the equation

$$\delta^2 u = 0.$$

This completes the analogy between a primitive function and an independent variable.

PROP. IX.

20. To find the second variation of the differential coefficient $\frac{d^n y}{dx^n}$.

We have seen already (Prop. I.) that

$$\delta.\frac{d^n y}{dx^n} = \frac{d^n\delta y}{dx^n}.$$

* It is evident, from the general equation of condition, that $F.0 = 0$. For

$$F.\phi = F.(\phi + 0) = F.\phi + F.0.$$

Hence

$$F.0 = F.\phi - F.\phi = 0.$$

Hence it is plain that

$$\delta^2 \frac{d^n y}{dx^n} = \frac{d^n \delta^2 y}{dx^n}.$$

But since y is a primitive function,

$$\delta^2 y = 0,$$

and, therefore,

$$\frac{d^n \delta^2 y}{dx^n} = 0.$$

Hence

$$\delta^2 . \frac{d^n y}{dx^n} = 0.$$

Prop. X.

21. To find the second variation of V, where

$$V = f\left(x, y, \frac{dy}{dx} \dots \frac{d^n y}{dx^n}\right).$$

We have seen (Prop. II.) that

$$\delta V = \frac{dV}{dy}\delta y + \frac{dV}{d . \frac{dy}{dx}} . \frac{d\delta y}{dx} + \&c. + \frac{dV}{d . \frac{d^n y}{dx^n}} . \frac{d^n \delta y}{dx^n};$$

hence

$$\delta^2 V = \delta\left(\frac{dV}{dy}\delta y\right) + \delta\left(\frac{dV}{d . \frac{dy}{dx}} . \frac{d\delta y}{dx}\right) + \&c.$$

But since

$$\delta^2 y = 0, \quad \frac{d\delta^2 y}{dx} = 0, \quad \&c.$$

$$\delta\left(\frac{dV}{dy}\delta y\right) = \delta y \delta . \frac{dV}{dy}.$$

Determining the value of $\delta . \frac{dV}{dy}$ in the same way as δV, we find

$$\delta . \frac{dV}{dy} = \frac{d^2 V}{dy^2}\delta y + \frac{d^2 V}{dy d . \frac{dy}{dx}} \frac{d\delta y}{dx} + \&c. + \frac{d^2 V}{dy d . \frac{d^n y}{dx^n}} \frac{d^n \delta y}{dx^n}.$$

Similarly

$$\delta\left(\frac{dV}{d\,.\,\frac{dy}{dx}}\frac{d\delta y}{dx}\right)=\frac{d\delta y}{dx}\,\delta\,.\,\frac{dV}{d\,.\,\frac{dy}{dx}},$$

and

$$\delta\,.\,\frac{dV}{d\,.\,\frac{dy}{dx}}=\frac{d^2V}{dyd\,.\,\frac{dy}{dx}}\delta y+\frac{d^2V}{\left(d\,.\,\frac{dy}{dx}\right)^2}\left(\frac{d\delta y}{dx}\right)^2+\&c.$$

$$\&c. = \&c.$$

Hence we find ultimately,

$$\delta^2V=\frac{d^2V}{dy^2}\delta y^2+2\frac{d^2V}{dyd\,.\,\frac{dy}{dx}}\delta y\frac{d\delta y}{dx}+\frac{d^2V}{\left(d\,.\,\frac{dy}{dx}\right)^2}\cdot\left(\frac{d\delta y}{dx}\right)^2+\&c.$$

Prop. XI.

22. To find the second variation of

$$\int Vdx,\ \text{where}\ V=f\left(x,y,\frac{dy}{dx}\dots\frac{d^ny}{dx^n}\right).$$

It has been already shown (Prop. III.) that

$$\delta\int Vdx=\int\delta Vdx;$$

and for the same reason it is evident that

$$\delta^2\int Vdx=\int\delta^2Vdx.$$

Substituting for δ^2V its value as found in the foregoing proposition, we have

$$\delta^2\int Vdx$$

$$=\int\left\{\frac{d^2V}{dy^2}\delta y^2+2\frac{d^2V}{dyd\,.\,\frac{dy}{dx}}\delta y\frac{d\delta y}{dx}+\frac{d^2V}{\left(d\,.\,\frac{dy}{dx}\right)^2}\cdot\left(\frac{d\delta y}{dx}\right)^2+\&c.\right\}dx.$$

It is easy to see that a similar method will give third, fourth, &c. variations; but it is unnecessary, for any practical purpose, to pursue this discussion any further.

CHAPTER III.

MAXIMA AND MINIMA OF INDETERMINATE FUNCTIONS OF ONE INDEPENDENT VARIABLE.

23. A *maximum* value of a function is one which exceeds any other value of that function which can be produced by an indefinitely small change in any of its varying elements.* Similarly, a *minimum* value is one which is less than any value which can be so produced.

In the mode of variation peculiar to the science with which we are at present concerned, a maximum or minimum may be defined to be a value of a derived function which exceeds or falls short of all other values which can be produced by an indefinitely small change in the form of its primitive.

The problem of maxima and minima, solved by the Differential Calculus, is, as is well known, as follows:

Let x be an independent variable, and $u\ \{=f(x)\}$ a function of that variable; find what *value* of x will render u a maximum or minimum.

The corresponding problem which the Calculus of Variations proposes to solve is this:

Let ϕ be a primitive indeterminate function, and let $u\ (=F.\phi)$ be a function derived therefrom; find what *form* of ϕ will render u a maximum or minimum.†

The general method of solution is as follows:—Substitute in the derived function u or $F.\phi$, $\phi + i\psi$ for ϕ. Let $F.(\phi + i\psi)$ be

* This definition is given in order to prevent the common mistake, that maximum and minimum values of functions are *the* greatest and *the* least values which those functions can have.

† I am aware that this problem (and indeed the science generally) is commonly defined with special reference to the case of definite integrals. But in stating the principles of a science it seems proper to give them all the generality of which they are susceptible, even though it be impossible, in the present state of mathematical knowledge, to give equally general applications of them.

expanded in powers of i, and it will appear by reasoning precisely the same as that employed in the Differential Calculus, that, if ϕ be a function of such a form as to render $F.\phi$ a maximum or minimum, the coefficient of i in the expansion must vanish, and that of i^2 preserve the same sign (negative for a maximum, and positive for a minimum) for all forms of the function ψ which the conditions of the question admit of. In other words, if the form only of ϕ be varied, we must have $\delta u = 0$; and if both form and independent variables be varied, we must have $Du = 0$.

We shall proceed to apply this theory to the case of functions derived by the processes of differentiation and integration.

Prop. I.

24. Let y be an indeterminate function of the single independent variable x, and let it be required to find what form of the function y will render

$$f\left(x, y, \frac{dy}{dx}, \frac{d^2y}{dx^2} \dots\right)$$

a maximum or minimum.

Assume

$$u = f\left(x, y, \frac{dy}{dx}, \frac{d^2y}{dx^2} \dots\right),$$

and let

$$du = Mdx + Ndy + P_1\, d\,.\frac{dy}{dx} + P_2\, d\,.\frac{d^2y}{dx^2} + \&c.$$

then (Chap. II. Prop. II.)

$$\delta u = N\delta y + P_1 \frac{d\delta y}{dx} + P_2 \frac{d^2\,\delta y}{dx^2} + \&c.$$

Now, if the form of the function y be such as to render u a maximum or minimum,

$$\delta u = 0,$$

or

$$N\delta y + P_1 \frac{d\delta y}{dx} + P_2 \frac{d^2\delta y}{dx^2} + \&c. = 0.$$

This equation it is manifestly impossible, in general, to satisfy, without destroying the independence of the form of the function ψ or δy, for unless the quantities N, P_1, P_2, &c., be separately zero, the equation

$$N\delta y + P_1 \frac{d\delta y}{dx} + P_2 \frac{d^2 \delta y}{dx^2} + \&c. = 0$$

will establish a relation between the form of the function δy or ψ, and the form of the function y or ϕ. Such a relation would, of course, render the entire proceeding nugatory. Nor is it in general possible to satisfy the equations

$$N = 0, \quad P_1 = 0, \quad \&c.,$$

inasmuch as each of these equations establishes a general relation between y and x, or, in other words, determines the form of the function y. Unless then it should so happen, that the forms of the function y, as determined by all these equations, should agree, it is plain that the equations

$$N = 0, \quad P_1 = 0, \quad \&c.,$$

cannot be satisfied; and as this will not happen generally, it is evident that the problem does not ordinarily admit of being solved.

If the function u contain but one of the quantities

$$y, \; \frac{dy}{dx}, \; \frac{d^2y}{dx^2}, \; \&c.,$$

or if, by the conditions of the question, the values of all but one of these quantities be fixed for each value of x, the equation

$$N\delta y + P_1 \frac{d\delta y}{dx} + \&c. = 0$$

will be reduced to a single term, and may therefore be satisfied. Thus, for example, let

$$u = f\left(x, \; y, \; \frac{dy}{dx}\right),$$

and let it be required to determine, among the forms of the function y which render that function of a given value for a given value of x, that form which renders u a maximum or minimum. In this case the equation

$$N\delta y + P_1 \frac{d\delta y}{dx} + \&c. = 0$$

is reduced to $P_1 = 0$, and may therefore be satisfied.

This will, perhaps, be most easily understood when stated geometrically. In that form the problem is as follows:

To determine a curve such that at every point the function

$$f\left(x,\ y,\ \frac{dy}{dx}\right)$$

may be a maximum or minimum, it being understood that the curve so found is, at each of its points, to be compared only with curves which pass through that point.

25. As an example of this case, let it be required to find a curve such that if at any point T (Fig. 1) a tangent be drawn, and produced to cut two given ordinates, CM, $C'M'$, the product $CM \,.\, C'M'$, may be a maximum or minimum.

Let O be the origin, and assume

$$a = OM, \quad a' = OM'.$$

We have, then,

$$CM = y + \frac{dy}{dx}(a - x),$$

$$C'M' = y + \frac{dy}{dx}(a' - x),$$

and, therefore,

$$V = CM \,.\, C'M' = \left(y + \frac{dy}{dx}(a - x)\right)\left(y + \frac{dy}{dx}(a' - x)\right).$$

Hence

$$\delta V = \left(2y + (a + a' - 2x)\frac{dy}{dx}\right)\delta y$$

$$+ \left\{\left(y + \frac{dy}{dx}(a - x)\right)(a' - x) + \left(y + \frac{dy}{dx}(a' - x)\right)(a - x)\right\}\frac{d\delta y}{dx}.$$

Now if it be one of the conditions of the problem that the curve is at each of its points to be compared only with curves passing through that point, it is evident that we must have

$$\delta y = 0,$$

which reduces the equation

$$\delta V = 0$$

to

$$\left(y + \frac{dy}{dx}(a - x)\right)(a' - x) + \left(y + \frac{dy}{dx}(a' - x)\right)(a - x) = 0\,;$$

or, by reduction,

$$2\frac{dy}{y} + \frac{dx}{a - x} + \frac{dx}{a' - x} = 0.$$

Integrating this, we find

$$y^2 = \pm e^2 (a - x)(a' - x), \qquad \text{(A)}$$

e^2 being an arbitrary constant. The curve is therefore an hyperbola or an ellipse, according as we use the upper or lower sign.

Let us next consider the second variation, $\delta^2 V$.

We have already seen (Art. 21) that, in general,

$$\delta^2 V = \frac{d^2 V}{dy^2}\delta y^2 + 2\frac{d^2 V}{dyd\,.\,\frac{dy}{dx}}\delta y\frac{d\delta y}{dx} + \frac{d^2 V}{\left(d\,.\,\frac{dy}{dx}\right)^2}\,.\left(\frac{d\delta y}{dx}\right)^2 + \&c.$$

Hence, in the present case, where

$$V = f\left(x, y, \frac{dy}{dx}\right),$$

and δy disappears by the conditions of the question,

$$\delta^2 V = \frac{d^2 V}{\left(d\,.\,\frac{dy}{dx}\right)^2}\left(\frac{d\delta y}{dx}\right)^2 = 2(a - x)(a' - x)\left(\frac{d\delta y}{dx}\right)^2.$$

Substituting for $2(a - x)(a' - x)$ its value, given by equation (A), we have

$$\delta^2 V = \pm\frac{2y^2}{e^2}\left(\frac{d\delta y}{dx}\right)^2.$$

The maximum value, corresponding to the negative sign, belongs therefore to the ellipse; and the minimum, corresponding to the positive sign, to the hyperbola. It is plain also that in the former case the curve lies wholly within the lines *CM*, *C'M'*, and in the latter wholly without them.

Prop. II.

26. Let

$$U = \int_{x_0}^{x_1} V dx, \text{ where } V = f\left(x, y, \frac{dy}{dx} \dots \frac{d^n y}{dx^n}\right),$$

and let it be required to determine the form of the function y, and the values of the limits x_0 and x_1, which will render U a maximum or minimum.

In this case we have (Chap. II. Prop. III.)

$$\begin{aligned} DU &= V_1 dx_1 - V_0 dx_0 \\ &+ \left(P_1 - \frac{dP_2}{dx} + \&c.\right)_1 \delta y_1 - \left(P_1 - \frac{dP_2}{dx} + \&c.\right)_0 \delta y_0 \\ &+ (P_2 - \&c.) \left(\frac{d\delta y}{dx}\right)_1 - (P_2 - \&c.) \left(\frac{d\delta y}{dx}\right)_0 \qquad \text{(A)} \\ &+ \&c. \\ &+ \int_{x_0}^{x_1} \left(N - \frac{dP_1}{dx} + \frac{d^2 P_2}{dx^2} - \&c. + (-1)^n \frac{d^n P_n}{dx^n}\right) \delta y \,.\, dx = 0. \end{aligned}$$

In seeking to determine the method of satisfying this equation, it is, in the first place, to be observed, that we may have to deal with one of two different cases:—1. The variation δy, or the form of the function ψ, may be wholly unrestricted (except by the considerations in Prop. II. of the preceding chapter). 2. The problem may be such as to render it necessary to assume only those forms of the function ψ which satisfy a certain condition or conditions.

In the first case we seek to determine, among all possible functions, that one which will render U a maximum or minimum. In the second case, we seek to determine the required function, not from among all possible functions, but from a certain class,* *sc.* from among those which satisfy certain conditions. Maxima and minima belonging to the first class are termed *absolute*, while those belonging to the second class are denominated *relative*. Problems involving maxima and minima of the latter species are

* This class may, of course, contain an infinite number of functions.

also frequently called (from a remarkable class among them) *isoperimetrical.*

We shall first consider the case of absolute maxima and minima.

27. Assume

$$a_1 = V_1 dx_1 + \left(P_1 - \frac{dP_2}{dx} + \&c.\right)_1 \delta y_1 + (P_2 - \&c.)\left(\frac{d\delta y}{dx}\right)_1 + \&c.$$

and

$$a_0 = V_0 dx_0 + \left(P_1 - \frac{dP_2}{dx} + \&c.\right)_0 \delta y_0 + \&c.$$

also

$$\beta = N - \frac{dP_1}{dx} + \frac{d^2P_2}{dx^2} - \&c. + (-1)^n \frac{d^nP_n}{dx^n}.$$

Equation (A) becomes then

$$a_1 - a_0 + \int_{x_0}^{x_1} \beta\delta y dx = 0.$$

Now this equation cannot be satisfied, without restricting the generality of δy or ψ, in any other way than by making

$$a_1 - a_0 = 0, \quad \beta = 0.$$

For if $a_1 - a_0$ be not $= 0$ we must have

$$\int_{x_0}^{x_1} \beta\delta y dx = a_0 - a_1,$$

an equation which, as is easily seen, implies that the integral of an arbitrary function may be expressed (without determining, or even restricting its general form) in terms of the limiting values of itself and a certain number of its differential coefficients. This is manifestly untrue. We have, therefore,

$$a_1 - a_0 = 0, \int_{x_0}^{x_1} \beta\delta y dx = 0.$$

Now it is known that, as the result of the process of integration depends in general on the form of the function to be integrated, it is impossible to determine the value of an expression such as $\int_{x_0}^{x_1} \beta\delta y dx$ without, at least, restricting* the generality of the func-

* I say *restricting*, because there are many cases of definite integrals, whose value can be determined without an absolute *determination* of the form of the function to be integrated.

tion δy. Hence it is evident that it is impossible to satisfy the equation

$$\int_{x_0}^{x_1} \beta \delta y dx = 0,$$

without either restricting the generality of the form of δy or making $\beta = 0$. As the former supposition is inadmissible, it is plain that we must have $\beta = 0$, or

$$N - \frac{dP_1}{dx} + \frac{d^2 P_2}{dx^2} - \&c. + (-1)^n \frac{d^n P_n}{dx^n} = 0,$$

a differential equation, by which the form of the function y is determined.

The equations

$$a_1 - a_0 = 0, \text{ and } \beta = 0,$$

differ essentially from one another, the second establishing a *general* relation between y and x, while the former is concerned merely with the values which these quantities have at the limits of integration. If this were not so, the solution of the problem would be impossible, as we could not have more than one *general* relation between y and x. The coefficients of the several increments in the equation

$$a_1 - a_0 = 0,$$

or

$$V_1 dx_1 - V_0 dx_0 + \left(P_1 - \frac{dP_2}{dx} + \&c.\right)_1 \delta y_1 - \left(P_1 - \frac{dP_2}{dx} + \&c.\right)_0 \delta y_0$$

$$+ (P_2 - \&c.)_1 \left(\frac{d\delta y}{dx}\right)_1 - (P_1 - \&c.) \left(\frac{d\delta y}{dx}\right)_0 \qquad \text{(B)}$$

$$+ \&c.$$

$$+ \left(P_n \frac{d^{n-1}\delta y}{dx^{n-1}}\right)_1 - \left(P_n \frac{d^{n-1}\delta y}{dx^{n-1}}\right)_0 = 0,$$

being constants, it is plain that this equation, if capable of being satisfied, can be satisfied only by means of the arbitrary constants which enter into the solution of the equation

$$N - \frac{dP_1}{dx} + \frac{d^2 P_2}{dx^2} - \&c. + (-1)^n \frac{d^n P_n}{dx^n} = 0.$$

This is, in general, a differential equation of the order $2n$, for since V, and, therefore, *in general*, P_n, contains $\frac{d^n y}{dx^n}$, it is plain that $\frac{d^n P_n}{dx^n}$ will contain $\frac{d^{2n} y}{dx^{2n}}$. The solution of this equation will, therefore, involve $2n$ arbitrary constants. Now if the limiting values of

$$x, y, \frac{dy}{dx} \dots \frac{d^{n-1}y}{dx^{n-1}}$$

be completely unrestricted, the equation

$$a_1 - a_0 = 0$$

will contain $2(n+1)$ arbitrary increments, viz.:

$$dx_1, \delta y_1, \delta . \left(\frac{dy}{dx}\right)_1 \dots \delta . \left(\frac{d^{n-1}y}{dx^{n-1}}\right)_1, \quad dx_0, \delta y_0, \dots \delta . \left(\frac{d^{n-1}y}{dx^{n-1}}\right)_0 .$$

In such a case it would be impossible to satisfy the equation

$$a_1 - a_0 = 0,$$

inasmuch as it would be necessary to equate to zero the coefficient of each of these arbitrary increments, which would furnish $2(n+1)$ equations between $2n$ quantities; and this was to be expected, for if there were no restriction on either the form of the function y or the limits of integration, it is easy to see that the integral might be made to have any value from zero to infinity, and would, therefore, be incapable of either maximum or minimum. The nature of the restriction upon the limits will, of course, be determined in each particular case by the conditions of the problem to be solved.

(1.) Let the limiting values of x, *sc.* x_0 and x_1, be given, i. e. let the question be, to determine such a form of the function y as will render $\int V dx$, when taken between *fixed* limits, a maximum or minimum.

In this case $dx_1 = 0$, $dx_0 = 0$, and the equation

$$a_1 - a_0 = 0$$

is equivalent to

$$\left(P_1 - \frac{dP_2}{dx} + \&c.\right)_1 = 0, \quad \left(P_1 - \frac{dP_2}{dx} + \&c.\right)_0 = 0,$$
$$(P_2 - \&c.)_1 = 0, \quad (P_2 - \&c.)_0 = 0, \qquad \text{(C)}$$
$$\&c. = 0, \quad \&c. = 0,$$
$$(P_n)_1 = 0, \quad (P_n)_0 = 0.$$

Now it is plain that the number of these equations ($2n$) is precisely the same as the number of arbitrary constants in the general solution of the equation

$$\beta = 0,$$

and that, therefore, the solution of the question is in this case complete.

(2.) Let the limiting values of both x and y be given.

In this case $dx_1 = 0$, $\delta y_1 = 0$, $dx_0 = 0$, $\delta y_0 = 0$; and the equation

$$a_1 - a_0 = 0$$

is equivalent to but $2n - 2$ equations, those, namely, which are formed by equating to zero the coefficients of the several variations,

$$\delta.\left(\frac{dy}{dx}\right)_1, \ \delta.\left(\frac{dy}{dx}\right)_0, \ \delta.\left(\frac{d^2y}{dx^2}\right) \ldots\ldots \delta.\left(\frac{d^{n-1}y}{dx^{n-1}}\right)_1, \ \delta.\left(\frac{d^{n-1}y}{dx^{n-1}}\right)_0.$$

But as two additional equations are furnished by the substitution, in the general solution, of the given limiting values of x and y, the number of equations remains still $2n$. Thus, if the integral of the equation

$$N - \frac{dP_1}{dx} + \frac{d^2P_2}{dx^2} - \&c. + (-1)^n \frac{d^nP_n}{dx^n} = 0$$

were

$$f(x, y, c_1, c_2 \ldots\ldots c_{2n}) = 0,$$

c_1, c_2, &c. being arbitrary constants, we should have for the determination of these constants the $2n$ equations,

$$f(x_1, y_1, c_1, c_2 \ldots\ldots c_{2n}) = 0, \quad f(x_0, y_0, c_1, c_2 \ldots\ldots c_{2n}) = 0,$$
$$\left(P_2 - \frac{dP_3}{dx} + \&c.\right)_1 = 0, \quad \left(P_2 - \frac{dP_3}{dx} + \&c.\right)_0 = 0,$$
$$(P_3 - \&c.)_1 = 0, \quad (P_3 - \&c.)_0 = 0, \qquad \text{(D)}$$
$$\&c. \quad \&c.$$
$$(P_n)_1 = 0, \quad (P_n)_0 = 0.$$

(3.) Similarly, if the limiting values of x, y, $\frac{dy}{dx}$ were given, the new datum would remove two equations from (D), namely,

$$\left(P_2 - \frac{dP_3}{dx} + \&c.\right)_1 = 0, \quad \left(P_2 - \frac{dP_3}{dx} + \&c.\right)_0 = 0,$$

But at the same time two new equations would be derived from the given limiting values of $\frac{dy}{dx}$, as follows:

Let the equation

$$f(x, y, c_1, c_2 \ldots . c_{2n}) = 0,$$

be differentiated with regard to x, and let the result be expressed by

$$f'\left(x, y, \frac{dy}{dx}, c_1, c_2 \ldots . c_{2n}\right) = 0,$$

then the two new equations will be

$$f'\left(x_1, y_1, \left(\frac{dy}{dx}\right)_1, c_1, c_2 \ldots . c_{2n}\right) = 0,$$

$$f'\left(x_0, y_0, \left(\frac{dy}{dx}\right)_0, c_1, c_2 \ldots . c_{2n}\right) = 0,$$

$\left(\frac{dy}{dx}\right)_1$ and $\left(\frac{dy}{dx}\right)_0$ being the given limiting values of $\frac{dy}{dx}$.

In the same way, if the limiting values of $\frac{d^2y}{dx^2}$ were given, the number of the equations (D) would be further diminished by two, and at the same time two new equations would be formed by differentiating twice the equation

$$f(x, y, c_1, c_2 \ldots . c_{2n}) = 0,$$

and substituting in the result the given limiting values of

$$x, y, \frac{dy}{dx}, \frac{d^2y}{dx^2}.$$

In all these cases, then, the number of equations is $2n$, the same as that of the arbitrary constants; and the same would be true if the limiting values of any number of the quantities

$$y, \frac{dy}{dx}, \frac{d^2y}{dx^2}, \frac{d^3y}{dx^3} \cdots \frac{d^{n-1}y}{dx^{n-1}},$$

were given. For it is evident that, by fixing the limiting values of any number of these quantities, as many new equations would be obtained as are removed from the equation

$$a_1 - a_0 = 0.$$

(4.) Instead of supposing the limiting values of x, y, $\frac{dy}{dx}$, &c. to be fixed, let any two or more of these limiting values be connected by a fixed equation or equations.

In this case it is evident that the variations of the quantities so connected would be no longer independent, and, therefore, that two or more of the terms of the equation

$$a_1 - a_0 = 0$$

would be reduced to one. This would, of course, reduce the number of equations to which that equation is equivalent, and, at the same time, introduce a sufficient number of new equations to make up the deficiency. Thus, for example, if the limiting values of y and x were connected by the determinate equations

$$y_1 = f_1(x_1), \quad y_0 = f_0(x_0),$$

we should have between δy_1, dx_1, δy_0, dx_0, the equations

$$\left(\frac{dy}{dx}\right)_1 dx_1 + \delta y_1 = f'_1(x_1) \,.\, dx_1, \quad \left(\frac{dy}{dx}\right)_0 dx_0 + \delta y_0 = f'_0(x_0) \,.\, dx_0.$$

If the values of δy_1, δy_0 derived from these equations be substituted in

$$a_1 - a_0 = 0\,;$$

and if the coefficients of the independent variations be equated to zero, it will be found that that equation is equivalent to the following:

$$V_1 + \left(P_1 - \frac{dP_2}{dx} + \&c.\right)_1 \left\{f'_1(x_1) - \left(\frac{dy}{dx}\right)_1\right\} = 0,$$

$$V_0 + \left(P_1 - \frac{dP_2}{dx} + \&c.\right)_0 \left\{f'_0(x_0) - \left(\frac{dy}{dx}\right)_0\right\} = 0,$$

$$(P_2 - \&c.)_1 = 0, \qquad (P_2 - \&c.)_0 = 0,$$

&c. &c.

the remaining equations being the same as before. These equations, which are in number $2n$, are, with the four equations,

$$y_1 = f_1(x_1), \qquad y_0 = f_0(x_0),$$
$$f(x_1, y_1, c_1, c_2 \ldots . c_{2n}) = 0, \quad f(x_0, y_0, c_1, c_2 \ldots . c_{2n}) = 0,$$

sufficient to determine the $2n + 4$ quantities,

$$x_1, y_1, x_0, y_0, c_1, c_2 \ldots . c_{2n}.$$

Similarly, if the limiting values of x and $\frac{dy}{dx}$ were connected by the equations

$$\left(\frac{dy}{dx}\right)_1 = f'_1(x_1), \quad \left(\frac{dy}{dx}\right)_0 = f'_0(x_0),$$

we should have

$$\left(\frac{d^2y}{dx^2}\right)_1 dx_1 + \delta . \left(\frac{dy}{dx}\right)_1 = f''_1(x_1) . dx_1, \quad \left(\frac{d^2y}{dx^2}\right)_0 dx_0 + \delta . \left(\frac{dy}{dx}\right)_0 = \&c.$$

hence the first three terms of each of the quantities a_1, a_0 would be reduced to one, and consequently the number of equations furnished by $a_1 - a_0 = 0$ to $2n - 2$.

To determine the $2n + 4$ quantities, y_1, &c., c_1, &c., c_n, there are, besides the $2n - 2$ equations just mentioned, and the four which occurred in the first part of this case, two others, namely,

$$f'(x_1, y_1, f'_1(x_1), c_1 \ldots . c_n) = 0, \quad f'(x_0, y_0, f'_0(x_0), c_1 \ldots . c_n) = 0,$$

formed by differentiating the general solution

$$f(x, y, c_1 \ldots . c_n) = 0,$$

and substituting in the result the limiting values of

$$x, y, \frac{dy}{dx}.$$

The same reasoning will show that in all cases in which the conditions of the question, by annulling or restricting any of the independent variations,

$$\delta y_1, \delta \left(\frac{dy}{dx}\right)_1, \&c.,$$

diminish the number of equations furnished by $a_1 - a_0 = 0$, they, at the same time, introduce the same number of new equations, thus leaving the total amount unchanged. Hence it appears that the equations furnished by the condition $DU = 0$ are, in general, necessary and sufficient for the solution of the problem, i. e. for the complete determination of the form of the function y.

28. Before giving any examples of the foregoing theory, it may be well to notice some exceptions to which it is liable.

(1.) If V be a linear function of the highest differential coefficient which it contains, it is manifest that P_n will not contain $\frac{d^n y}{dx^n}$, and, therefore, that $\frac{d^n P_n}{dx^n}$ will be, at most, of the order $2n - 1$.

Therefore the equation

$$N - \frac{dP_1}{dx} + \&c. + (-1)^n \frac{d^n P_n}{dx^n} = 0$$

cannot be of an order higher than $2n - 1$, and its solution will only contain $2n - 1$ disposable constants. In this case, then, the equation

$$a_1 - a_0 = 0,$$

which is, as we have seen, equivalent to $2n$ equations, cannot, in general, be satisfied.

It is, indeed, easy to show that if V be a linear function of the highest differential coefficient which it contains, the equation

$$\beta = 0$$

can never rise above the order $2n - 2$. To prove this, let

$$\frac{d^n y}{dx^n} = v,$$

then $V = \theta v + \theta'$, where θ and θ' are functions of

$$x, y, \frac{dy}{dx} \dots \frac{d^{n-1}y}{dx^{n-1}}.$$

As we have seen already that the equation

$$\beta = 0$$

does not contain $\frac{d^{2n}y}{dx^{2n}}$, it only remains to show that it does not contain $\frac{d^{2n-1}y}{dx^{2n-1}}$. Now this coefficient, if it occur at all, can be found only in one of the terms $\frac{d^{n-1}P_{n-1}}{dx^{n-1}}$, $\frac{d^n P_n}{dx^n}$. But since

$$V = \theta v + \theta', \quad P_n = \theta, \quad \frac{d^n P_n}{dx^n} = \frac{d^n \theta}{dx^n}.$$

In order, then, to find the coefficient of $\frac{d^{2n-1}y}{dx^{2n-1}}$ in $\frac{d^n P_n}{dx^n}$ or in $\frac{d^n\theta}{dx^n}$, it will be necessary to form the values of $\frac{d\theta}{dx}, \frac{d^2\theta}{dx^2}$, &c., $\frac{d^n\theta}{dx^n}$, rejecting at each step every term except that of the highest order. Assuming $u = \frac{d^{n-1}y}{dx^{n-1}}$, we have

$$\frac{d\theta}{dx} = \left(\frac{d\theta}{dx}\right) + \left(\frac{d\theta}{dy}\right) \cdot \frac{dy}{dx} + \&c. + \left(\frac{d\theta}{du}\right) \cdot \frac{du}{dx}.$$

In this expression the term $\left(\frac{d\theta}{du}\right) \cdot \frac{du}{dx}$, or $\left(\frac{d\theta}{du}\right) \cdot \frac{d^n y}{dx^n}$, is the only one to be retained, as it is evident that all the other terms are of an order less than n.

Similarly in $\frac{d^2\theta}{dx^2}$ the only term of the order $n+1$ is $\left(\frac{d\theta}{du}\right) \cdot \frac{d^2u}{dx^2}$, and in the same way it can be shown that the only term in $\frac{d^n\theta}{dx^n}$, which is of the order $2n-1$, is $\left(\frac{d\theta}{du}\right) \cdot \frac{d^n u}{dx^n}$, or $\left(\frac{d\theta}{du}\right) \cdot \frac{d^{2n-1}y}{dx^{2n-1}}$.

But since

$$V = \theta v + \theta', \quad P_{n-1} = v \cdot \left(\frac{d\theta}{du}\right) + \left(\frac{d\theta'}{du}\right),$$

if then we proceed as before to form the values of

$$\frac{dP_{n-1}}{dx}, \frac{d^2 P_{n-1}}{dx^2} \cdots\cdots \frac{d^{n-1} P_{n-1}}{dx^{n-1}},$$

retaining after each differentiation only the term of the highest order, it will appear, as before, that the only term of the order $2n-1$ in $\frac{d^{n-1}P_{n-1}}{dx^{n-1}}$ is $\left(\frac{d\theta}{du}\right) \cdot \frac{d^{n-1}v}{dx^{n-1}}$, or $\left(\frac{d\theta}{du}\right) \cdot \frac{d^{2n-1}y}{dx^{2n-1}}$.

As this result is the same with that found for the term of the highest order in $\frac{d^n P_n}{dx^n}$, it is evident that that term will disappear from

$$\frac{d^{n-1}P_{n-1}}{dx^{n-1}} - \frac{d^n P_n}{dx^n},$$

and therefore from the equation

$$\beta = 0.$$

Hence it is evident that this equation cannot be of an order higher than $2n - 2$.

This reduction of the equation

$$\beta = 0$$

to another whose order is less by *two* than that of the first, is not noticed by Lacroix, who merely shows* that when V is a linear function of the highest differential coefficient, this equation will be depressed by *one* order. It was first remarked by the illustrious Euler,† whose demonstration, however, as far as it is general, proves rather that such a reduction was to be expected, than that it actually takes place. His demonstration is nearly as follows:

Assume, as above,

$$u = \frac{d^{n-1}y}{dx^{n-1}}, \quad v = \frac{d^n y}{dx^n};$$

then, since V is a linear function of v, we must have

$$V = \theta v + \theta',$$

θ and θ' not containing differential coefficients of an order higher than $n - 1$. Hence

$$\int V dx = \int \theta v dx + \int \theta' dx.$$

Now it is easy to see that the integral $\int \theta v dx$ can be reduced to the form

$$\theta_1 + \int \theta_2 dx,$$

θ_1, θ_2 not containing any differential coefficient higher than u. For if we assume

$$\int \theta v dx = \theta_1 + \int \theta_2 dx,$$

and differentiate, we find

$$\theta v = \left(\frac{d\theta_1}{dx}\right) + \left(\frac{d\theta_1}{dy}\right) \cdot \frac{dy}{dx} + \&c. + \left(\frac{d\theta_1}{du}\right) \cdot \frac{du}{dx} + \theta_2.$$

* Traitè de Calc., tom. ii. p. 740.

† Meth. Inven., pp. 62, 75.

Hence as $v = \dfrac{du}{dx}$, it is evident that the equation

$$\int \theta v dx = \theta_1 + \int \theta_2 dx$$

is satisfied by making

$$\left(\frac{d\theta_1}{du}\right) = \theta, \text{ and } \theta_2 = -\left\{\left(\frac{d\theta_1}{dx}\right) + \left(\frac{d\theta_1}{dy}\right) \cdot \frac{dy}{dx} + \&c.\right\},$$

the first of which conditions shows that θ_1 is found by integrating θ with regard to u solely. From this it appears that when V is a linear function of the highest differential coefficient which it contains, $\int V dx$ can always be reduced to another, $\int V_1 dx$, in which V_1 contains no differential coefficient of an order higher than $n-1$. If then this reduction be made previous to the application of the Calculus of Variations, it is evident that the order of the differential equation arrived at will be $2n-2$.

This mode of demonstration is important, as showing that the case which we have been considering cannot properly be regarded in the light of an *exception* to the generality of the method. For, if the given integral be reduced to its lowest terms, previous to the application of the Calculus of Variations, it will assume the form

$$\phi + \int \phi' dx,$$

or, when taken between limits,

$$\phi_1 - \phi_0 + \int_{x_0}^{x_1} \phi' dx,$$

in which ϕ_1, ϕ_0 contain differential coefficients of an order equal to that of the highest coefficient contained in ϕ'. This case is considered in a subsequent proposition, and is evidently distinct from that which we have been considering in the present proposition, in which the quantity which is to be made a maximum or minimum contains no terms free from the sign of integration. Hence it appears that the existence of a case such as that which we have been here discussing merely denotes the necessity of reducing an integral to its lowest terms, previously to the application of the Calculus of Variations.

(2.) Another exception to this theory is as follows:

Let $V = yf(x) + F(x, p)$; then $N = f(x)$, and $P = \frac{dF}{dp}$. The equation

$$N - \frac{dP}{dx} = 0$$

becomes in this case immediately integrable, giving

$$P = \int f(x)\, dx + c = f_1(x) + c.$$

This equation, being solved for p, gives a result of the form

$$p = F(x, c),$$

and, therefore,

$$y = F_1(x, c) + c'.$$

Now let it be supposed that the limiting values of x are given, those of y remaining indeterminate. The equation

$$a_1 - a_0 = 0$$

is equivalent to

$$P_1 = 0, \quad P_0 = 0,$$

or

$$f_1(x_1) + c = 0, \quad f_1(x_0) + c = 0.$$

These equations, containing but one indeterminate quantity, c, cannot, of course, in general be satisfied, and, therefore, although the general solution contains the requisite number of constants, the problem in this shape does not admit of a solution. If, however, $f(x) = 0$, so that

$$V = F(x, p),$$

the two equations

$$P_0 = 0, \quad P_1 = 0,$$

become identical, and the problem becomes soluble, but indeterminate, containing one arbitrary constant.

The conclusion here arrived at is not peculiar to functions of the first order. If V be of such a form as to satisfy the equation

$$N = \frac{dV}{dy} = f(x),$$

and if the limiting values of x only are given, it is easy to see that the same reasoning will apply. The equation

$$\beta = 0$$

will be, as in the case of functions of the first order, immediately integrable, giving

$$P - \frac{dQ}{dx} + \&c. = f_1(x) + c;$$

and the first two equations furnished by $a_1 - a_0 = 0$, will be

$$f_1(x_1) + c = 0, \quad f_1(x_0) + c = 0,$$

which cannot be satisfied generally unless $f_1(x) = 0$, i. e. unless y disappear altogether from V. And in general let $\frac{d^p y}{dx^p}$ be the lowest differential coefficient contained in V. Then if V be such as to satisfy the equation

$$\frac{dV}{d\,.\,\frac{d^p y}{dx^p}} = f(x),$$

and if only the limiting values of x and of coefficients higher than the pth be given, similar reasoning will show that the problem does not admit of a solution.

(3.) Let $N = 0$, and let the limiting values of x only be given. In this case the equation

$$\beta = 0$$

becomes

$$\frac{dP_1}{dx} - \frac{d^2 P_2}{dx^2} + \&c. = 0;$$

or, integrating,

$$P_1 - \frac{dP_2}{dx} + \&c. = c.$$

Hence it is evident that the two conditions furnished by equating to zero the coefficients of δy_1, δy_0, namely,

$$\left(P_1 - \frac{dP_2}{dx} + \&c.\right)_1 = 0, \quad \left(P_1 - \frac{dP_2}{dx} + \&c.\right)_0 = 0,$$

are equivalent to but one, namely, $c = 0$, and that, therefore, the equation

$$a_1 - a_0 = 0$$

is equivalent to but $2n - 1$ equations, instead of $2n$, as before. The problem is, therefore, in this case, indeterminate.

This might have been anticipated, for as y does not enter either into the function V, or into the conditions which are supposed to be fulfilled at the limits, we might have taken $\frac{dy}{dx}$ for the primitive function, and it would then have required all the equations which $DU=0$ affords, to determine an equation without arbitrary constants between x and $\frac{dy}{dx}$. The relation between x and y, derived by integration from that between x and $\frac{dy}{dx}$, will, therefore, contain one arbitrary constant. But if either of the limiting values of y were given, the problem would become determinate as before. Similarly, if $N=0$, $P_1=0$, and if the limiting values of y and $\frac{dy}{dx}$ be both indeterminate, the solution will involve two arbitrary constants, and would be rendered determinate only by fixing at least one of the limiting values for both y and $\frac{dy}{dx}$. And, in general, if the first m terms of the equation

$$N-\frac{dP_1}{dx}+\&c.=0$$

disappear, and if no limitation of the extreme values of y, $\frac{dy}{dx}$ . . . $\frac{d^{m-1}y}{dx^{m-1}}$ be given, the solution will involve m arbitrary constants.

29. As in certain cases the equation

$$N-\frac{dP_1}{dx}+\&c.=0$$

admits of being integrated one or more times without a determination of the form of the function V, we shall proceed to consider some of these cases, inasmuch as, by so doing, we shall be enabled to establish some general equations, which will greatly facilitate the application of the foregoing theory to particular examples. These cases may be arranged under two general classes, one of which has been already alluded to.

(1.) Let the first m of the quantities $y, \frac{dy}{dx}, \frac{d^2y}{dx^2} \dots$ be wanting in V, i. e. let

$$V=f\left(x, \frac{d^{m+1}y}{dx^{m+1}} \dots \frac{d^ny}{dx^n}\right).$$

In this case the first m terms of the equation

$$N - \frac{dP_1}{dx} + \&c. = 0$$

are wanting, and that equation becomes

$$\frac{d^m P_m}{dx^m} - \frac{d^{m+1} P_{m+1}}{dx^{m+1}} + \&c. = 0;$$

which equation, being integrated m times, becomes

$$P_m - \frac{dP_{m+1}}{dx} + \&c. = c_0 + c_1 x + c_2 x^2 + \&c. + c_{m-1} x^{m-1},$$

a differential equation of the order $2n - m$.

(2.) Let

$$V = f\left(y, \frac{dy}{dx}, \frac{d^2y}{dx^2} \cdots \frac{d^ny}{dx^n}\right),$$

not containing the independent variable.

In this case

$$dV = Ndy + P_1 d.\frac{dy}{dx} + P_2 d.\frac{d^2y}{dx^2} + \&c.$$

$$= \left(N\frac{dy}{dx} + P_1\frac{d^2y}{dx^2} + P_2\frac{d^3y}{dx^3} + \&c. + P_n\frac{d^{n+1}y}{dx^{n+1}}\right) dx;$$

or, if we substitute for N its value derived from the equation

$$N - \frac{dP_1}{dx} + \&c. = 0,$$

$$dV = \left(P_1\frac{d^2y}{dx^2} + \frac{dy}{dx}.\frac{dP_1}{dx}\right) dx + \left(P_2\frac{d^3y}{dx^3} - \frac{dy}{dx}.\frac{d^2P_2}{dx^2}\right) dx + \&c.$$

$$+ \left(P_n\frac{d^{n+1}y}{dx^{n+1}} - (-1)^n.\frac{dy}{dx}.\frac{d^nP_n}{dx^n}\right) dx;$$

and, therefore,

$$V = c + \int\left(P_1\frac{d^2y}{dx^2} + \frac{dy}{dx}.\frac{dP_1}{dx}\right) dx$$

$$+ \int\left(P_2\frac{d^3y}{dx^3} - \frac{dy}{dx}.\frac{d^2P_2}{dx^2}\right) dx + \&c.$$

$$+ \int\left(P_n\frac{d^{n+1}y}{dx^{n+1}} - (-1)^n\frac{dy}{dx}.\frac{d^nP_n}{dx^n}\right) dx.$$

But if the quantity $\int P_n \frac{d^{n+1}y}{dx^{n+1}}$ be integrated by parts, its value will be found to be

$$\int P_n \frac{d^{n+1}y}{dx^{n+1}} \cdot dx = P_n \frac{d^n y}{dx^n} - \frac{dP_n}{dx} \cdot \frac{d^{n-1}y}{dx^{n-1}} + \&c. + (-1)^n \int \frac{dy}{dx} \cdot \frac{d^n P_n}{dx^n};$$

hence

$$\int \left(P_n \frac{d^{n+1}y}{dx^{n+1}} - (-1)^n \frac{dy}{dx} \cdot \frac{d^n P_n}{dx^n}\right) dx = P_n \frac{d^n y}{dx^n} - \frac{dP_n}{dx} \cdot \frac{d^{n-1}y}{dx^{n-1}} + \&c.$$

$$+ (-1)^{n-1} \frac{dy}{dx} \cdot \frac{d^{n-1}P_n}{dx^{n-1}};$$

and, therefore,

$$V = c + P_1 \frac{dy}{dx} + \left(P_2 \frac{d^2y}{dx^2} - \frac{dy}{dx} \cdot \frac{dP_2}{dx}\right)$$

$$+ \left(P_3 \frac{d^3y}{dx^3} - \frac{dP_3}{dx} \cdot \frac{d^2y}{dx^2} + \frac{d^2P_3}{dx^2} \cdot \frac{dy}{dx}\right)$$

$$+ \&c.$$

$$+ P_n \frac{d^n y}{dx^n} - \frac{dP_n}{dx} \cdot \frac{d^{n-1}y}{dx^{n-1}} + \&c. + (-1)^{n-1} \frac{d^{n-1}P_n}{dx^{n-1}} \cdot \frac{dy}{dx}, \qquad \text{(E)}$$

a differential equation, which is, at most, of the order $2n - 1$.

Hence it appears that if the function V do not contain the independent variable, the equation

$$\beta = 0$$

can be reduced, at least, one order.

For the sake of convenience in the application of this formula, we shall give a few of the most important cases:

(a) $$V = f\left(\frac{dy}{dx}\right).$$

Here $V = c + P_1 \frac{dy}{dx}$; but as V, and therefore P_1, are functions of $\frac{dy}{dx}$, this equation is equivalent to $\frac{dy}{dx} = c_1$, or $y = c_1 x + c_2$. Hence it appears that linear functions have the property of giving a maximum or minimum value to every function of $\frac{dy}{dx}$ which admits of such a value.

(b) $$\text{If } V = f\left(y, \frac{dy}{dx}\right), \text{ then } V = c + P_1 \frac{dy}{dx}.$$

(c) $$\text{If } V = f\left(y, \frac{d^2y}{dx^2}\right), \text{ then } V = c + P_2 \frac{d^2y}{dx^2} - \frac{dy}{dx} \cdot \frac{dP_2}{dx}.$$

(3.) Let the simplifications of the form of V in both these classes be combined, i. e. let neither the independent variable nor the first m of the quantities $y, \frac{dy}{dx}$, &c. be found in V.

The equation

$$\beta = 0$$

becomes, as in case 1, by integration,

$$P_m - \frac{dP_{m+1}}{dx} + \&c. = c_0 + c_1 x + c_2 x^2 + \&c. + c_{m-1} x^{m-1};$$

hence

$$P_m = \frac{dP_{m+1}}{dx} - \&c. + c_0 + c_1 x + c_2 x^2 + \&c. + c_{m-1} x^{m-1}.$$

Substituting this value in the equation

$$dV = \left(P_m \frac{d^{m+1}y}{dx^{m+1}} + P_{m+1} \frac{d^{m+2}y}{dx^{m+2}} + \&c. + P_n \frac{d^{n+1}y}{dx^{n+1}}\right) dx,$$

$$dV = \left(P_{m+1} \frac{d^{m+2}y}{dx^{m+2}} + \frac{dP_{m+1}}{dx} \cdot \frac{d^{m+1}y}{dx^{m+1}}\right) dx$$

$$+ \left(P_{m+2} \frac{d^{m+3}y}{dx^{m+3}} - \frac{d^2 P_{m+2}}{dx^2} \cdot \frac{d^{m+1}y}{dx^{m+1}}\right) dx$$

$$+ \&c.$$

$$+ \left(P_n \frac{d^{n+1}y}{dx^{n+1}} - (-1)^{n-1} \frac{d^{n-m}P_n}{dx^{n-m}} \cdot \frac{d^{m+1}y}{dx^{m+1}}\right) dx$$

$$+ (c_0 + c_1 x + c_2 x^2 + \&c. + c_{m-1} x^{m-1}) \frac{d^{m+1}y}{dx^{m+1}}.$$

Integrating by parts, we find

$$V = c + P_{m+1} \frac{d^{m+1}y}{dx^{m+1}} + \left(P_{m+2} \frac{d^{m+2}y}{dx^{m+2}} - \frac{dP_{m+2}}{dx} \cdot \frac{d^{m+1}y}{dx^{m+1}}\right) + \&c.$$

$$+ P_n \frac{d^n y}{dx^n} - \frac{dP_n}{dx} \cdot \frac{d^{n-1}y}{dx^{n-1}} + \&c. + (-1)^{n-m-1} \frac{d^{n-m-1}P_n}{dx^{n-m-1}} \cdot \frac{d^{m+1}y}{dx^{m+1}}$$

$$+ \int (c_0 + c_1 x + c_2 x^2 + \&c. + c_{m-1} x^{m-1}) \frac{d^{m+1}y}{dx^{m+1}} \cdot dx. \qquad \text{(F)}$$

But as in general

$$\int x^k \frac{d^{m+1}y}{dx^{m+1}} = x^k \frac{d^m y}{dx^m} - kx^{k-1}\frac{d^{m-1}y}{dx^{m-1}}$$

$$+ k.k-1.x^{k-2}\frac{d^{m-2}y}{dx^{m-2}} - \&c. + (-1)^k . k . k-1.k-2 \ldots 1. \frac{d^{m-k}y}{dx^{m-k}},$$

if k be made successively 1, 2, ... $m-1$, and the values of the integrals,

$$\int x \frac{d^{m+1}y}{dx^{m+1}}, \int x^2 \frac{d^{m+1}y}{dx^{m+1}} \ldots \int x^{m-1} \frac{d^{m+1}y}{dx^{m+1}},$$

be substituted in equation (F), the resulting equation will evidently be of the order $2n-m-1$, i. e. of an order $m+1$ degrees lower than that of the original differential equation.

Thus, for example, let

$$V = f\left(\frac{dy}{dx}, \frac{d^2y}{dx^2}\right).$$

Here the equation $\beta = 0$ is

$$\frac{dP_1}{dx} - \frac{d^2P_2}{dx^2} = 0,$$

which becomes, by integration,

$$P_1 = \frac{dP_2}{dx} + c.$$

Substituting this value in the equation

$$dV = \left(P_1 \frac{d^2y}{dx^2} + P_2 \frac{d^3y}{dx^3}\right) dx,$$

and integrating, we find

$$V = c' + c\frac{dy}{dx} + P_2\frac{d^2y}{dx^2}, \qquad \text{(d)}$$

a differential equation of the second order.

We shall next proceed to give some examples of the application of the foregoing formulæ.

Example 1.

30. To find the form of the function y which will render

$$U = \int_{x_0}^{x_1} \sqrt{\left(1 + \frac{dy^2}{dx^2}\right)} \, . \, dx$$

a maximum or minimum.

Here $V = f\left(\frac{dy}{dx}\right)$, and, therefore, ((a), p. 52),

$$y = cx + c'.$$

To determine the arbitrary constants c and c',

(1.) If the limiting values of both y and x be given, we have the two equations,

$$y_0 = cx_0 + c', \quad y_1 = cx_1 + c',$$

the equation

$$a_1 - a_0 = 0$$

disappearing, inasmuch as

$$\delta y_1 = 0, \quad \delta y_0 = 0.$$

(2.) If the limiting values of x only be given, the equation

$$a_1 - a_0 = 0$$

becomes

$$c\,(\delta y_1 - \delta y_0) = 0, \text{ or } c = 0;$$

hence, in this case, the other constant, c', is indeterminate. This is an example of exception 2.

(3.) If the limiting values of x and y be connected by the equations

$$y_1 = f_1(x_1), \quad y_0 = f_0(x_0),$$

the increments δy_1, dx_1, δy_0, dx_0, are connected by the equations

$$\delta y_1 + \left(\frac{dy}{dx}\right)_1 dx_1 = m_1 dx_1,$$

$$\delta y_0 + \left(\frac{dy}{dx}\right)_0 dx_0 = m_0 dx_0,$$

putting m_1, m_0 for $f'_1(x_1)$, $f'_0(x_0)$. Hence the equation

$$a_1 - a_0 = 0$$

is equivalent to

$$\sqrt{(1+c^2)} + \frac{c}{\sqrt{(1+c^2)}}(m_1 - c) = 0,$$

$$\sqrt{(1+c^2)} + \frac{c}{\sqrt{(1+c^2)}}(m_0 - c) = 0,$$

or

$$1 + m_1 c = 0, \quad 1 + m_0 c = 0;$$

which equations, with the four others,

$$y_1 = cx_1 + c', \quad y_0 = cx_0 + c', \quad y_1 = f_1(x_1), \quad y_0 = f_0(x_0),$$

are sufficient to determine the six unknown quantities,

$$y_1, x_1, y_0, x_0, c, c'.$$

Example 2.

31. To find the form of the function y, which will make

$$U = \int_{x_0}^{x_1} y^n \sqrt{\left(1 + \frac{dy^2}{dx^2}\right)} dx$$

a maximum or minimum.

In this case $V = f\left(y, \frac{dy}{dx}\right)$, hence ((b), p. 53),

$$V = P_1 \frac{dy}{dx} + c;$$

or since

$$P_1 = \frac{y^n \frac{dy}{dx}}{\sqrt{\left(1 + \frac{dy^2}{dx^2}\right)}},$$

$$y^n \sqrt{\left(1 + \frac{dy^2}{dx^2}\right)} = \frac{y^n \frac{dy^2}{dx^2}}{\sqrt{\left(1 + \frac{dy^2}{dx^2}\right)}} + c,$$

therefore

$$y^n = c\sqrt{\left(1 + \frac{dy^2}{dx^2}\right)};$$

or making, for the sake of symmetry, $c = a^n$,

$$dx = \frac{a^n\,dy}{\sqrt{(y^{2n} - a^{2n})}}.$$

This comes under the general class of binomial differentials included in the expression

$$y^{m'-1}(a + by^{n'})^{\frac{p'}{q'}}.$$

In the present instance

$$m' = 1, \quad n' = 2n, \quad p' = -1, \quad q' = 2.$$

Now it is known that the expression

$$y^{m'-1}(a + by^{n'})^{\frac{p'}{q'}}$$

is integrable if either

$$\frac{m'}{n'} = i, \text{ or } \frac{m'}{n'} + \frac{p'}{q'} = i;^*$$

hence it is easy to see that the equation

$$dx = \frac{a^n dy}{\sqrt{(y^{2n} - a^{2n})}}$$

is integrable if the value of n be any term of the series

$$1, \tfrac{1}{2}, \tfrac{1}{3}, \tfrac{1}{4}, \text{\&c.}$$

As the cases $n = 1$ and $n = \frac{1}{2}$ will occur again, we shall not at present dwell further upon them.

Let

$$n = \tfrac{1}{3}, \quad dx = \frac{a^{\frac{1}{3}}\,dy}{\sqrt{(y^{\frac{2}{3}} - a^{\frac{2}{3}})}};$$

assume

$$z = y^{\frac{1}{3}}, \quad b = a^{\frac{1}{3}},$$

then

$$dx = \frac{3bz^2dz}{\sqrt{(z^2 - b^2)}}.$$

Hence

$$x + c = 3b.\int \frac{z^2dz}{\sqrt{(z^2 - b^2)}}.$$

* Vid. Lacroix Traitè Elem., Art. 192.

But

$$\int \frac{z^2 dz}{\sqrt{(z^2 - b^2)}} = z\sqrt{(z^2 - b^2)} - \int dz \sqrt{(z^2 - b^2)}$$

$$= z\sqrt{(z^2 - b^2)} - \int \frac{z^2 dz}{\sqrt{(z^2 - b^2)}} + b^2 \int \frac{dz}{\sqrt{(z^2 - b^2)}}.$$

Hence, finally,

$$\int \frac{z^2 dz}{\sqrt{(z^2 - b^2)}} = \tfrac{1}{2}\left[z\sqrt{(z^2 - b^2)} + b^2 l.\{z + \sqrt{(z^2 - b^2)}\}\right];$$

therefore

$$x + c = \frac{3a^{\frac{1}{3}}}{2}\cdot\left[y^{\frac{1}{3}}\sqrt{(y^{\frac{2}{3}} - a^{\frac{2}{3}})} + a^{\frac{2}{3}} l.\{y^{\frac{1}{3}} + \sqrt{(y^{\frac{2}{3}} - a^{\frac{2}{3}})}\}\right].$$

To determine the constants:

(1). If the limiting values of x and y be both given, the constants a and c are determined by the two equations,

$$x_1 + c = \frac{3a^{\frac{1}{3}}}{2}\cdot\left[y_1^{\frac{1}{3}}\sqrt{(y_1^{\frac{2}{3}} - a^{\frac{2}{3}})} + a^{\frac{2}{3}} l\{y_1^{\frac{1}{3}} + \sqrt{(y_1^{\frac{2}{3}} - a^{\frac{2}{3}})}\}\right],$$

$$x_0 + c = \frac{3a^{\frac{1}{3}}}{2}\cdot\left[y_0^{\frac{1}{3}}\sqrt{(y_0^{\frac{2}{3}} - a^{\frac{2}{3}})} + a^{\frac{2}{3}} l\{y_0^{\frac{1}{3}} + \sqrt{(y_0^{\frac{2}{3}} - a^{\frac{2}{3}})}\}\right].$$

(2.) If the limiting values of x only be given, the equation

$$a_1 - a_0 = 0$$

is equivalent to

$$\left(\frac{y^n \frac{dy}{dx}}{\sqrt{\left(1 + \frac{dy^2}{dx^2}\right)}}\right)_1 = 0, \quad \left(\frac{y^n \frac{dy}{dx}}{\sqrt{\left(1 + \frac{dy^2}{dx^2}\right)}}\right)_0 = 0.$$

These equations are satisfied by making

$$y_1 = 0, \; y_0 = 0, \text{ or } \left(\frac{dy}{dx}\right)_1 = 0, \; \left(\frac{dy}{dx}\right)_0 = 0.$$

The first supposition gives evidently $a = 0$, as the general solution would otherwise be impossible. But if $a = 0$, it is easy to see that the only solution which the equation admits of is $y = 0$. For if y be not $= 0$, we must have $x + c = 0$, and therefore

$$x_1 + c = 0, \quad x_0 + c = 0, \quad \text{or } x_1 = x_0,$$

which is impossible.* The second supposition gives $y_1 = y_0 = a$, the value $-a$ being evidently inadmissible; hence, from the general solution,

$$x_1 + c = \tfrac{1}{2}\, ala, \quad x_0 + c = \tfrac{1}{2}\, ala,$$

and therefore $x_1 = x_0$, which is impossible, as before. Hence, in this case, the problem does not admit of a finite solution.

(3.) If the limiting values of both x and y be given for one limit, and only that of x for the other, it is evident that the equation

$$\alpha_1 - \alpha_0 = 0$$

will become

$$\alpha_0 = 0, \text{ or } x_0 + c = 0,$$

which, with the equation

$$x_1 + c = \frac{3a^{\frac{1}{3}}}{2} \cdot [y_1^{\frac{1}{3}} \sqrt{(y_1^{\frac{2}{3}} - a^{\frac{2}{3}})} + a^{\frac{2}{3}} l \{ y_1^{\frac{1}{3}} + \sqrt{(y_1^{\frac{2}{3}} - a^{\frac{2}{3}})} \}],$$

will suffice for the determination of the constants a and c.

(4.) If the limiting values of x and y be connected by the equations

$$y_1 = f_1(x_1), \quad y_0 = f_0(x_0), \quad \text{or} \quad dy_1 = m_1 dx_1, \quad dy_0 = m_0 dx_0,$$

the equation

$$\alpha_1 - \alpha_0 = 0$$

will be equivalent to the following, viz.:

$$\left\{ V + P_1 \left(m_1 - \frac{dy}{dx} \right) \right\}_1 = 0, \quad \left\{ V + P_1 \left(m_0 - \frac{dy}{dx} \right) \right\}_0 = 0;$$

or, putting for V and P_1 their values, and reducing,

$$1 + m_1 \left(\frac{dy}{dx} \right)_1 = 0, \quad 1 + m_0 \left(\frac{dy}{dx} \right)_0 = 0.†$$

* It is unnecessary to consider either of the suppositions,

$$y_1 = 0, \left(\frac{dy}{dx} \right)_0 = 0, \text{ or } y_0 = 0, \left(\frac{dy}{dx} \right)_1 = 0,$$

inasmuch as the foregoing reasoning shows that neither y_1 nor y_0 can vanish.

† For the geometrical interpretation of these equations, vid. Chapter IV., where a more general problem is discussed.

Example 3.

32. To find the form of the function y, which will render

$$\int_{x_0}^{x_1} y^n \frac{\frac{d^2y}{dx^2}}{\frac{dy}{dx}} dx$$

a maximum or minimum.

Since V does not contain the independent variable, this problem is a case of the general one discussed in pp. 51, 52. We find, therefore, by applying equation (E) to this case,

$$V = c + P_1 \frac{dy}{dx} + P_2 \frac{d^2y}{dx^2} - \frac{dy}{dx} \cdot \frac{dP_2}{dx};$$

which, since

$$V = y^n \frac{\frac{d^2y}{dx^2}}{\frac{dy}{dx}}, \quad P_1 = -y^n \frac{\frac{d^2y}{dx^2}}{\frac{dy^2}{dx^2}}, \quad P_2 = \frac{y^n}{\frac{dy}{dx}},$$

$$\frac{dP_2}{dx} = ny^{n-1} - y^n \frac{\frac{d^2y}{dx^2}}{\frac{dy^2}{dx^2}},$$

becomes

$$ny^{n-1} \frac{dy}{dx} = c;$$

hence, by integration,

$$y^n = cx + c'.$$

This equation containing but two arbitrary constants, it is impossible, in general, to satisfy by means of these constants more than two equations. Hence as the equation

$$a_1 - a_0 = 0$$

is in general equivalent to four, it is impossible, by means of the two arbitrary constants, to satisfy this equation. The present problem is, in fact, a case of Excep. 1, p. 44. And although in certain cases the equation

$$a_1 - a_0 = 0$$

is equivalent to but two, it will be found that the impossibility of satisfying it by any finite value of y_1 remains. This will appear by an enumeration of the several cases.

(1.) Let the limiting values of x only be given, and it is evident that the equation

$$a_1 - a_0 = 0$$

is equivalent to

$$\left(P_1 - \frac{dP_2}{dx}\right)_1 = 0, \quad \left(P_1 - \frac{dP_2}{dx}\right)_0 = 0, \quad (P_2)_1 = 0, \quad (P_2)_0 = 0.$$

If the values of P_1, P_2 be substituted, it will be found that these equations are equivalent to but two, namely,

$$y_1 = 0, \quad y_0 = 0.$$

Hence, from the general solution, we find

$$cx_1 + c' = 0, \quad cx_0 + c' = 0,$$

which can only be satisfied by

$$c = 0, \quad c' = 0;$$

and, therefore, the only possible solution of the problem is

$$y = 0,$$

which would, of course, render the given definite integral zero also.

(2.) If the limiting values of both x and y be given, the equation

$$a_1 - a_0 = 0$$

is reduced to

$$(P_2)_1 = 0, \quad (P_2)_0 = 0, \quad \text{or} \quad y_1 = 0, \quad y_0 = 0,$$

which cannot, of course, be satisfied, unless the given limiting values of y be each zero, in which case the solution becomes

$$y = 0.$$

(3.) If the limiting values of x and y be connected by the given equations,

$$dy_1 = m_1 dx_1, \quad dy_0 = m_0 dx_0,$$

the equations

$$(P_2)_1 = 0, \quad (P_2)_0 = 0,$$

will give, as before,

$$y_1 = 0, \quad y_0 = 0.$$

This case is, therefore, reduced to the preceding. It appears, then, that the problem admits of no solution except

$$y = 0.$$

Example 4.

33. To determine the function y, such as to render

$$\int_{x_0}^{x_1} \left\{ y - a \sqrt{\left(1 + \frac{dy^2}{dx^2}\right)} \right\} dx$$

a maximum or minimum.

In this case

$$V = y - a \sqrt{\left(1 + \frac{dy^2}{dx^2}\right)},$$

$$P_1 = - \frac{a \dfrac{dy}{dx}}{\sqrt{\left(1 + \dfrac{dy^2}{dx^2}\right)}}.$$

As V does not contain the independent variable, we have (p. 53, (b)),

$$V = P_1 \frac{dy}{dx} + c,$$

or

$$y - a \sqrt{\left(1 + \frac{dy^2}{dx^2}\right)} = - \frac{a \dfrac{dy^2}{dx^2}}{\sqrt{\left(1 + \dfrac{dy^2}{dx^2}\right)}} + c.$$

Hence, by reduction,

$$(y - c) \sqrt{\left(1 + \frac{dy^2}{dx^2}\right)} = a,$$

and, therefore,

$$dx = \frac{(y - c)\, dy}{\sqrt{\{a^2 - (y - c)^2\}}}.$$

Integrating this, we have

$$x - b = - \sqrt{\{a^2 - (y - c)^2\}},$$

or

$$(y - c)^2 + (x - b)^2 = a^2.$$

This equation determines the form of the function y. To determine the constants c, c', we shall consider the several cases which may arise.

(1.) Let the limiting values of both x and y be given.

In this case we have

$$dx_0 = 0, \quad dx_1 = 0, \quad \delta y_0 = 0, \quad \delta y_1 = 0.$$

The equation

$$a_1 - a_0 = 0,$$

therefore, disappears, and the arbitrary constants, c, c', are determined by the equations,

$$(y_0 - c)^2 + (x_0 - b)^2 = a^2,$$
$$(y_1 - c)^2 + (x_1 - b)^2 = a^2,$$

formed by the substitution of the given limiting values,

$$y_0, \; x_0, \; y_1, \; x_1,$$

in the general solution.

(2.) Let the limiting values of x be given, those of y remaining indeterminate.

The equation

$$a_1 - a_0 = 0$$

is, in this case, equivalent to

$$(P_1)_0 = 0, \quad (P_1)_1 = 0.$$

Substituting for $(P_1)_0$, $(P_1)_1$, their values

$$-a\left(\frac{\frac{dy}{dx}}{\sqrt{\left(1+\frac{dy^2}{dx^2}\right)}}\right)_0, \quad -a\left(\frac{\frac{dy}{dx}}{\sqrt{\left(1+\frac{dy^2}{dx^2}\right)}}\right)_1,$$

these equations give

$$x_0 - b = 0,$$
$$x_1 - b = 0.$$

As these latter contain but one indeterminate quantity, b, it is plain that they cannot be satisfied, and that, therefore, the given integral does not, in this case, admit of a maximum or minimum value. This is an example of Excep. 2, pp. 47, 48. The geometrical interpretation will be found in the chapter on the applications of the Calculus of Variations to Geometry.

(3.) Finally, let it be supposed that the limiting values of x and y are connected by the equations

$$y_0 = f_0(x_0),$$
$$y_1 = f_1(x_1);$$

or,

$$dy_0 = m_0 dx_0,$$
$$dy_1 = m_1 dx_1;$$

f_0, f_1, denoting two given functions.

Substituting for V_1, V_0, &c., in the equations of Art. 27 (4), we find

$$y_1 - a\sqrt{\left(1 + \frac{dy_1^2}{dx_1^2}\right)} - a\left(m_1 - \frac{dy_1}{dx_1}\right)\frac{\frac{dy_1}{dx_1}}{\sqrt{\left(1 + \frac{dy_1^2}{dx_1^2}\right)}} = 0,$$

$$y_0 - a\sqrt{\left(1 + \frac{dy_0^2}{dx_0^2}\right)} - a\left(m_0 - \frac{dy_0}{dx_0}\right)\frac{\frac{dy_0}{dx_0}}{\sqrt{\left(1 + \frac{dy_0^2}{dx_0^2}\right)}} = 0;$$

or, by reduction,

$$y_1\sqrt{\left(1 + \frac{dy_1^2}{dx_1^2}\right)} - a\left(1 + m_1\frac{dy_1}{dx_1}\right) = 0,$$

$$y_0\sqrt{\left(1 + \frac{dy_0^2}{dx_0^2}\right)} - a\left(1 + m_0\frac{dy_0}{dx_0}\right) = 0.$$

Substituting for $\frac{dy_1}{dx_1}, \frac{dy_0}{dx_0}$, their values found by differentiating the general solution, and reducing, we have ultimately

$$c + (x_1 - b)\, m_1 = 0,$$
$$c + (x_0 - b)\, m_0 = 0,$$

which, with the four equations,

$$(y_1 - c)^2 + (x_1 - b)^2 = a^2, \quad y_1 = f_1(x_1),$$
$$(y_0 - c)^2 + (x_0 - b)^2 = a^2, \quad y_0 = f_0(x_0),$$

are sufficient to determine the six quantities,

$$x_1,\ y_1,\ x_0,\ y_0,\ b,\ c.$$

Prop. III.

34. Let

$$U = V' + \int_{x_0}^{x_1} V dx,$$

where V has the same form as before, and

$$V' = f'\left\{ x_0, y_0, \left(\frac{dy}{dx}\right)_0 \ldots \left(\frac{d^{n'}y}{dx^{n'}}\right)_0, \; x_1, y_1, \left(\frac{dy}{dx}\right)_1 \ldots \left(\frac{d^{n''}y}{dx^{n''}}\right)_1 \right\};$$

and let it be required to determine the form of the function y and the values of the limits x_0, x_1, such as to render U a maximum or minimum.

The general equation

$$\beta = 0,$$

being derived solely from terms under the sign of integration, will evidently be the same as in the foregoing Proposition. Hence it will only be necessary to consider the terms which refer to the limits. Assume

$$dV' = M'dx_0 + N'dy_0 + P'_1 d\left(\frac{dy}{dx}\right)_0 + \&c. + P'_{n'} d\left(\frac{d^{n'}y}{dx^{n'}}\right)_0$$

$$+ M''dx_1 + N''dy_1 + P''_1 d\left(\frac{dy}{dx}\right)_1 + \&c. + P''_{n''} d\left(\frac{d^{n''}y}{dx^{n''}}\right)_1.$$

Then it is easily seen that the terms added to DU by V' will be

$$M'dx_0 + N'\delta y_0 + P'_1 \left(\frac{d\delta y}{dx}\right)_0 + \&c. + P'_{n'} \left(\frac{d^{n'}\delta y}{dx^{n'}}\right)_0$$

$$+ M''dx_1 + N''\delta y_1 + P''_1 \left(\frac{d\delta y}{dx}\right)_1 + \&c. + P''_{n''} \left(\frac{d^{n''}\delta y}{dx^{n''}}\right)_1.$$

The left hand member of the equation

$$a_1 - a_0 = 0$$

will therefore be augmented by these terms; and as they are obviously of the same form with those which are already found in that equation, the mode of treating it, in the several cases which may arise, is precisely the same as before.

One important remark, however, must not be omitted.

It will readily appear, from the reasoning of pp. 38–44, that the possibility of satisfying the condition

$$DU = 0$$

depends upon the fact that the number of independent increments which enter into the equation

$$a_1 - a_0 = 0$$

does not, in general, exceed that of the arbitrary constants which are found in the integral of the equation

$$\beta = 0.$$

If, then, any case should arise in which the number of these increments does exceed that of the constants, it is plain that the condition

$$DU = 0$$

would no longer be capable of being satisfied. Now, in the case before us, the number of the increments

$$dx_0,\ \delta y_0,\ \left(\frac{d\delta y}{dx}\right)_0 \dots \left(\frac{d^{n'}\delta y}{dx^{n'}}\right)_0,$$

relative to the inferior limit, which are found in the terms introduced by V', is evidently $n' + 2$. The number of increments, relative to the same limit, which already exist in a_0, is, as has been before observed, $n + 1$. If, then,

$$n' + 2 > n + 1,$$

or, in other words, if

$$n' > n - 1,$$

the problem does not admit of a solution. Similar conclusions apply, of course, to the superior limit. Hence it appears that, if the new quantity V contain any coefficient of an order higher than $n - 1$, the given function does not admit of either a maximum or a minimum.

One instance of this has already been alluded to (p. 47), where it has been shown that if V be a linear function of the highest differential coefficient which it contains, the given integral

$$\int_{x_0}^{x_1} V dx$$

can be reduced to the form

$$\phi_1 - \phi_0 + \int \phi' dx,$$

in which ϕ_1, ϕ_0 contain coefficients of an order equal to that of the highest coefficient in ϕ'. In this case, therefore,

$$n' = n'' = n > n - 1;$$

and, as we have before seen, it is impossible to satisfy the equation

$$DU = 0.$$

Examples of this Proposition will be found in Chapter IV. As the mode of treating any case which may arise is perfectly similar to that of Prop. II., we shall content ourselves with giving here a single example of the excepted case just alluded to, principally for the purpose of pointing out its identity with a case formerly noticed.

Example.

35. Let

$$U = y_1{}^n l \left(\frac{dy}{dx}\right)_1 - y_0{}^n l \left(\frac{dy}{dx}\right)_0 - n \int_{x_0}^{x_1} y^{n-1} \frac{dy}{dx}\, l \frac{dy}{dx}\, dx.$$

Here

$$V' = y_1{}^n l \left(\frac{dy}{dx}\right)_1 - y_0{}^n l \left(\frac{dy}{dx}\right)_0,$$

$$V = -ny^{n-1} \frac{dy}{dx}\, l \frac{dy}{dx};$$

whence

$$P_1 = -ny^{n-1} l \frac{dy}{dx} - ny^{n-1}.$$

Substituting the values of V and P_1 in the equation

$$V = c + P_1 \frac{dy}{dx},$$

we find, by reduction,

$$ny^{n-1} \frac{dy}{dx} = c;$$

or, integrating,

$$y^n = cx + c'.$$

With regard to the terms outside the sign of integration, it is evident that

$$a_1 = V_1 dx_1 + \{(P_1)_1 + N'\}\,\delta y_1 + P'_1 \left(\frac{d\delta y}{dx}\right)_1,$$

$$a_0 = V_0 dx_0 + \{(P_1)_0 - N''\}\,\delta y_0 - P''_1 \left(\frac{d\delta y}{dx}\right)_0.$$

But

$$N' = \frac{dV'}{dy_1} = ny_1^{n-1} l \left(\frac{dy}{dx}\right)_1;$$

hence

$$(P_1)_1 + N' = -\,ny_1^{n-1}.$$

Also

$$P'_1 = \frac{dV'}{d\,.\left(\frac{dy}{dx}\right)_1} = \frac{y_1^n}{\left(\frac{dy}{dx}\right)_1};$$

hence

$$a_1 = V_1 dx_1 - ny_1^{n-1}\,\delta y_1 + \frac{y_1^n}{\left(\frac{dy}{dx}\right)_1} \cdot \left(\frac{d\delta y}{dx}\right)_1.$$

Similarly,

$$a_0 = V_0 dx_0 - ny_0^{n-1}\,\delta y_0 + \frac{y_0^n}{\left(\frac{dy}{dx}\right)_0} \cdot \left(\frac{d\delta y}{dx}\right)_0.$$

Now if we compare these values of a_1, a_0, with those which occur in Ex. 3, pp. 60, 61, it is easily seen that the coefficients of

$$\delta y_1, \left(\frac{d\delta y}{dx}\right)_1, \quad \delta y_0, \left(\frac{d\delta y}{dx}\right)_0,$$

in the two expressions, are identical.

The general solution,

$$y^n = cx + c',$$

being also identical with that of Ex. 3, the conclusions there deduced applying to the present case, showing that the problem admits of no solution except

$$y = 0.$$

The identity of the two problems may also be shown by proving, according to the method of Euler (pp. 46, 47), that the value of U in p. 60, namely,

$$\int_{x_0}^{x_1} y^n \frac{\frac{d^2y}{dx^2}}{\frac{dy}{dx}}\, dx,$$

may be reduced to the form

$$y_1^n\, l\left(\frac{dy}{dx}\right)_1 - y_0^n\, l\left(\frac{dy}{dx}\right)_0 - n\int y^{n-1}\frac{dy}{dx}\, l\frac{dy}{dx}\, dx.$$

Assume

$$\int y^n \frac{\frac{d^2y}{dx^2}}{\frac{dy}{dx}}\, dx = \phi + \int \phi' dx.$$

Differentiating both sides, we have

$$y^n \frac{\frac{d^2y}{dx^2}}{\frac{dy}{dx}} = \frac{d\phi}{dx} + \frac{d\phi}{dy}\frac{dy}{dx} + \frac{d\phi}{d\,.\,\frac{dy}{dx}}\frac{d^2y}{dx^2} + \phi'. \tag{G}$$

Equating the coefficients of $\frac{d^2y}{dx^2}$, we have

$$\frac{d\phi}{d\,.\,\frac{dy}{dx}} = \frac{y^n}{\frac{dy}{dx}};$$

whence, by integration,

$$\phi = y^n l\frac{dy}{dx} + f(x, y),$$

f denoting an arbitrary function.

Substituting this value in the remaining terms of equation (G), we have

$$\phi' = -ny^{n-1} l\frac{dy}{dx} - \frac{df}{dx} - p\frac{df}{dy};$$

whence

$$\int_{x_0}^{x_1} \phi' dx = -n\int_{x_0}^{x_1} y^{n-1}\, l\frac{dy}{dx} - f(x_1, y_1) + f(x_0, y_0).$$

But since

$$\phi = y^n l \frac{dy}{dx} + f(x, y),$$

we have, by substituting successively, x_1, y_1, and x_0, y_0,

$$\phi_1 - \phi_0 = y_1{}^n l \left(\frac{dy}{dx}\right)_1 - y_0{}^n l \left(\frac{dy}{dx}\right)_0 + f(x_1, y_1) - f(x_0, y_0).$$

Hence, therefore,

$$\int_{x_0}^{x_1} y^n \frac{\frac{d^2y}{dx^2}}{\frac{dy}{dx}} dx = \phi_1 - \phi_0 + \int_{x_0}^{x_1} \phi' dx$$

$$= y_1{}^n l \left(\frac{dy}{dx}\right)_1 - y_0{}^n l \left(\frac{dy}{dx}\right)_0 - n \int_{x_0}^{x_1} y^{n-1} \frac{dy}{dx} l \frac{dy}{dx} dx.$$

Q. E. D.

Prop. IV.

36. Let

$$U = \int_{x_0}^{x_1} V dx,$$

where

$$V = f\left\{x, y, \frac{dy}{dx} \ldots \frac{d^n y}{dx^n}, x_0, y_0, \left(\frac{dy}{dx}\right)_0 \ldots \left(\frac{d^{n'} y}{dx^{n'}}\right)_0,\right.$$

$$\left. x_1, y_1, \left(\frac{dy}{dx}\right)_1 \ldots \left(\frac{d^{n''} y}{dx^{n''}}\right)_1\right\},$$

and let it be required to determine the form of the function y, and the values of the limits x_0, x_1, which will render U a maximum or minimum.

The general equation,

$$DU = 0$$

becomes in this case (Chap. II. Prop. IV.),

$$\left[V_1 + \int_{x_0}^{x_1} \left\{\mu_1 + \nu_1 \left(\frac{dy}{dx}\right)_1 + \pi_1 \left(\frac{d^2y}{dx^2}\right)_1 + \&c.\right\} dx\right] dx_1$$

$$+ \left[- V_0 + \int_{x_0}^{x_1} \left\{\mu_0 + \nu_0 \left(\frac{dy}{dx}\right)_0 + \pi_0 \left(\frac{d^2y}{dx^2}\right)_0 + \&c.\right\} dx\right] dx_0$$

$$+\left\{\left(P_1-\frac{dP_2}{dx}+\&c.\right)_1+\int_{x_0}^{x_1}\nu_1 dx\right\}\delta y_1$$

$$-\left\{\left(P_1-\frac{dP_2}{dx}+\&c.\right)_0-\int_{x_0}^{x_1}\nu_0 dx\right\}\delta y_0$$

$$+\left\{(P_2-\&c.)_1+\int_{x_0}^{x_1}\pi_1 dx\right\}\left(\frac{d\delta y}{dx}\right)_1-\left\{(P_2-\&c.)_0-\int_{x_0}^{x_1}\pi_0 dx\right\}\left(\frac{d\delta y}{dx}\right)_0$$

$$+\ \&c.$$

$$+\int_{x_0}^{x_1}\left(N-\frac{dP_1}{dx}+\frac{d^2P_2}{dx^2}-\&c.\right)\delta y=0.$$

Putting this equation, as in Prop. II., under the form

$$a_1-a_0+\int_{x_0}^{x_1}\beta\delta y dx=0,$$

it is, in the first place, evident that β is the same as before, and that, therefore, the form of the general differential equation by which the form of the function y is determined, is not altered by the supposition that V contains the limiting values of

$$x,\ y,\ \frac{dy}{dx},\ \&c.$$

The terms contained in a_1-a_0 are also of the same nature as in Prop. II., constituting, in fact, a series of the form

$$A_1dx_1+B_1\delta y_1+C_1\left(\frac{d\delta y}{dx}\right)_1+\&c.+A_0dx_0+B_0\delta y_0+C_0\left(\frac{d\delta y}{dx}\right)_0+\&c.$$

A_1, B_1, C_1, &c., A_0, B_0, C_0, &c. being constants.

For in the expressions

$$\int_{x_0}^{x_1}\mu_1 dx,\ \&c.,\ \int_{x_0}^{x_1}\mu_0 dx,\ \&c.,$$

the same supposition is made as in the case of

$$\left(P_1-\frac{dP_2}{dx}+\&c.\right)_1,\ \left(P_1-\frac{dP_2}{dx}+\&c.\right)_0,$$

and the other coefficients of the several increments which enter into $a_1 - a_0$ in Prop. II.; the supposition, namely, that the value of y derived from the equation

$$\beta = 0$$

has been substituted in μ_1, μ_0, &c. If this substitution be made, and the operation of definite integration be then performed, the terms

$$\int_{x_0}^{x_1} \mu_1 dx, \quad \int_{x_0}^{x_1} \mu_0 dx,$$

will become constants.

Hence it is evident that the mode of treating the equation

$$DU = 0,$$

in the several cases which may occur, is precisely similar to that given in Prop. II.

The restriction on the possibility of the problem is analogous to that which we found to hold in Prop. III. If either n' or n'' exceed $n - 1$, it is easily seen that the number of arbitrary increments in $a_1 - a_0$ is greater than the number of arbitrary constants in the integral of the equation

$$\beta = 0,$$

and that, therefore, it is not, in general, possible to satisfy the condition

$$DU = 0.$$

From this and the preceding Proposition we infer that the quantity

$$V' + \int_{x_0}^{x_1} V dx$$

does not, in general, admit of a maximum or minimum, if either V or V' contain either of the *limiting* values of a differential coefficient of an order equal to or greater than the order of the highest *general* coefficient,

$$\frac{d^n y}{dx^n},$$

which is found in V.

Example.

37. Let

$$U = \int_{x_0}^{x_1} \frac{\sqrt{\left(1 + \frac{dy^2}{dx^2}\right)}}{\sqrt{(a-y)}} dx,$$

in which a is a given function of x_1, y_1; and let it be required to determine y so as to make U a minimum.

In this case, V being a function of y and $\frac{dy}{dx}$, not containing the independent variable, we have, (p. 53) (b),

$$V = c + P_1 \frac{dy}{dx}.$$

But since

$$V = \frac{\sqrt{\left(1 + \frac{dy^2}{dx^2}\right)}}{\sqrt{(a-y)}}, \text{ and } P_1 = \frac{\frac{dy}{dx}}{\sqrt{(a-y)} \cdot \sqrt{\left(1 + \frac{dy^2}{dx^2}\right)}},$$

we have, by substituting these values,

$$\frac{\sqrt{\left(1 + \frac{dy^2}{dx^2}\right)}}{\sqrt{(a-y)}} = c + \frac{\frac{dy^2}{dx^2}}{\sqrt{(a-y)} \cdot \sqrt{\left(1 + \frac{dy^2}{dx^2}\right)}}.$$

Whence, by reduction,

$$1 = c\sqrt{(a-y)} \cdot \sqrt{\left(1 + \frac{dy^2}{dx^2}\right)},$$

or, putting $\frac{1}{c^2} = b$, and solving for $\frac{dy}{dx}$,

$$\frac{dy}{dx} = \sqrt{\left(\frac{b-a+y}{a-y}\right)}. \tag{H}$$

This equation, which is easily integrated by putting

$$a - y = b \cos^2\theta,$$

gives

$$x + b' = b \cos^{-1} \sqrt{\left(\frac{a-y}{b}\right)} + \sqrt{\{b(a-y) - (a-y)^2\}}. \tag{I}$$

To determine the arbitrary constants b and b':

1. If the limiting values of both x and y be given, it is plain that a will be a given constant, and that, therefore, the problem differs in no respect from that of Prop. II. All the terms of $a_1 - a_0$ will in this case disappear of themselves, and the constants b, b' will be determined, as in Prop. II., by the substitution of the given limiting values of x and y in the general solution.

2. Let the limiting values of x be given, those of y remaining indeterminate.

Since a is only a function of x_1, y_1, we have, in the formula of Art. 13, p. 18,

$$a_1 = \left[V_1 + \int_{x_0}^{x_1}\left\{\mu_1 + \nu_1\left(\frac{dy}{dx}\right)_1\right\}dx\right]dx_1 + \left\{(P_1)_1 + \int_{x_0}^{x_1}\nu_1 dx\right\}\delta y_1,$$

$$a_0 = V_0 dx_0 + (P_1)_0 \delta y_0;$$

or, since the limiting values, x_1, x_0, are fixed, and, therefore,

$$dx_1 = 0, \quad dx_0 = 0,$$

$$a_1 = \left\{(P_1)_1 + \int_{x_0}^{x_1}\nu_1 dx\right\}\delta y_1, \quad a_0 = (P_1)_0 \delta y_0.$$

The equation

$$a_1 - a_0 = 0$$

is, therefore, equivalent to

$$(P_1)_1 + \int_{x_0}^{x_1}\nu_1 dx = 0, \quad (P_1)_0 = 0.$$

But

$$\nu_1 = \frac{dV}{dy_1} = \frac{dV}{da}\,\frac{da}{dy_1} = -\tfrac{1}{2}\frac{da}{dy_1}\int_{x_0}^{x_1}\frac{\sqrt{\left(1 + \frac{dy^2}{dx^2}\right)}}{(a-y)^{\frac{3}{2}}}\,dx.$$

Substituting the values of $(P_1)_1$, $(P_1)_0$, ν_1, in the above equations, we easily find that the second is equivalent to

$$\left(\frac{dy}{dx}\right)_0 = 0, \text{ or } y_0 - a + b = 0;$$

and the first to

$$\left(\frac{\frac{dy}{dx}}{\sqrt{(a-y)}\cdot\sqrt{\left(1+\frac{dy^2}{dx^2}\right)}}\right)_1 - \tfrac{1}{2}\frac{da}{dy_1}\int_{x_0}^{x_1}\frac{\sqrt{\left(1+\frac{dy^2}{dx^2}\right)}}{(a-y)^{\frac{3}{2}}}dx = 0;$$

or, substituting for $\frac{dy}{dx}$ its value derived from (H),

$$\sqrt{\left(\frac{b-a+y_1}{a-y_1}\right)} - \tfrac{1}{2}\frac{bda}{dy_1}\int_{x_0}^{x_1}\frac{dx}{(a-y)^2} = 0. \qquad \text{(K)}$$

If we put for dx its value,

$$dy\sqrt{\left(\frac{a-y}{b-a+y}\right)},$$

we have

$$\int\frac{dx}{(a-y)^2} = \int\frac{dy}{(a-y)^{\frac{3}{2}}\cdot\sqrt{(b-a+y)}} = \frac{2}{b}\sqrt{\left(\frac{b-a+y}{a-y}\right)}.$$

Substituting this value in (K), and recollecting that

$$y_0 - a + b = 0,$$

we have

$$\sqrt{\left(\frac{b-a+y_1}{a-y_1}\right)}\left(1-\frac{da}{dy_1}\right) = 0. \qquad \text{(L)}$$

This equation may be satisfied by making either

$$b - a + y_1 = 0,$$

or

$$\frac{da}{dy_1} = 1.$$

The first of these gives

$$y_1 = y_0 = a - b.$$

Substituting this value of y_1 in the general solution (I), we find

$$x_1 + b' = b\cos^{-1}1 = 2i\pi b.$$

Similarly,

$$x_0 + b' = b\cos^{-1}1 = 2i''\pi b.$$

Hence

$$b = \frac{x_1 - x_0}{2(i-i'')\pi}, \quad b' = \frac{i''x_1 - ix_0}{i-i''}.$$

The form of the function a being given, its value is found from the equation

$$a = f(x_1, y_1) = f(x_1, a - b).$$

If we suppose the second factor of equation (L) to vanish, it is easy to see that we shall have for the determination of the arbitrary constants the equations

$$x_0 + b' = 2i''\pi b,$$

$$x_1 + b' = b\cos^{-1}\sqrt{\left(\frac{a - y_1}{b}\right)} + \sqrt{\{b(a - y_1) - (a - y_1)\}^2},$$

y_1 being given by the equation

$$\frac{da}{dy_1} = 1.$$

It is unnecessary to dwell further upon this example, which will occur again in the applications of the Calculus of Variations to Mechanics.

Prop. V.

38. If y be a function of x which satisfies the equation

$$D\int_{x_0}^{x_1} V dx = 0,$$

to determine what further conditions are requisite, in order that the corresponding value of the given integral should be a real maximum or minimum, and to distinguish between maxima and minima.

The solution of this problem, which is one of the most difficult in the whole science, may be justly said to be due to Jacobi. For although Legendre, and, subsequently, Lagrange, have examined and partially solved it, the methods which they have given are encumbered with difficulties, which render them, in practice, comparatively useless. Both these writers have, it is true, succeeded in pointing out the criterion by which a maximum is distinguished from a minimum value, as well as in establishing *one* of the conditions essential to the existence of any such value.

But the condition thus established, although necessary, is easily seen to be insufficient; and the methods by which they have endeavoured to remedy this deficiency depend upon differential equations, or (in the latter case) *inequalities*, which they have given no means of integrating. This important defect in the theory of maxima and minima has been fully supplied by the method of M. Jacobi, which we shall now proceed to state.*

39. We have already seen that the form of the function y, which corresponds to a maximum or minimum value of a given integral, is determined by a differential equation of an order double of that of the highest differential coefficient which V contains. The value of y so determined is, therefore, of the form

$$f(x, c_1, c_2, \dots c_{2n}).$$

If, then, this value of y be substituted in the given integral, it is evident that the value of that integral will depend partly upon the *form* of the function f, and partly upon the *values* of the several constants, $c_1, c_2, \dots c_{2n}$, which are introduced in the expression for y. This consideration naturally suggests the division of the question, with which we are at present concerned, into two, namely,

1. Whether the general *form* of the function y, as determined by the equation

$$N - \frac{dP}{dx} + \&c. = 0,$$

be such as to give a real maximum or minimum value to $\int V dx$, for *given* values of the constants which enter into the solution of that equation.

2. Whether the *values* so assigned to these constants in any particular case be such as to render the value of the given integral greater or less than that of any other which might be deduced from it by an indefinitely small change in any of these constants.

* For the original Memoir of M. Jacobi, the reader is referred to Crelle, B. 17, S. 68, or to Liouville, Journal de Math., tom. iii. p. 44. The author of the present work has also been largely indebted to a very able Memoir by M. Delaunay (Liouville, tom. vi. p. 209), containing a development of M. Jacobi's discoveries. Such a development was the more to be desired, on account of the very succinct nature of M. Jacobi's memoir.

These two questions are evidently perfectly distinct, and therefore admit of being considered separately from each other. Thus, for example, if we can show that the value of the integral $\int V dx$, corresponding to

$$y = f(x, c_1, c_2, \ldots . c_{2n}),$$

cannot be augmented either by an indefinitely small change in the form of the function f, the values of $c_1, c_2, \ldots . c_{2n}$ remaining unchanged, or by a similar change in any one or more of these constants, the form of the function remaining unchanged, it is evident that the value so assigned to the integral is a real maximum. Now we have already seen (Prop. II.) that if the limiting values of

$$x, y, \frac{dy}{dx}, \ldots . \frac{d^{n-1}y}{dx^{n-1}}$$

be *given*, the values of the constants $c_1, c_2, \ldots . c_{2n}$ will be found as functions of these limiting values; and it is easy to see, conversely, that if the values of $c_1, c_2, \ldots . c_{2n}$ be *given*, the limiting values x, y, &c., must have been given also.

Hence it is evident that the foregoing subdivision of the question may be stated as follows:

1. Whether the form of the function y, as deduced from the equation

$$\beta = 0,$$

be such as to render the given integral a maximum or minimum for *given* values of

$$x_0, y_0, \left(\frac{dy}{dx}\right)_0 \ldots . \left(\frac{d^{n-1}y}{dx^{n-1}}\right)_0, \quad x_1, y_1, \left(\frac{dy}{dx}\right)_1 \ldots . \left(\frac{d^{n-1}y}{dx^{n-1}}\right)_1.$$

2. Whether the *particular* values of these quantities, derived from the equation

$$a_1 - a_0 = 0,$$

be such as to render the corresponding value of the given integral greater or less than any other which might be deduced from it by an indefinitely small change in any one or more of them.

It is with the former of these questions only that the Calculus of Variations is, properly speaking, concerned. The second belongs to the ordinary problem of maxima and minima of functions of more than one variable, and is completely solved by the rules which the Differential Calculus furnishes for such cases. We shall, therefore, confine our attention to the former question.

The essential condition of the existence of a maximum or minimum value of a given derived function is, as has been stated at the commencement of the present Chapter, that the value of y, which causes the *first* variation to vanish, should render the *sign* of the *second* independent of the form of δy or ψ, provided that that form be not inconsistent with any of the conditions of the question. Now in the case which we are here considering, in which the limiting values of y and its first $n-1$ differential coefficients, are supposed to be fixed, it is plain that no form of δy is admissible which does not satisfy the conditions

$$\delta y_0 = 0, \left(\frac{d\delta y}{dx}\right)_0 = 0, \text{ \&c. } \left(\frac{d^{n-1}\delta y}{dx^{n-1}}\right)_0 = 0,$$

$$\delta y_1 = 0, \left(\frac{d\delta y}{dx}\right)_1 = 0, \text{ \&c. } \left(\frac{d^{n-1}\delta y}{dx^{n-1}}\right)_1 = 0.$$

It is, therefore, essential to the existence of a maximum or minimum, that the sign of the second variation of the given integral should be the same for *every* form of δy which does satisfy these conditions, in addition to those noticed in Art. 9.

It is, moreover, necessary that no such form of δy should render the quantity under the integral sign in the second variation infinite, for any value of x within the limits of integration.

Lastly, it is, *in general*, necessary that no form of δy which satisfies the foregoing conditions should cause the second variation to vanish. For if any such form of δy cause the second variation to vanish, without, at the same time, annulling the *third*, it is plain (as in the Differential Calculus) that there will be neither maximum nor minimum.

40. It will contribute to the clearness of the subsequent discussion, which is necessarily somewhat tedious, to premise that the object of the method which we are now about to state is to prove that the second variation of the given integral may be reduced (by the method of integration by parts) to the form

$$\int_{x_0}^{x_1} Q_n \left(A\delta y + A_1 \frac{d\delta y}{dx} + A_2 \frac{d^2\delta y}{dx^2} + \&c. + A_n \frac{d^n\delta y}{dx^n}\right)^2 dx,$$

in which

$$Q_n = \frac{d^2 V}{\left(d.\frac{d^n y}{dx^n}\right)^2},$$

and A, A_1, A_2, ... A_n are functions of x.

From this transformation it will immediately appear—

1. That the *sign* of the second variation depends upon the sign of Q_n, the other factor being essentially positive, and that, therefore, one of the required conditions is

$$Q_n > \text{ or } < 0,$$

for all values of x from x_0 to x_1. For, if Q_n *change* its sign within the limits of integration, it is possible, as we shall hereafter show, to render the sign of the second variation either positive or negative. In such a case there would be neither maximum nor minimum.*

2. That if this condition be fulfilled, the value of the integral will be a maximum or minimum (according as Q_n is negative or positive), provided that the quantity

$$Q_n \left(A\delta y + A_1 \frac{d\delta y}{dx} + A_2 \frac{d^2\delta y}{dx^2} + \&c. + A_n \frac{d^n\delta y}{dx^n}\right)^2,$$

do not become infinite for any value of x within the limits of integration, and provided also that no admissible value of δy satisfy the equation

$$A\delta y + A_1 \frac{d\delta y}{dx} + \&c. = 0.$$

3. We shall further show that, in order to determine whether or not this condition be satisfied, it is only necessary to have the complete integral of the equation

$$\beta = 0;$$

and that, therefore, in every case in which we are enabled to solve the problem of Prop. II., we shall also have the means of deciding whether or not the solution so found give a real maximum or minimum.

* This condition has been noticed both by Legendre and Lagrange.

We shall commence with the following Lemmas, which are necessary to this investigation:

LEMMA I.

41. If β be the coefficient of δy under the sign of integration in the variation of a given integral, the variation $\delta\beta$ may be put under the form

$$\delta\beta = A_0\delta y + \frac{d\,.\,A_1\dfrac{d\delta y}{dx}}{dx} + \frac{d^2.\,A_2\dfrac{d^2\,\delta y}{dx^2}}{dx^2} + \&c. + \frac{d^n.\,A_n\dfrac{d^n\,\delta y}{dx^n}}{dx^n}.$$

Putting for β its value,

$$N - \frac{dP_1}{dx} + \frac{d^2 P_2}{dx^2} - \&c.,$$

and taking the variation, we have

$$\delta\beta = \delta N - \delta.\frac{dP_1}{dx} + \delta\,.\,\frac{d^2P_2}{dx^2} - \&c.$$

$$= \delta N - \frac{d\delta P_1}{dx} + \frac{d^2\delta P_2}{dx^2} - \&c.$$

Consider any term of this series, e. g.,

$$\frac{d^m\delta P_m}{dx^m}.$$

Forming the variation of P_m according to the usual rules, we have

$$\delta P_m = \frac{dP_m}{dy}\delta y + \frac{dP_m}{d.\dfrac{dy}{dx}}\frac{d\delta y}{dx} + \&c. + \frac{dP_m}{d\,.\,\dfrac{d^{m'}y}{dx^{m'}}}\frac{d^{m'}\delta y}{dx^{m'}} + \&c.$$

But since

$$P_m = \frac{dV}{d.\dfrac{d^m y}{dx^m}},$$

we have

$$\frac{dP_m}{d.\dfrac{d^{m'}y}{dx^{m'}}} = \frac{d^2 V}{d\,.\,\dfrac{d^m y}{dx^m}\,d\,.\,\dfrac{d^{m'}y}{dx^{m'}}}.$$

Hence, if we assume, for the sake of brevity,

$$y^{(m)} = \frac{d^m y}{dx^m}, \quad y^{(m')} = \frac{d^{m'} y}{dx^{m'}}, \quad u = \delta y,$$

the general term of the series for

$$\frac{d^m \delta P_m}{dx^m}$$

may be written

$$\frac{d^m}{dx^m}\left(\frac{d^2 V}{dy^{(m)}\, dy^{(m')}} \cdot \frac{d^{m'} u}{dx^{m'}}\right). \tag{1}$$

Now if m be equal to m', this term has already the required form. If m be not equal to m', it is plain that in the expansion of

$$\frac{d^{m'} \delta P_{m'}}{dx^{m'}}$$

there will be found a term

$$\frac{d^{m'}}{dx^{m'}}\left(\frac{d^2 V}{dy^{(m)}\, dy^{(m')}} \cdot \frac{d^m u}{dx^m}\right). \tag{2}$$

And, as the terms in the original series for $\delta\beta$ are alternately positive and negative, the terms (1) and (2) will have the same or contrary signs, according as $m - m'$ is even or odd.

Hence it appears that the entire series for $\delta\beta$, with the exception of those terms which have already the required form, may be arranged in groups of the form

$$\frac{d^m . K \dfrac{d^{m'} u}{dx^{m'}}}{dx^m} \pm \frac{d^{m'} . K \dfrac{d^m u}{dx^m}}{dx^{m'}},$$

where

$$K = \frac{d^2 V}{dy^{(m)}\, dy^{(m')}}.$$

But, by a theorem of the Differential Calculus,* a group such as the foregoing may always be represented by a series of the form

$$\frac{d^{m'} . C_{m'} \dfrac{d^{m'} u}{dx^{m'}}}{dx^{m'}} + \frac{d^{m'+1} . C_{m'+1} \dfrac{d^{m'+1} u}{dx^{m'+1}}}{dx^{m'+1}} + \&c.;$$

* For a proof of this theorem, vid. note upon p. 82.

m' being supposed less than m. Hence the proposition is evident, as it is plain that no term can occur of an order higher than

$$\frac{d^n . A_n \frac{d^n \delta y}{dx^n}}{dx^n}.$$

It is easily seen that

$$A_n = \pm \frac{d^2 V}{(dy^{(n)})^2},$$

the upper sign corresponding to the case in which n is even, and the lower to that in which n is odd.

For the coefficient of

$$\left\{\frac{d^n \delta y}{dx^n}\right\}^2$$

in $\delta\beta$ is evidently

$$\pm \frac{dP_n}{dy^{(n)}} = \pm \frac{d^2 V}{(dy^{(n)})^2}.$$

The coefficient of the same term in the series just given is A_n. Hence, as these coefficients must be identical, we have

$$A_n = \pm \frac{d^2 V}{(dy^n)^2}.$$

Lemma II.

42. If u be a function of x which satisfies the differential equation

$$Au + \frac{d . A_1 \frac{du}{dx}}{dx} + \&c. + \frac{d^n . A_n \frac{d^n u}{dx^n}}{dx^n} = 0,$$

(A, A_1, &c., being functions of x), the expression

$$u\left(Ay + \frac{d . A_1 \frac{dy}{dx}}{dx} + \&c. + \frac{d^n . A_n \frac{d^n y}{dx^n}}{dx^n}\right) = U$$

is integrable independently of y, and the integral may be expressed as follows:

$$\int U dx = B_1 \frac{dt}{dx} + \frac{d \, . \, B_2 \frac{d^2 t}{dx^2}}{dx} + \&c. + \frac{d^{n-1} \, . \, B_n \frac{d^n t}{dx^n}}{dx^{n-1}},$$

where $t = \frac{y}{u}$, and B_1, B_2, &c., are functions of x depending upon A_1, A_2, &c. The value of the last of these, B_n, is

$$B_n = A_n u^2.^*$$

Lemma III.

43. Let

$$y = f(x, c_1, c_2, \ldots . c_{2n})$$

be the integral of the equation

$$\beta = 0,$$

and let

$$\frac{dy}{dc_1}, \frac{dy}{dc_2}, \&c. \ldots . \frac{dy}{dc_{2n}}$$

be the differential coefficients of y with regard to the several arbitrary constants which it contains. Then if

$$C_1, C_2, \ldots . C_{2n}\dagger$$

be a new system of arbitrary constants, the complete integral of the equation

$$\delta\beta = 0$$

will be

$$\delta y = C_1 \frac{dy}{dc_1} + C_2 \frac{dy}{dc_2} + \&c. + C_{2n} \frac{dy}{dc_{2n}}.$$

For, if in the equation

$$y = f(x, c_1, c_2 \ldots . c_{2n}),$$

we suppose the several constants, $c_1, c_2 \ldots . c_{2n}$, to receive the indefinitely small increments $dc_1, dc_2, \ldots . dc_{2n}$, the corresponding increment of y will be

$$\frac{dy}{dc_1} dc_1 + \frac{dy}{dc_2} dc_2 + \&c. + \frac{dy}{dc_{2n}} dc_{2n}.$$

* The proof of this theorem of M. Jacobi will be found in the note on p. 84. As it properly belongs to the Integral Calculus, I have, in the text, contented myself with enunciating it.

† These constants are, of course, indefinitely small.

But as the new value of y is derived from the former one by an indefinitely small change in the *values* of the arbitrary constants which it contains, it is plain that this value will still continue to satisfy the equation

$$\beta = 0.$$

If, therefore, we make

$$\delta y = \frac{dy}{dc_1} dc_1 + \frac{dy}{dc_2} dc_2 + \&c. + \frac{dy}{dc_{2n}} dc_{2n},$$

it is evident that this value of δy will satisfy the equation

$$\delta\beta = 0.$$

But since dc_1, dc_2, ... dc_{2n} are arbitrary increments, we may replace them by the arbitrary constants C_1, C_2, ... C_{2n}. Hence it appears that the expression

$$\delta y = C_1 \frac{dy}{dc_1} + C_2 \frac{dy}{dc_2} + \&c. + C_{2n} \frac{dy}{dc_{2n}}$$

will *satisfy* the equation

$$\delta\beta = 0.$$

But as β contains, in general,

$$y, \frac{dy}{dx} \cdots \frac{d^{2n}y}{dx^{2n}},$$

it is plain that $\delta\beta$ will contain

$$\delta y, \frac{d\delta y}{dx} \cdots \frac{d^{2n}\delta y}{dx^{2n}}.$$

The equation

$$\delta\beta = 0$$

will therefore be of the order $2n$ in δy. Hence the value given above for δy contains a number of arbitrary constants equal to the order of the equation. It is, therefore, the complete integral.

Lemma IV.

44. If β be the coefficient of δy under the sign of integration in δU, the second variation will be given by the equation

$$\delta^2 U = \int_{x_0}^{x_1} \delta\beta \delta y dx.$$

For in the case which we are now considering, that, namely, in which the limiting values of

$$y, \frac{dy}{dx}, \frac{d^2y}{dx^2} \cdots \frac{d^{n-1}y}{dx^{n-1}}$$

are given, the first variation becomes

$$\delta U = \int \beta \delta y dx.$$

Taking the variation of both sides of this equation, and recollecting that

$$\delta^2 y = 0,$$

we have

$$\delta^2 U = \int_{x_0}^{x_1} \delta\beta\delta y dx.$$

45. These Lemmas being premised, we shall next proceed to consider the mode of effecting the transformation indicated in Art. 40; and, to render the reasoning more easily intelligible, we shall commence with the simplest cases.

(1.) Let

$$V = f\left(x, y, \frac{dy}{dx}\right).$$

In this case (Art. 11)

$$\beta = N - \frac{dP_1}{dx},$$

and therefore (Lemma I.),

$$\delta\beta = A\delta y + \frac{d \,.\, A_1 \dfrac{d\delta y}{dx}}{dx}.$$

Let c_1, c_2 be the arbitrary constants contained in the integral of the equation

$$\beta = 0,$$

and assume

$$u = C_1 \frac{dy}{dc_1} + C_2 \frac{dy}{dc_2}.$$

Then (Lemma III.) the equation

$$\delta\beta = 0$$

is satisfied by making

$$\delta y = u,$$

and therefore (Lemma II.) $u\delta\beta$ is a perfect differential, whose integral is of the form

$$B_1 \frac{d}{dx}\left(\frac{\delta y}{u}\right).$$

Assume

$$\delta y = u\delta' y,$$

and substitute this value in the second variation,

$$\int_{x_0}^{x_1} \delta\beta\,\delta y\,dx.$$

This gives

$$\delta^2 U = \int_{x_0}^{x_1} u\delta\beta\,\delta' y\,dx. \qquad \text{(a)}$$

But since

$$\int u\delta\beta dx = B_1 \frac{d}{dx}\left(\frac{\delta y}{u}\right) = B_1 \frac{d\delta' y}{dx},$$

if we integrate the right hand member of equation (a) by parts, we have

$$\delta^2 U = \left(B_1 \frac{d\delta' y}{dx}\right)_1 \delta' y_1 - \left(B_1 \frac{d\delta' y}{dx}\right)_0 \delta' y_0 - \int_{x_0}^{x_1} B_1 \left(\frac{d\delta' y}{dx}\right)^2 dx.$$

Now since the limiting values of y are fixed, we must have

$$\delta y_1 = 0, \quad \delta y_0 = 0,$$

and, therefore,

$$u_1\delta' y_1 = 0, \quad u_0\delta' y_0 = 0.$$

If, therefore, it be possible to determine the arbitrary constants which enter into the value of u in such a way that neither u_0 nor u_1 may vanish, we have, necessarily,

$$\delta' y_1 = 0, \quad \delta' y_0 = 0;$$

and, therefore,

$$\delta^2 U = -\int_{x_0}^{x_1} B_1 \left(\frac{d\delta' y}{dx}\right)^2 dx.$$

Moreover, we have seen (Lemmas I. II.) that

$$B_1 = u^2 A_1 = -u^2 \frac{d^2 V}{\left(d \cdot \frac{dy}{dx}\right)^2};$$

also,

$$\delta' y = \frac{\delta y}{u}, \quad \frac{d\delta' y}{dx} = u \frac{\frac{d\delta y}{dx} - \frac{du}{dx}\delta y}{u^2}.$$

Making these substitutions in the value of $\delta^2 U$, we find

$$\delta^2 U = \int_{x_0}^{x_1} \frac{d^2 V}{\left(d \cdot \frac{dy}{dx}\right)^2} \left(\frac{1}{u}\frac{du}{dx}\delta y - \frac{d\delta y}{dx}\right)^2 dx. \quad \text{(b)}$$

46. From this expression it immediately appears that the given integral U will not admit of either maximum or minimum if

$$\frac{d^2 V}{\left(d \cdot \frac{dy}{dx}\right)^2} \ (= Q_1)$$

change its sign between the limits of integration. For since we are at liberty to assume for δy $\{= i\psi(x)\}$ any function which vanishes at the limits of integration, and satisfies the conditions

$$\psi(x) < \infty, \quad \frac{d\psi(x)}{dx} < \infty,$$

for all values of x lying between the limiting values, it is evident that we may, consistently with these conditions, suppose

$$\delta y = 0, \quad \frac{d\delta y}{dx} = 0,$$

for all values of x which make Q_1 negative. This supposition would render $\delta^2 U$ positive. But if we suppose δy to satisfy the same conditions for all values of x which render Q_1 positive, it is plain that we shall have $\delta^2 U$ negative. As, therefore, the second variation, $\delta^2 U$, may, in this case, be made either positive or negative, the given integral will not admit of either maximum or minimum.

Again, it is necessary to the existence of either maximum or minimum, that the quantity under the integral sign in $\delta^2 U$ should not become infinite for any value of x lying between the limiting values. Whether or not this condition be fulfilled, will be seen immediately when the complete integral of the equation

$$\beta = 0$$

is known. For in that case the values of the several quantities,

$$Q_1,\ u,\ \frac{du}{dx},$$

are found by simple differentiation in terms of x. Let the values so found be substituted in the expression

$$\frac{d^2 V}{\left(d \cdot \frac{dy}{dx}\right)^2} \left(\frac{1}{u}\frac{du}{dx}\delta y - \frac{d\delta y}{dx}\right)^2,$$

and it will at once appear whether or not it be possible to determine the arbitrary constants which enter into u, so as to render this expression finite for all values of x lying between x_0 and x_1.

Lastly, it is, in general, necessary to the existence of either maximum or minimum that no admissible value of δy satisfy the equation

$$\frac{1}{u}\frac{du}{dx}\delta y - \frac{d\delta y}{dx} = 0,$$

or

$$\delta y = u.\text{*}$$

Hence it is easily seen that the integral will not, in general, admit of either maximum or minimum, if it be possible to determine u so as to satisfy the conditions

$$u_1 = 0, \quad u_0 = 0.$$

This is also evident from the equation

$$\delta^2 U = \int_{x_0}^{x_1} \delta\beta\, \delta y\, dx;$$

for we have seen (Lemma III.) that the supposition

$$\delta y = u$$

satisfies the equation

$$\delta\beta = 0,$$

* It is plain from the form of u that the arbitrary constant may be neglected.

and therefore reduces the second variation to zero. If, therefore, u be an *admissible* value for δy, i. e. if it be possible to determine the arbitrary constants which enter into u, so as to cause its limiting values to vanish, there will be, in general, neither maximum nor minimum.

47. If Q_1 remain finite for all values of x between x_0 and x_1, it is evident, from the form to which the second variation has been reduced, that the quantity under the sign of integration can be rendered infinite only by one of the suppositions,

$$\frac{du}{dx} = \infty, \text{ or } u = 0.$$

We have before seen (Art. 46) that the given integral will not, in general, admit of either maximum or minimum, if it be possible so to determine C_1 and C_2 as to satisfy the equations

$$u_1 = 0, \quad u_0 = 0.$$

Again, it appears from the latter of the two foregoing suppositions, that it is further necessary that it be possible to determine these constants in such a way that u may not vanish for any value of x between x_0 and x_1. These cases admit of important geometrical interpretations.

Let AB (fig. 2) be a curve whose equation is

$$y = f(x, c_1, c_2),$$

the integral of the equation

$$\beta = 0;$$

and let A, B be the points corresponding to the limiting values of y and x.

Then, in the first of the above-mentioned cases it is evidently possible to draw between A and B a *second* curve, satisfying the same general equation, and indefinitely near the first. For it has been already shown that the curve whose equation is

$$y = f(x, c_1, c_2) + u,$$

satisfies the general equation

$$\beta = 0.$$

And since

$$u_1 = 0, \quad u_0 = 0,$$

it is plain that this curve passes through the points *A*, *B*.

In the second case, it is possible to draw through *A* and some intermediate point *C* a curve satisfying the general differential equation, and indefinitely near to the first. For if we make

$$\frac{C_1}{C_2} = -\frac{\left(\frac{dy}{dc_2}\right)_0}{\left(\frac{dy}{dc_1}\right)_0},$$

it is plain that we shall have

$$u_0 = 0.$$

Now if u vanish for some intermediate value of x, it is easily seen, as in the former case, that there must be some intermediate point *C* at which the curve

$$y = f(x, c_1, c_2) + u$$

will intersect the curve *AB*. Hence it is evidently possible to draw the curve as above described.

We have, therefore, the following rule:

Let *AB* be the curve whose differential equation is

$$\beta = 0,$$

and which passes through the given points *A*, *B*. Draw through *A* a curve satisfying the same general equation, and indefinitely near to *AB*. This curve will, in general, meet the curve *AB* at some other point *C*. Then it appears from the foregoing discussion that the given integral will not, in general, admit of a maximum or minimum unless *C* fall *beyond B*.

48. As an example of the foregoing theory, let us consider the case discussed in Ex. 1.

Here we have

$$V = \sqrt{\left(1 + \frac{dy^2}{dx^2}\right)}$$

whence

$$Q_1 = \frac{d^2 V}{\left(d \cdot \frac{dy}{dx}\right)^2} = \frac{1}{\left(1 + \frac{dy^2}{dx^2}\right)^{\frac{3}{2}}}.$$

Now as the radical in V is supposed to be taken *positively* throughout the integration, it is plain that Q_1 will continue always *positive.* This is the first condition requisite to the existence of a *minimum* value of the integral.

Let us now consider whether the other conditions be satisfied. The complete integral of the equation

$$\beta = 0$$

being, in this case, of the form

$$y = c_1 x + c_2,$$

we have

$$\frac{dy}{dc_1} = x, \quad \frac{dy}{dc_2} = 1,$$

and, therefore,

$$u = C_1 x + C_2, \quad \frac{du}{dx} = C_1.$$

Substituting these values in equation (b), p. 88, we have

$$\delta^2 U = \frac{1}{\sqrt{(1 + C_1{}^2)^3}} \int_{x_0}^{x_1} \left(\frac{C_1}{C_1 x + C_2} \delta y - \frac{d\delta y}{dx} \right)^2 dx.$$

Now if we assume values for C_1 and C_2, such that $C_1 x + C_2$ may not vanish for any value of x between x_0 and x_1, it is plain that the quantity under the sign of integration cannot become infinite.

Lastly, it is easily seen that no admissible value of δy will cause the second variation to vanish. For we have before seen (p. 89) that the only value of δy which satisfies this condition is

$$\delta y = u,$$

and that, in order to render this an *admissible* value, we should have

$$u_1 = 0, \quad u_0 = 0.$$

These equations become, in the present case,

$$C_1 x_1 + C_2 = 0, \quad C_1 x_0 + C_2 = 0,$$

which are manifestly impossible for any finite values of C_1, C_2. Hence it is evident that the value of y found in Ex. 1 renders the given integral a *minimum.*

The conclusion here arrived at may be extended beyond the individual case. For we have seen, p. 52 (a), that, in general, if

$$V = f\left(\frac{dy}{dx}\right),$$

the complete integral of the equation

$$\beta = 0$$

will be

$$y = c_1 x + c_2;$$

we, have, therefore, as before,

$$u = C_1 x + C_2.$$

Hence it will readily appear, as in the preceding article, that such values may be assigned to C_1 and C_2 as to render the quantity under the sign of integration in $\delta^2 U$ *finite* for all values of x between x_0 and x_1; and further, that no admissible value of δy will cause $\delta^2 U$ to vanish.

Again, the value of Q_1 being

$$Q_1 = \frac{d^2 V}{\left(d.\frac{dy}{dx}\right)^2} = \frac{d^2 V}{dc_1^2};$$

and therefore constant, cannot change its sign. Hence, and from p. 52, we infer—

(1.) That every integral of the form

$$\int_{x_0}^{x_1} f\left(\frac{dy}{dx}\right) dx$$

(which is not in itself integrable) admits of a maximum or minimum value when the limiting values of y and x are fixed.

(2.) That this value is found by making

$$\frac{dy}{dx} = c_1,$$

and is, therefore,

$$(x_1 - x_0) f(c_1).$$

(3.) That, as in all other cases, the maximum and minimum

values are distinguished from each other by the sign of Q_1, or (in this case) by that of

$$\frac{d^2 f(c_1)}{dc_1^2}.*$$

49. Let

$$V = f\left(x, y, \frac{dy}{dx}, \frac{d^2y}{dx^2}\right).$$

In this case (Art. 11)

$$\beta = N - \frac{dP_1}{dx} + \frac{d^2 P_2}{dx^2},$$

and therefore (Lemma I.)

$$\delta\beta = A\delta y + \frac{d.A_1\dfrac{d\delta y}{dx}}{dx} + \frac{d^2.A_2\dfrac{d^2\delta y}{dx^2}}{dx^2}.$$

Denoting by c_1, c_2, c_3, c_4 the arbitrary constants which enter into the integral of the equation

$$\beta = 0,$$

and assuming

$$u = C_1\frac{dy}{dc_1} + C_2\frac{dy}{dc_2} + C_3\frac{dy}{dc_3} + C_4\frac{dy}{dc_4},$$

it appears, as before, that the equation

$$\delta\beta = 0$$

is satisfied by making

$$\delta y = u;$$

and that, therefore,

$$u\left(A\delta y + \frac{d.A_1\dfrac{d\delta y}{dx}}{dx} + \frac{d^2.A_2\dfrac{d^2\delta y}{dx^2}}{dx^2}\right)dx$$

is integrable.

Assuming, as before,

$$\delta y = u\delta' y,$$

* As other examples of this theory will be given in the following chapter, I do not think it necessary here to multiply them further.

and integrating by parts, we have

$$\int_{x_0}^{x_1} \delta\beta\delta y dx = \int_{x_0}^{x_1} u\delta\beta\delta' y dx = -\int_{x_0}^{x_1} \frac{d\delta' y}{dx}\left(B_1 \frac{d\delta' y}{dx} + \frac{d\,.\,B_2 \frac{d^2\delta' y}{dx^2}}{dx}\right) dx\,; \quad \text{(c)}$$

the parts outside the sign of integration vanishing, as before, at the limits.

Now if C'_1, C'_2, C'_3, C'_4 be a new system of arbitrary constants, and if we make

$$u' = C'_1 \frac{dy}{dc_1} + C'_2 \frac{dy}{dc_2} + C'_3 \frac{dy}{dc_3} + C'_4 \frac{dy}{dc_4},$$

it is easily seen that the supposition

$$\delta' y = \frac{u'}{u}$$

will satisfy the equation

$$B_1 \frac{d\delta' y}{dx} + \frac{d\,.\,B_2 \frac{d^2\delta' y}{dx^2}}{dx} = 0. \quad \text{(d)}$$

For, since

$$\int u\delta\beta dx = B_1 \frac{d\delta' y}{dx} + \frac{d\,.\,B_2 \frac{d^2\delta' y}{dx^2}}{dx},$$

it is plain that every value of δy which satisfies the equation

$$\delta\beta = 0$$

will also satisfy equation (d). But the supposition

$$\delta' y = \frac{u'}{u}$$

gives

$$\delta y = u',$$

and therefore $\delta\beta = 0$ (Lemma III.)

Hence it appears that this supposition satisfies equation (d), and therefore renders the expression

$$\frac{d\,.\left(\frac{u'}{u}\right)}{dx}\left(B_1 \frac{d\delta' y}{dx} + \frac{d\,.\,B_2 \frac{d^2\delta' y}{dx^2}}{dx}\right)$$

integrable.

Proceeding as before, assume

$$\frac{d\delta y}{dx} = \frac{d\left(\frac{u'}{u}\right)}{dx}\delta'' y,$$

and integrate by parts the second member of equation (c).

Then, since, by the principle of Lemma II.,

$$\int \frac{d\left(\frac{u'}{u}\right)}{dx}\left(B_1\frac{d\delta y}{dx} + \frac{d \, . \, B_2\frac{d^2\delta y}{dx^2}}{dx}\right)dx = E_1\frac{d\delta'' y}{dx},$$

we shall have, ultimately,

$$\int_{x_0}^{x_1} \delta\beta\delta y dx = \int_{x_0}^{x_1} E_1\left(\frac{d\delta'' y}{dx}\right)^2 dx.$$

The conclusions derived from this transformation are analogous to those obtained in the preceding case:

(1.) It is essential to the existence of a maximum or minimum value that E_1 should not change sign within the limits of integration. But since

$$E_1 = \left(\frac{d \, . \, \frac{u'}{u}}{dx}\right)^2 . \, B_2 = \left(\frac{d \, . \, \frac{u'}{u}}{dx}\right)^2 . u^2 A_2 = \left(\frac{d \, . \, \frac{u'}{u}}{dx}\right)^2 \frac{u^2 d^2 V}{\left(d \, . \, \frac{d^2 y}{dx^2}\right)^2},$$

it is plain that the sign of E_1 is the same as that of

$$\frac{d^2 V}{\left(d \, . \, \frac{d^2 y}{dx^2}\right)^2} = (Q_2).$$

Hence, in the case of a maximum, we must have

$$Q_2 < 0,$$

and in that of a minimum,

$$Q_2 > 0,$$

for all values of x between x_0 and x_1.

(2.) It is, moreover, necessary that the quantity under the sign of integration,

$$E_1\left(\frac{d\delta'' y}{dx}\right),$$

should not become infinite for any value of x within the limiting values.

But since

$$\frac{d.\frac{u'}{u}}{dx}\delta''y = \frac{d\delta'y}{dx} = \frac{d.\frac{\delta y}{u}}{dx},$$

if we assume, for the sake of brevity,

$$v = \frac{u'}{u},$$

and differentiate, we shall find

$$\frac{d\delta''y}{dx} = \frac{1}{\frac{dv}{dx}} \frac{d^2.\frac{\delta y}{u}}{dx^2} - \frac{1}{\frac{dv^2}{dx^2}} \frac{d^2v}{dx^2} \frac{d.\frac{\delta y}{u}}{dx}.$$

Substituting this value, we find

$$\delta^2 U = \int_{x_0}^{x_1} Q_2 \left(A\delta y + A_1 \frac{d\delta y}{dx} + \frac{d^2\delta y}{dx^2}\right)^2 dx,$$

where

$$A = \frac{\frac{du}{dx}\frac{d^2u'}{dx^2} - \frac{du'}{dx}\frac{d^2u}{dx^2}}{u\frac{du'}{dx} - u'\frac{du}{dx}},$$

$$A_1 = \frac{u\frac{d^2u'}{dx^2} - u'\frac{d^2u}{dx^2}}{u\frac{du'}{dx} - u'\frac{du}{dx}}.$$

Hence we have, in the case of a maximum,

(1.) $Q_2 < 0$ for all values of x between x_0 and x_1.

(2.) $Q_2\left(A\delta y + A_1 \frac{d\delta y}{dx} + \frac{d^2\delta y}{dx^2}\right)^2$ not infinite.

(3.) The equation

$$A\delta y + A_1 \frac{d\delta y}{dx} + \frac{d^2\delta y}{dx^2} = 0$$

impossible for any admissible value of δy.

For a minimum the conditions are:

(1.) $Q_2 > 0$ for all values of x between x_0 and x_1.

(2.) }
(3.) } As before.

Now if the general solution of the equation

$$\beta = 0$$

be known, the quantities u, u' are known also in terms of x and arbitrary constants. If then from these values we form the values of A and A_1, it will immediately appear whether or not these conditions be satisfied.

50. Conclusions similar to those of the preceding article may be drawn as to the geometrical interpretation of the foregoing conditions.

Thus if the equation

$$A\delta y + A_1 \frac{d\delta y}{dx} + \frac{d^2\delta y}{dx^2} = 0$$

be satisfied by any admissible value of δy, it is easy to see that if AB (fig. 3) be a curve whose equation is

$$\beta = 0,$$

and which has, at the points A, B, the given limiting values of

$$x,\ y,\ \frac{dy}{dx},$$

it is possible to draw a second curve indefinitely near to the first, satisfying the same differential equation, and touching the curve AB at the points A, B.

For it will readily appear, as in the preceding case, that the general solution of

$$A\delta y + A_1 \frac{d\delta y}{dx} + \frac{d^2\delta y}{dx^2} = 0$$

is

$$\delta y = u.$$

If, then, this be an admissible value of δy, we must have

$$u_1 = 0,\quad u_0 = 0,\quad \left(\frac{du}{dx}\right)_1 = 0,\quad \left(\frac{du}{dx}\right)_0 = 0. \qquad \text{(c)}$$

Now the equation of a curve indefinitely near to the curve AB, and satisfying the same differential equation, will evidently be of the form

$$y' = y + u,$$

y being the ordinate of the curve AB. And it is plain that if u satisfy the four equations (c), this second curve will touch the curve AB in the points A, B.

And it may be readily shown, as in the foregoing case, that the given integral will not admit of either maximum or minimum, if a second curve can be drawn, as above described, between A and any intermediate point C.

51. As an example of the foregoing theory, let

$$V = \left(\frac{d^2y}{dx^2}\right)^m,$$

and let it be required to determine y, such as to make

$$\int_{x_0}^{x_1} V dx$$

a maximum or minimum, the limiting values of x, y, and $\frac{dy}{dx}$ being given.

In this case we have

$$N = 0, \quad P_1 = 0, \quad P_2 = m\left(\frac{d^2y}{dx^2}\right)^{m-1}.$$

The equation

$$\beta = 0$$

becomes, therefore,

$$\frac{d^2P_2}{dx^2} = 0;$$

or, integrating twice, and substituting for P_2 its value,

$$\frac{d^2y}{dx^2} = (ax + b)^{\frac{1}{m-1}}.$$

Hence it is easy to see that the complete integral will be

$$y = (c_1x + c_2)^{\frac{2m-1}{m-1}} + c_3x + c_4. \qquad \text{(a)}$$

From this we find

$$Q_2 = m.m-1.\left(\frac{d^2y}{dx^2}\right)^{m-2} = m.m-1.(ax+b)^{\frac{m-2}{m-1}}.$$

Let us suppose m to be an integer. Now it is evident that if m be even, the factor

$$(ax+b)^{\frac{m-2}{m-1}}$$

is always positive. If m be odd, this factor is positive or negative according as we agree to take the positive or negative sign of the radical

$$\pm\,(ax+b)^{\frac{1}{m-1}}.$$

In either case Q_2 cannot *change* its sign. The remaining factor, $m.m-1$, is evidently always positive, while m is any integer different from unity. But, as the effect of giving a negative sign to

$$(ax+b)^{\frac{1}{m-1}}$$

would merely be to render $\frac{d^2y}{dx^2}$, and therefore V, negative, it is plain that, if we understand by a *minimum* that value which is nearest to zero, the value here found will in all cases be a minimum value.

Let us next consider whether the other conditions stated in the previous Article be fulfilled.

Putting, for the sake of brevity,

$$\frac{2m-1}{m-1} = n,$$

and differentiating equation (a), with respect to the arbitrary constants which it contains, we find

$$\frac{dy}{dc_1} = n(c_1x+c_2)^{n-1}x, \quad \frac{dy}{dc_2} = n(c_1x+c_2)^{n-1},$$

$$\frac{dy}{dc_3} = x, \quad \frac{dy}{dc_4} = 1.$$

Hence

$$u = n(c_1x + c_2)^{n-1}(C_1x + C_2) + C_3x + C_4;$$

and, similarly,

$$u' = n(c_1x + c_2)^{n-1}(C'_1x + C'_2) + C'_3x + C'_4.$$

Now it is easy to see that if we make

$$C_1 = 0, \quad C_2 = 0, \quad C'_1 = 0, \quad C'_2 = 0,$$

the second variation will become

$$\delta^2 U = \int_{x_0}^{x_1} Q_2 \left(\frac{d^2\delta y}{dx^2}\right)^2 dx.$$

Hence it is evident that the quantity under the sign of integration in $\delta^2 U$ may always be made finite, provided that Q_2 does not become infinite within the limits of integration. But we have already seen that

$$Q_2 = m.m - 1.(ax + b)^{\frac{m-2}{m-1}}.$$

Now since m is supposed to be an integer, it is plain that

$$\frac{m-2}{m-1}$$

cannot be negative, and therefore that Q_2 cannot become infinite for any finite value of x.

It remains then to consider whether the second variation can be made to vanish by any admissible form of δy. Now we have seen that the most general value of δy, which causes the second variation to vanish, is

$$\delta y = u.$$

In order, therefore, that this may be an admissible value of δy, it is plain that it must be possible to determine the constants which u contains, in such a way as to satisfy the equations

$$u_0 = 0, \quad u_1 = 0, \quad \left(\frac{du}{dx}\right)_0 = 0, \quad \left(\frac{du}{dx}\right)_1 = 0. \qquad \text{(b)}$$

Let us suppose

$$x_0 = 0, \quad y_0 = 0, \quad \left(\frac{dy}{dx}\right)_0 = 0.$$

These conditions reduce the complete integral to

$$y = (ax + b)^n - nab^{n-1}\,x - b^n, \qquad \text{(c)}$$

putting a, b, for c_1, c_2. Hence

$$u = n\,(ax + b)^{n-1}\,(C_1x + C_2) + C_3x + C_4,$$

$$\frac{du}{dx} = nC_1\,(ax + b)^{n-1} + n\,.\,\overline{n-1}\,(C_1x + C_2)\,(ax + b)^{n-2}\,a + C_3.$$

The four conditions (b) give, therefore,

$$nb^{n-1}\,C_2 + C_4 = 0;$$
$$nC_1b^{n-1} + n\,.\,\overline{n-1}\;C_2b^{n-2}\,a + C_3 = 0;$$
$$n\,(ax_1 + b)^{n-1}\,(C_1x_1 + C_2) + C_3x_1 + C_4 = 0;$$
$$nC_1(ax_1 + b)^{n-1} + n\,.\,\overline{n-1}\,(C_1x_1 + C_2)\,(ax_1 + b)^{n-2}\,a + C_3 = 0.$$

Eliminating the several quantities,

$$\frac{C_2}{C_1}, \quad \frac{C_3}{C_1}, \quad \frac{C_4}{C_1}, \quad a, \; b,$$

between these equations and the equations

$$\left.\begin{aligned} y_1 &= (ax_1 + b)^n - nab^{n-1}\,x_1 - b^n \\ \left(\frac{dy}{dx}\right)_1 &= na\,(ax_1 + b)^{n-1} - nab^{n-1}, \end{aligned}\right\} \qquad \text{(d)}$$

formed by the substitution of the given limiting values,

$$x_1, \; y_1, \; \left(\frac{dy}{dx}\right)_1,$$

in (c) and its first differential, we shall finally arrive at an equation containing these limiting values only. If this equation be satisfied either by these limiting values, or by any other system of values of

$$x, \; y, \; \frac{dy}{dx},$$

which is found within the limits of integration, the given integral will not, in general, admit of either maximum or minimum.

52. The question may, however, be treated in somewhat a simpler manner, by the geometrical considerations of p. 98. For we have there seen that, if it be possible to make the second variation vanish, a curve may be drawn between the extreme points A, B, indefinitely near to the first, satisfying the same general equation, and having the same limiting values of the differential coefficient. Now it is evident that if the extreme points be such as to satisfy this condition, we must have

$$\left(\frac{dy}{da}\right)_0 + \left(\frac{dy}{db}\right)_0 \frac{db}{da} = 0,$$

$$\left(\frac{d^2y}{dxda}\right)_0 + \left(\frac{d^2y}{dxdb}\right)_0 \frac{db}{da} = 0,$$

$$\left(\frac{dy}{da}\right)_1 + \left(\frac{dy}{db}\right)_1 \frac{db}{da} = 0,$$

$$\left(\frac{d^2y}{dxda}\right)_1 + \left(\frac{d^2y}{dxdb}\right)_1 \frac{db}{da} = 0.$$

Substituting the values of these differential coefficients derived from (c), it is easily seen that the first two are identical, as they ought to be.

The third and fourth become, by the same process,

$$\{(ax_1 + b)^{n-1} - b^{n-1}\}x_1 + \{(ax_1 + b)^{n-1} - b^{n-1} - (n-1)ab^{n-2}x_1\}\frac{db}{da} = 0,$$

$$(ax_1 + b)^{n-1} - b^{n-1} + (n-1)(ax_1 + b)^{n-2}ax_1$$
$$+ (n-1)a\{(ax_1 + b)^{n-2} - b^{n-2}\}\frac{db}{da} = 0.$$

Eliminating $\frac{db}{da}$ between these equations, and reducing, we have

$$\{(ax_1 + b)^{n-1} - b^{n-1}\}^2 = (n-1)^2 a^2 x_1^2 (ax_1 + b)^{n-2} b^{n-2}. \qquad \text{(e)}$$

Eliminating a and b between the equations (e) and (d), we find, as before, an equation between

$$x_1, y_1, \left(\frac{dy}{dx}\right)_1.$$

If this equation be satisfied either by these quantities or by any system of values of

$$x, y, \frac{dy}{dx},$$

which is found within the limits of integration, we infer, as before, that the given integral does not admit of either maximum or minimum.

For example, suppose

$$m = 2, \quad n = \frac{2m - 1}{m - 1} = 3,$$

equation (e) becomes, in this case,

$$\{(ax_1 + b)^2 - b^2\}^2 = 4a^2x_1^2(ax_1 + b)\,b,$$

or, by reduction,

$$a^4x_1^4 = 0\,;$$

or, since x_1 cannot vanish,

$$a = 0,$$

an impossible equation, since it would reduce equation (c) to

$$y_1 = 0.$$

It is, therefore, in the present case, impossible to draw a curve in the manner stated at the commencement of this article.

Hence, and from p. 100, we infer that the value of the given integral is in this case a real *minimum*.

53. The mode of applying the theory of M. Jacobi to functions involving differential coefficients of an order higher than the second, being strictly analogous to that which has been stated in the preceding pages, it may be sufficient here to give a brief sketch of this extension.

(1.) The second variation being, in general, of the form

$$\int_{x_0}^{x_1}\left(A\delta y + \frac{d.A_1\dfrac{d\delta y}{dx}}{dx} + \&c. + \frac{d^n.A_n\dfrac{d^n\,\delta y}{dx^n}}{dx^n}\right)dx,$$

assume, as before,

$$u = C_1\frac{dy}{dc_1} + C_2\frac{dy}{dc_2} + \&c. + C_{2n}\frac{dy}{dc_{2n}},$$

$$\delta y = u\delta' y,$$

and integrate the expression by parts. It will thus be reduced to the form

$$\delta^2 U = -\int_{x_0}^{x_1} \left(B_1 \frac{d\delta' y}{dx} + \&c. + \frac{d^{n-1}.B_n \frac{d^n \delta' y}{dx^n}}{dx^{n-1}} \right) \frac{d\delta' y}{dx}\, dx.$$

(2.) Assume

$$u' = C'_1 \frac{dy}{dc_1} + C'_2 \frac{dy}{dc_2} + \&c. + C'_{2n} \frac{dy}{dc_{2n}},$$

$$\frac{d\delta' y}{dx} = \frac{d\,.\frac{u'}{u}}{dx} \delta'' y,$$

and integrate again by parts. The second variation will thus be reduced a second order.

(3.) Assuming, for the sake of brevity,

$$v = \frac{d\,.\,\frac{u'}{u}}{dx},$$

let v' be a quantity formed from v in the same way in which u' was formed from u, namely, by a change in the arbitrary constants which it contains, and make

$$\frac{d\delta'' y}{dx} = \frac{d\,.\frac{v'}{v}}{dx} \delta''' y.$$

Integrate by parts as before, and proceed with similar assumptions and integrations, until the second variation is reduced to the form

$$\pm \int_{x_0}^{x_1} E_n \left(\frac{d\delta^{(n)} y}{dx} \right)^2 dx,$$

the upper sign being taken if n be even, and the lower, if n be odd.

(4.) It is evident, from Lemmas I. and II., that this quantity has the same sign as Q_n. For it appears, from the concluding remark of Lemma I., that

$$B_n = A_n u^2.$$

Similarly, if D_n be the coefficient of

$$\frac{d^{n-1}\delta''y}{dx^{n-1}}$$

in the quantity which remains after the next integration,

$$D_n = B_n\left(\frac{d\,.\,\frac{u'}{u}}{dx}\right)^2 = A_n u^2\left(\frac{d\,.\,\frac{u'}{u}}{dx}\right)^2;$$

and, finally,

$$E_n = A_n u^2\left(\frac{d\,.\,\frac{u'}{u}}{dx}\right)^2\ldots\ldots$$

But from Lemma I. we have

$$A_n = \pm\, Q_n.$$

Hence in all cases the form to which the second variation has been reduced is

$$\int_{x_0}^{x_1} Q_n u^2\left(\frac{d\,.\,\frac{u'}{u}}{dx}\right)^2\ldots\ldots\left(\frac{d\delta^{(n)}y}{dx}\right)^2.$$

It appears, therefore, as in the preceding cases, that for a maximum we must have

$$Q_n < 0,$$

and for a minimum,

$$Q_n > 0,$$

for all values of x from x_0 to x_1.

(5.) It is easily seen that the form to which we have reduced $\delta^2 U$ is identical with that indicated in Art. 40. For

$$\delta' y = \frac{\delta y}{u},$$

$$\frac{d\delta' y}{dx} = \frac{d}{dx}\left(\frac{\delta y}{u}\right),$$

$$\frac{d\delta'' y}{dx} = \frac{d}{dx}\left(\frac{1}{v}\,\frac{d\delta' y}{dx}\right) = \frac{d}{dx}\left\{\frac{1}{v}\,\frac{d}{dx}\left(\frac{\delta y}{u}\right)\right\},$$

$$\frac{d\delta''' y}{dx} = \&c.$$

&c. &c. &c.

If these several differentiations be actually performed, it is evident that the form of $\delta^2 U$ will be as stated in Art. 40.

(6.) If A, B be the given extreme points of the curve satisfying the several conditions of the question, the given integral will not, in general, admit of either maximum or minimum, if between A and B there be found two points, C, D, such that a *second* curve may be drawn through them indefinitely near to the first, satisfying the same differential equation, and having at the points C, D the same values of

$$\frac{dy}{dx}, \frac{d^2y}{dx^2}, \dots \frac{d^{n-1}y}{dx^{n-1}}.$$

(7.) As the determination of the curve which satisfies all the conditions of the question depends upon the solution of the system of equations formed by the substitution of the given limiting values of x, y, &c., in

$$\beta = 0,$$

it is plain that there will be, in general, more than one such curve. Now if two of these curves become coincident, i. e., if the equations by which they are determined have one or more pairs of equal roots, it appears from (6) that the given integral does not admit of either maximum or minimum.

Prop. VI.

53. To determine the functions y, z, which will render $\int_{x_0}^{x_1} V dx$ a maximum or minimum, where

$$V = f\left(x, y, \frac{dy}{dx} \dots \frac{d^n y}{dx^n}, z, \frac{dz}{dx} \dots \frac{d^m z}{dx^m}\right).$$

The equation

$$DU = 0$$

becomes, in this case (Chap. II. Prop. V.),

$$V_1 dx_1 - V_0 dx_0 + \left(P_1 - \frac{dP_2}{dx} + \&c.\right)_1 \delta y_1 - \left(P_1 - \frac{dP_2}{dx} + \&c.\right)_0 \delta y_0$$

$$+ (P_2 - \&c.)_1 \left(\frac{d\delta y}{dx}\right)_1 - (P_2 - \&c.)_0 \left(\frac{d\delta y}{dx}\right)_0$$

$$+ \&c. \qquad \text{(A)}$$

$$+ \left(P_n \frac{d^{n-1}\delta y}{dx^{n-1}}\right)_1 - \left(P_n \frac{d^{n-1}\delta y}{dx^{n-1}}\right)_0$$

$$+ \int_{x_0}^{x_1} \left(N - \frac{dP_1}{dx} + \frac{d^2P_2}{dx^2} - \&c. + (-1)^n \frac{d^nP_n}{dx^n}\right) \delta y dx$$

$$+ \left(P'_1 - \frac{dP'_2}{dx} + \&c.\right)_1 \delta z_1 - \left(P'_1 - \frac{dP'_2}{dx} + \&c.\right)_0 \delta z_0$$

$$+ (P'_2 - \&c.)_1 \left(\frac{d\delta z}{dx}\right)_1 - (P'_2 - \&c.)_0 \left(\frac{d\delta z}{dx}\right)_0$$

$$+ \&c.$$

$$+ \left(P'_m \frac{d^{m-1}\delta z}{dx^{m-1}}\right)_1 - \left(P'_m \frac{d^{m-1}\delta z}{dx^{m-1}}\right)_0$$

$$+ \int_{x_0}^{x_1} \left(N' - \frac{dP'_1}{dx} + \frac{d^2P'_1}{dx^2} - \&c. + (-1)^m \frac{d^mP'_m}{dx^m}\right) \delta z dx$$

$$= 0.$$

If the functions y, z be independent of each other, their variations, δy, δz, will also be independent; and, therefore, by the same reasoning as that employed in Prop. II., we must have the general equations,

$$\left.\begin{aligned} N - \frac{dP_1}{dx} + \frac{d^2P_2}{dx^2} - \&c. + (-1)^n \frac{d^nP_n}{dx^n} = 0, \\ N' - \frac{dP'_1}{dx} + \frac{d^2P'_2}{dx^2} - \&c. + (-1)^m \frac{d^mP'_m}{dx^m} = 0; \end{aligned}\right\} \qquad \text{(B)}$$

and the equation at the limits,

$$V_1dx_1 - V_0dx_0 + \left(P_1 - \frac{dP_2}{dx} + \&c.\right)_1 \delta y_1 - \left(P_1 - \frac{dP_2}{dx} + \&c.\right)_0 \delta y_0$$

$$+ (P_2 - \&c.)_1 \left(\frac{d\delta y}{dx}\right)_1 - (P_2 - \&c.)_0 \left(\frac{d\delta y}{dx}\right)_0$$

$$+ \&c. \qquad \text{(C)}$$

$$+\left(P'_1 - \frac{dP'_2}{dx} + \&c.\right)_1 \delta z_1 - \left(P'_1 - \frac{dP'_2}{dx} + \&c.\right)_0 \delta z_0$$

$$+ (P'_2 - \&c.)_1 \left(\frac{d\delta z}{dx}\right)_1 - (P'_2 - \&c.)_0 \left(\frac{d\delta z}{dx}\right)_0$$

$$+ \&c. = 0.$$

The mode of treating these equations is perfectly analogous to that used in Prop. II., and it appears conclusively, from the reasoning there adopted, that the number of equations is not, in general, affected by any auxiliary equations which may be introduced. For every such equation will either remove a term from (C), by annulling the variation which enters into that term, or it will consolidate two terms into one; thus in all cases diminishing by one the number of equations to which (C) is equivalent, and, at the same time, increasing the number by one, namely, itself. In proving, therefore, that (B) and (C) furnish a sufficient number of equations for the complete solution of the problem, it will be sufficient to show it for a single case, namely, that in which the limiting values of x only are fixed. Now it is easy to see that the equation

$$N - \frac{dP_1}{dx} + \&c. + (-1)^n \frac{d^n P_n}{dx^n} = 0$$

is of the order $2n$ in y, and $m + n$ in z; and the equation

$$N' - \frac{dP'_1}{dx} + \&c. + (-1) \frac{d^m P'_m}{dx^m} = 0$$

of the order $m + n$ in y, and $2m$ in z. These equations are, therefore, of the form

$$\phi\left(x, y, \frac{dy}{dx} \cdots \frac{d^{2n}y}{dx^{2n}}, z, \frac{dz}{dx} \cdots \frac{d^{m+n}z}{dx^{m+n}}\right) = 0,$$

$$\psi\left(x, y, \frac{dy}{dx} \cdots \frac{d^{m+n}y}{dx^{m+n}}, z, \frac{dz}{dx} \cdots \frac{d^{2m}z}{dx^{2m}}\right) = 0.$$

If, according to the ordinary method, we differentiate the first of these $2m$ times, and the second $m + n$ times, we shall thus have $3m + n + 2$ equations, between which, if we eliminate

$$z, \frac{dz}{dx} \cdots \frac{d^{3m+n}z}{dx^{3m+n}},$$

we shall have a resulting equation in y of the order $2(m+n)$, whose solution, therefore, will contain $2(m+n)$ arbitrary constants. Now it is evident that the equations at the limits are precisely $2(m+n)$ in number, namely, the $2n$ equations,

$$\left(P_1 - \frac{dP_2}{dx} + \&c.\right)_0 = 0, \quad \left(P_1 - \frac{dP_2}{dx} + \&c.\right)_1 = 0,$$

$$(P_2 - \&c.)_0 = 0, \quad (P_2 - \&c.)_1 = 0,$$

$$\&c. \qquad \&c.$$

$$(P_n)_0 = 0, \qquad (P_n)_1 = 0;$$

and the $2m$ equations,

$$\left(P'_1 - \frac{dP'_2}{dx} + \&c.\right)_0 = 0, \quad \left(P'_1 - \frac{dP'_2}{dx} + \&c.\right)_1 = 0,$$

$$(P'_2 - \&c.)_0 = 0, \qquad (P'_2 - \&c.)_1 = 0,$$

$$(P'_m)_0 = 0, \qquad (P'_m)_1 = 0.$$

The problem is, therefore, in general, perfectly determinate. Similar exceptions to those noticed in the case of a single dependent variable will of course occur. Thus, for example, if $N = 0$, and neither of the limiting values of y be given, the problem is indeterminate, and the explanation is precisely the same as before.

54. If V do not contain the independent variable, we may find a single equation of the order $2(m+n)-1$ without any particular determination of the form of the function V.

For since $M = 0$, we have

$$dV = Ndy + P_1 d.\frac{dy}{dx} + P_2 d.\frac{d^2y}{dx^2} + \&c. + P_n d.\frac{d^ny}{dx^n}$$

$$+ N'dz + P'_1 d.\frac{dz}{dx} + P'_2 d.\frac{d^2z}{dx^2} + \&c. + P'_m d.\frac{d^mz}{dx^m}.$$

Substituting for N and N' their values, derived from the equations

$$N - \frac{dP_1}{dx} + \&c. = 0,$$

$$N' - \frac{dP'_1}{dx} + \&c. = 0,$$

and proceeding exactly as in Art. (29), we find

$$V = c + P_1 \frac{dy}{dx} + \left(P_2 \frac{d^2y}{dx^2} - \frac{dy}{dx} \cdot \frac{dP_2}{dx}\right) + \&c.$$

$$+ P'_1 \frac{dz}{dx} + \left(P'_2 \frac{d^2z}{dx^2} - \frac{dz}{dx} \cdot \frac{dP'_2}{dx}\right) + \&c.$$

A more particular enumeration of the several cases which may occur is obviously unnecessary.

Example.

55. To find the forms of the functions y and z which will render

$$\int \sqrt{\left(1 + \frac{dy^2}{dx^2} + \frac{dz^2}{dx^2}\right)}\, dx$$

a maximum or minimum.

Here

$$N = 0, \quad N' = 0,$$

$$P_1 = \frac{\frac{dy}{dx}}{\sqrt{\left(1 + \frac{dy^2}{dx^2} + \frac{dz^2}{dx^2}\right)}}, \quad P'_1 = \frac{\frac{dz}{dx}}{\sqrt{\left(1 + \frac{dy^2}{dx^2} + \frac{dz^2}{dx^2}\right)}},$$

$$P_2 = 0, \ \&c. \qquad P'_2 = 0, \ \&c.$$

The equations

$$N - \frac{dP_1}{dx} + \&c. = 0, \quad N' - \frac{dP'_1}{dx} + \&c. = 0,$$

become, therefore,

$$\frac{dP_1}{dx} = 0, \quad \frac{dP'_1}{dx} = 0;$$

or,

$$P_1 = \frac{\frac{dy}{dx}}{\sqrt{\left(1 + \frac{dy^2}{dx^2} + \frac{dz^2}{dx^2}\right)}} = c, \quad P'_1 = \frac{\frac{dz}{dx}}{\sqrt{\left(1 + \frac{dy^2}{dx^2} + \frac{dz^2}{dx^2}\right)}} = c'.$$

These equations, if solved for the two differential coefficients, will evidently give results of the form

$$\frac{dy}{dx} = m, \qquad \frac{dz}{dx} = n;$$

or, finally,

$$y = mx + p, \qquad z = nx + q.$$

To determine the arbitrary constants:

(1.) If the limiting values of x, y, z be given, the constants m, n, p, q are determined by the four equations,

$$y_1 = mx_1 + p, \qquad z_1 = nx_1 + q,$$
$$y_0 = mx_0 + p, \qquad z_0 = nx_0 + q.$$

(2.) Let the limiting values of x only be given, those of y and z remaining indeterminate.

Here the terms without the sign of integration give, as is easily seen, the equations

$$(P_1)_1 = 0, \qquad (P'_1)_1 = 0,$$
$$(P_1)_0 = 0, \qquad (P'_1)_0 = 0.$$

These are, however, only equivalent to the two equations,

$$m = 0, \qquad n = 0.$$

The remaining constants, p, q, are therefore indeterminate. This case is analogous to that of Ex. (3), Art. 28.

(3.) Let the limiting values of x, y, z be connected by the equations,

$$f_1(x_1, y_1, z_1) = 0, \quad f_0(x_0, y_0, z_0) = 0.$$

Taking the complete increments of these equations, as in Art. 27, (4), we have

$$\left(\frac{df_1}{dx_1} + \frac{df_1}{dy_1}\frac{dy_1}{dx_1} + \frac{df_1}{dz_1}\frac{dz_1}{dx_1}\right) dx_1 + \frac{df_1}{dy_1}\delta y_1 + \frac{df_1}{dz_1}\delta z_1 = 0,$$

$$\left(\frac{df_0}{dx_0} + \frac{df_0}{dy_0}\frac{dy_0}{dx_0} + \frac{df_0}{dz_0}\frac{dz_0}{dx_0}\right) dx_0 + \frac{df_0}{dy_0}\delta y_0 + \frac{df_0}{dz_0}\delta z_0 = 0. \qquad \text{(a)}$$

Assume

$$m_1 = \frac{\dfrac{df_1}{dy_1}}{\dfrac{df_1}{dx_1}}, \qquad m_0 = \frac{\dfrac{df_0}{dy_0}}{\dfrac{df_0}{dx_0}},$$

$$n_1 = \frac{\dfrac{df_1}{dz_1}}{\dfrac{df_1}{dx_1}}, \qquad n_0 = \frac{\dfrac{df_0}{dz_0}}{\dfrac{df_0}{dx_0}};$$

and substitute for

$$\frac{dy_1}{dx_1}, \quad \frac{dy_0}{dx_0}, \quad \frac{dz_1}{dx_1}, \quad \frac{dz_0}{dx_0},$$

their values derived from equation (a). We have then

$$(1 + mm_1 + nn_1)\,dx_1 + m_1\delta y_1 + n_1\delta z_1 = 0,$$
$$(1 + mm_0 + nn_0)\,dx_0 + m_0\delta y_0 + n_0\delta z_0 = 0.$$

Eliminating, by means of these equations, dx_1, dx_0, from the equations

$$V_1 dx_1 + (P_1)_1\,\delta y_1 + (P'_1)_1\,\delta z_1 = 0,$$
$$V_0 dx_0 + (P_1)_0\,\delta y_0 + (P'_0)_0\,\delta z_0 = 0,$$

and equating to zero the coefficients of the variations

$$\delta y_1,\ \delta z_1,\ \delta y_0,\ \delta z_0,$$

we have

$$m_1 V_1 - (P_1)_1\,(1 + mm_1 + nn_1) = 0,$$
$$n_1\, V_1 - (P'_1)_1\,(1 + mm_1 + nn_1) = 0,$$
$$m_0 V_0 - (P_1)_0\,(1 + mm_0 + nn_0) = 0,$$
$$n_0\, V_0 - (P'_1)_0\,(1 + mm_0 + nn_0) = 0.$$

Replacing V_1, (P_1), &c., by their values, and solving the first two equations for m, n, we easily find

$$m = m_1, \quad n = n_1. \tag{b}$$

Similarly, from the third and fourth equations we find

$$m = m_0, \quad n = n_0. \tag{c}$$

Eliminating $x_1, y_1, z_1, x_0, y_0, z_0$, by the equations

$$y_1 = mx_1 + p, \qquad y_0 = mx_0 + p,$$
$$z_1 = nx_1 + q, \qquad z_0 = nx_0 + q,$$
$$f_1(x_1, y_1, z_1) = 0, \quad f_0(x_0, y_0, z_0) = 0,$$

we have the four equations (b) and (c) for the determination of the four unknown quantities, m, n, p, q.

56. We may here extend to functions containing two or more dependent variables a remark made in Art. 29.

Let

$$V = f\left(\frac{dy}{dx}, \frac{dz}{dx}\right),$$

and let it be required to determine the functions y, z, so as to make

$$\int_{x_0}^{x_1} V dx$$

a maximum or minimum.

The equations

$$N - \frac{dP_1}{dx} + \&c. = 0, \quad N' - \frac{dP'_1}{dx} + \&c. = 0,$$

become, as in the foregoing example,

$$\frac{dP_1}{dx} = 0, \qquad \frac{dP'_1}{dx} = 0;$$

or,

$$P_1 = c, \qquad P'_1 = c'.$$

But since V, and therefore P_1, P'_1, are functions of

$$\frac{dy}{dx}, \frac{dz}{dx}$$

only, it is plain that these equations are equivalent to

$$\frac{dy}{dx} = m, \quad \frac{dz}{dx} = n;$$

or, as before,

$$y = mx + p, \quad z = nx + q.$$

We infer, therefore, as in Art. 29, that if V contain only differential coefficients of the first order, the given integral will receive a maximum or minimum value, by making the several dependent variables which it contains *linear* functions of the independent variable. If

$$V = f\left(\frac{d^n y}{dx^n}, \frac{d^m z}{dx^m}, \ldots\right),$$

containing but one differential coefficient of each dependent variable, it is easy to see that the solution will be obtained from the system of differential equations,

$$\frac{d^n y}{dx^n} = X, \quad \frac{d^m z}{dx^m} = X', \text{ \&c.},$$

X, X', &c., being given functions of x.

Prop. VII.

57. To determine the functions y and z which will render

$$\int_{x_0}^{x_1} V dx$$

a maximum or minimum; V having the same form as in the foregoing Proposition, and the functions y and z being connected by the equation

$$L = 0.$$

It has been already shown (Prop. VI., Chap. II.) that, if the equation

$$L = 0$$

be integrable with regard to either of the dependent variables which it contains, so as to furnish a result of the form

$$z = f\left(x, y, \frac{dy}{dx}, \text{ \&c.}\right),$$

the given integral, and therefore its variation, will have the same form with that found in Prop. III. of the same Chapter. In such a case, then, the present problem would differ in no respect from that of Prop. II. But as it is in general impossible to effect this integration, we are obliged, for the solution of the present problem, to have recourse to the method of Lagrange, which has been already partly explained in Art. 16. The spirit of this method, as will be seen by referring to the Article quoted, consists in the introduction of a new indeterminate quantity, λ, by a suitable determination of which we are enabled to reduce the variation of the given integral to a form in which but one of the arbitrary variations, δy or δz, shall appear under the sign of integration.

In fact, if we denote by Θ the aggregate of terms without the sign of integration in δU, the general formula of p. 23 gives

$$\delta U = \Theta + \int_{x_0}^{x_1}\left(N + \lambda\alpha - \frac{d(P_1 + \lambda\beta)}{dx} + \&\text{c.}\right)\delta y\,dx$$

$$+ \int_{x_0}^{x_1}\left(N' + \lambda\alpha' - \frac{d(P'_1 + \lambda\beta')}{dx} + \&\text{c.}\right)\delta z\,dx.$$

Suppose now that the indeterminate quantity λ is so assumed as to satisfy the equation

$$N' + \lambda\alpha' - \frac{d(\mathrm{P}'_1 + \lambda\beta')}{dx} + \&\text{c.} = 0.$$

Then it will appear, by reasoning precisely analogous to that of Prop. II., that the equation

$$\delta U = 0,$$

or

$$\Theta + \int_x^{x_1}{}_0\left(N + \lambda\alpha - \frac{d(P_1 + \lambda\beta)}{dx} + \&\text{c.}\right)\delta y\,dx = 0,$$

cannot be satisfied (without restricting the variation δy) in any other way than by making

$$\Theta = 0,$$

$$N + \lambda\alpha - \frac{d(P_1 + \lambda\beta)}{dx} + \&\text{c.} = 0.$$

Hence it is evident that we have, in general, for the solution of the problem, the three general equations,

$$L = 0,$$

$$N + \lambda\alpha - \frac{d(P_1 + \lambda\beta)}{dx} + \&c. = 0,$$

$$N' + \lambda\alpha' - \frac{d(P'_1 + \lambda\beta')}{dx} + \&c. = 0, \qquad \text{(A)}$$

being the same in number with the unknown quantities y, z, λ.

58. The whole method may be otherwise stated as follows: The general value of δV is

$$\delta V = N\delta y + P_1 \frac{d\delta y}{dx} + \&c. + N'\delta z + P'_1 \frac{d\delta z}{dx} + \&c.,$$

being the same as when y and z are actually independent. Now it is plain that the defect of this expression, in its present form, is that δy, δz cannot be treated as independent variations, and, therefore, that the rules of the foregoing Proposition cannot be applied. It is necessary, therefore, in the first place, to determine such a modification of the value of δV as will enable us to regard these variations as independent of each other. Now the equation by which they are connected, namely,

$$\alpha\delta y + \beta \frac{d\delta y}{dx} + \&c. + \alpha'\delta z + \beta' \frac{d\delta z}{dx} + \&c. = 0,$$

being a *linear* equation, it follows, from the theory of such equations, that the condition which it furnishes is satisfied by multiplying its left hand member by an indeterminate quantity λ, adding it to the value of δV, and then treating δy, δz as independent of each other.

This will evidently give, as before,

$$N + \lambda\alpha - \frac{d(P_1 + \lambda\beta)}{dx} + \&c. = 0,$$

$$N' + \lambda\alpha' - \frac{d(P'_1 + \lambda\beta')}{dx} + \&c. = 0,$$

which, with the given equation,

$$L = 0,$$

are sufficient to determine the three functions, y, z, λ, in terms of x and arbitrary constants.

For the determination of these constants we have, as before, the several equations which are formed by equating to zero the several coefficients of δy_1, δz_1, δy_0, δz_0, &c., which occur in the terms freed from the integral sign; i. e., we have as many equations as there are *independent* variations among these terms. But as the equation

$$L = 0$$

is supposed to hold between the functions y and z, however these functions may be varied, it is plain that the variations δy_1, δz_1, &c., δy_0, δz_0, &c., are not necessarily independent of each other, and, therefore, that the number of the ancillary equations may not be equal to the number of these variations.

In order, then, to determine how far the question admits of a definite solution, we must consider, firstly, what is the number of arbitrary constants contained in the values of y, z, λ; and, secondly, what is the number of the ancillary equations which the conditions of the question furnish for their determination.

In discussing the general theory of the number of equations furnished by the terms outside the sign of integration, it will be sufficient to consider *one* out of the many cases which may arise from the various data of the problems to be solved. For it is plain, from what has been said, p. 43, that this variety does not, *in general*, affect the *number* of the ancillary equations. Suppose, then, that the limiting values of x are given, those of

$$y, \frac{dy}{dx}, \text{ \&c.}, \quad z, \frac{dz}{dx}, \text{ \&c.},$$

remaining indeterminate. Suppose also that V contains y and its differential coefficients, as far as the nth order, inclusive, and z and its coefficients, as far as the mth order, also inclusive. Suppose, lastly, that the equation

$$L = 0$$

is of the order n' in y, and m' in z.

59. We shall consider successively the three cases:

1. $m > m'$, $n > n'$. 2. $m > m'$, $n < n'$. 3. $m < m'$, $n < n'$.

(1.) Let $m > m'$ and $n > n'$. In this case it is evident that the equations for the determination of y, z, λ are of the following orders:

$2n$ in y,	$m+n$ in z,	n' in λ,	(1)
$m+n$ . . .	$2m$	m'	(2)
n'	m'	0	(3)

Differentiating, according to the usual method, equation (1) m' times, and equation (2) n' times, and eliminating λ, we have two equations of the orders

$m+n+n'$ in y,	$2m+n'$ in z,	(1′)
n'	m'	(2′)

it being supposed that $m - m' > n - n'$.

Again, differentiating (1′) m' times, and (2′) $2m+n'$ times, and eliminating z, we have a final equation in y of the order $2(m+n')$, whose solution, therefore, will contain $2(m+n')$ arbitrary constants. As in the process of elimination, λ and z have been successively determined in terms of the remaining variables, no new arbitrary constants will be introduced in the values of these quantities. The total number of these constants will therefore be $2(m+n')$.

Let us next consider the number of equations which are furnished for the determination of these constants, by the terms which are free from the sign of integration. The number of distinct variations which enter into these terms is evidently $2(m+n)$, namely:

$$\delta y_1,\ \left(\frac{d\delta y}{dx}\right)_1 \cdots\cdots \left(\frac{d^{n-1}\delta y}{dx^{n-1}}\right)_1,$$

$$\delta y_0,\ \left(\frac{d\delta y}{dx}\right)_0 \cdots\cdots \left(\frac{d^{n-1}\delta y}{dx^{n-1}}\right)_0,$$

$$\delta z_1,\ \left(\frac{d\delta z}{dx}\right)_1 \cdots\cdots \left(\frac{d^{m-1}\delta z}{dx^{m-1}}\right)_1,$$

$$\delta z_0,\ \left(\frac{d\delta z}{dx}\right)_0 \cdots\cdots \left(\frac{d^{m-1}\delta z}{dx^{m-1}}\right)_0.$$

If, then, these variations were independent of each other, we should have, by equating to zero their several coefficients, $2(m+n)$ equations. But as the equation

$$L = 0$$

is supposed to hold among all the admissible values of y and z, it is plain that the variations δy_1, &c., are not independent, but are connected by the $2(n-n')$ equations

$$\begin{aligned} &\delta L_1 = 0, \quad \delta L_0 = 0, \quad \left(\frac{d\delta L}{dx}\right)_1 = 0, \quad \left(\frac{d\delta L}{dx}\right)_0 = 0, \ldots \\ &\ldots \left(\frac{d^{n-n'-1}\delta L}{dx^{n-n'-1}}\right)_1 = 0, \quad \left(\frac{d^{n-n'-1}\delta L}{dx^{n-n'-1}}\right)_0 = 0. \end{aligned} \qquad \text{(B)}$$

These equations being $2(n-n')$ in number, it is evident that the number of independent variations contained in the terms which are free from the sign of integration is only $2(m+n) - 2(n-n') = 2(m+n')$, the same as that of the arbitrary constants. The differentiation of δL cannot be pushed beyond the order $n - n' - 1$, inasmuch as the next step would introduce the new variations,

$$\left(\frac{d^n \delta y}{dx^n}\right)_1, \quad \left(\frac{d^n \delta y}{dx^n}\right)_0.$$

(2.) Let $m > m'$ and $n < n'$. Then, if $m + n > m' + n'$, the three equations are of the orders,

$2n'$ in y,	$m+n$ in z,	n' in λ,	(1)
$m+n$	$2m$	m'	(2)
n'	m'	0	(3)

Eliminating λ, as before, between (1) and (2), we have in y and z two equations of the orders

$m+n+n'$ in y,	$2m+n'$ in z,	(1′)
n'	m'	(2′)

Eliminating z between these equations, we have a final equation in y of the order $2(m+n')$, whose solution will therefore, as before, contain $2(m+n')$ arbitrary constants. And it is easy to see that a similar result would be obtained by supposing

$$m + n < m' + n'.$$

No relations between the variations δy_1, δy_0, &c., are furnished in this case by the equation

$$L = 0,$$

for it is evident that δL_1, δL_0 will contain

$$\left(\frac{d^{n'}\delta y}{dx^{n'}}\right)_1, \quad \left(\frac{d^{n'}\delta y}{dx^{n'}}\right)_0,$$

respectively, while the highest which occur in the terms free from the sign of integration are only

$$\left(\frac{d^{n'-1}\delta y}{dx^{n'-1}}\right)_1, \quad \left(\frac{d^{n'-1}\delta y}{dx^{n'-1}}\right)_0.$$

It is therefore impossible to eliminate any of the quantities δy_1, δy_0, &c., by means of the equations

$$\delta L_1 = 0, \quad \delta L_0 = 0;$$

and, *à fortiori*, no such elimination can be effected by any of the equations

$$\left(\frac{d\delta L}{dx}\right)_1 = 0, \quad \left(\frac{d\delta L}{dx}\right)_0 = 0, \text{ \&c.}$$

Each of these equations, in fact, introduces a new variation. The number of equations which are furnished by the terms free from the sign of integration is therefore, in this case, equal to the number of the variations,

$$\delta y_1, \delta y_0, \left(\frac{d\delta y}{dx}\right)_1, \text{ \&c., } \delta z_1, \delta z_0, \left(\frac{d\delta z}{dx}\right)_1, \text{ \&c.,}$$

which they contain. But of these quantities there are evidently $2(m + n')$, namely, the $2n'$ quantities,

$$\delta y_1, \left(\frac{d\delta y}{dx}\right)_1 \cdots \left(\frac{d^{n'-1}\delta y}{dx^{n'-1}}\right)_1, \quad \delta y_0, \left(\frac{d\delta y}{dx}\right)_0 \cdots \left(\frac{d^{n'-1}\delta y}{dx^{n'-1}}\right)_0,$$

and the $2m$ quantities,

$$\delta z_1, \left(\frac{d\delta z}{dx}\right)_1 \cdots \left(\frac{d^{m-1}\delta z}{dx^{m-1}}\right)_1, \quad \delta z_0, \left(\frac{d\delta z}{dx}\right)_0 \cdots \left(\frac{d^{m-1}\delta z}{dx^{m-1}}\right)_0.$$

The number of ancillary equations is therefore again the same as that of the arbitrary constants.

(3.) Let $m' > m$ and $n' > n$. The equations in y, z, λ are, in this case, of the orders

$$\begin{array}{llll} 2n' \text{ in } y, & m'+n' \text{ in } z, & n' \text{ in } \lambda, & (1) \\ m'+n' \ldots & 2m' \ldots\ldots & m' \ldots. & (2) \\ n' \ldots\ldots & m' \ldots\ldots & 0 \ldots.. & (3) \end{array}$$

Eliminating λ, as before, between equations (1) and (2), we have, for the determination of y and z, two equations of the orders

$$\begin{array}{lll} 2n'+m' \text{ in } y, & 2m'+n' \text{ in } z, & (1') \\ n' \ldots\ldots & m' \ldots\ldots & (2') \end{array}$$

Again, eliminating z, we have a final result in y of the order $2(m'+n')$. The number of arbitrary constants is, therefore, in this case, $2(m'+n')$.

With regard to the number of the ancillary equations which are furnished by the terms free from the sign of integration, it is, in the first place, evident that these terms contain $2(m'+n')$ independent variations, sc.

$$\delta y_1, \left(\frac{d\delta y}{dx}\right)_1 \ldots \left(\frac{d^{n'-1}\delta y}{dx^{n'-1}}\right)_1,$$

$$\delta y_0, \left(\frac{d\delta y}{dx}\right)_0 \ldots \left(\frac{d^{n'-1}\delta y}{dx^{n'-1}}\right)_0,$$

$$\delta z_1, \left(\frac{d\delta z}{dx}\right)_1 \ldots \left(\frac{d^{m'-1}\delta z}{dx^{m'-1}}\right)_1,$$

$$\delta z_0, \left(\frac{d\delta z}{dx}\right)_0 \ldots \left(\frac{d^{m'-1}\delta z}{dx^{m'-1}}\right)_0.$$

It would seem, therefore, at first sight, that the number of equations should be $2(m'+n')$, the same as that of the arbitrary constants. But this case requires a little more consideration.

If we assume

$$\beta_{n'} = \frac{dL}{d\,.\,\frac{d^{n'}y}{dx^{n'}}}, \quad \beta_{n'-1} = \frac{dL}{d\,.\,\frac{d^{n'-1}y}{dx^{n'-1}}}, \quad \&c.,$$

and

$$\beta'_{m'} = \frac{dL}{d\,.\,\frac{d^{m'}z}{dx^{m'}}}, \quad \beta'_{m'-1} = \frac{dL}{d\,.\,\frac{d^{m'-1}z}{dx^{m'-1}}}, \text{ \&c.,}$$

it is easy to see that the coefficient of

$$\left(\frac{d^{n'-1}\delta y}{dx^{n'-1}}\right)_1$$

will be

$$\lambda_1(\beta_{n'})_1,$$

and that the coefficients of

$$\left(\frac{d^{n'-2}\delta y}{dx^{n'-2}}\right)_1, \quad \left(\frac{d^{n'-3}\delta y}{dx^{n'-3}}\right)_1, \text{ \&c.,}$$

will be

$$\lambda_1(\beta_{n'-1})_1 - \left(\frac{d.\lambda\beta_{n'}}{dx}\right)_1, \quad \lambda_1\beta_{n'-2} - \left(\frac{d.\lambda\beta_{n'-1}}{dx}\right)_1 + \left(\frac{d^2.\lambda\beta_{n'}}{dx^2}\right)_1, \text{ \&c.}$$

Equating these coefficients to zero, we have the several conditions,

$$\left.\begin{aligned}
&\lambda_1(\beta_{n'})_1 = 0,\\
&\lambda_1(\beta_{n'-1})_1 - \left(\frac{d.\lambda\beta_{n'}}{dx}\right)_1 = 0,\\
&\lambda_1(\beta_{n'-2})_1 - \left(\frac{d.\lambda\beta_{n'-1}}{dx}\right)_1 + \left(\frac{d^2.\lambda\beta_{n'}}{dx^2}\right)_1 = 0,\\
&\text{\&c. \&c.}\\
&\lambda_1(\beta_{n'-(n'-n-1)})_1 - \text{\&c.} + (-1)^{n'-n-1}\left(\frac{d^{n'-n-1}\lambda\beta_{n'}}{dx^{n'-n-1}}\right)_1 = 0.
\end{aligned}\right\} \quad (C)$$

These equations, which are $n' - n$ in number, being derived from the coefficients of

$$\left(\frac{d^{n'-1}\delta y}{dx^{n'-1}}\right)_1, \quad \left(\frac{d^{n'-2}\delta y}{dx^{n'-2}}\right)_1 \cdots\cdots \left(\frac{d^{n'-(n'-n)}\delta y}{dx^{n'-(n'-n)}}\right)_1,$$

are, as is evident, wholly free from any terms derived from V.

The first of these equations may be satisfied by making either $\lambda_1 = 0$ or $(\beta_{n'})_1 = 0$. We shall examine these suppositions separately.

(1.) Let $\lambda_1 = 0$, and $(\beta_{n'})_1 >$ or < 0. This supposition reduces the second of the preceding equations to

$$(\beta_{n'})_1 \cdot \left(\frac{d\lambda}{dx}\right)_1 = 0;$$

or, since $(\beta_{n'})_1$ is not supposed to vanish,

$$\left(\frac{d\lambda}{dx}\right)_1 = 0;$$

and it is easily seen that the suppositions,

$$\lambda_1 = 0, \quad \left(\frac{d\lambda}{dx}\right)_1 = 0, \quad \beta_{n'} > \text{or} < 0,$$

reduce the third of these equations to

$$\left(\frac{d^2\lambda}{dx^2}\right)_1 = 0,$$

and so on for the other equations.

Hence it is evident that if $(\beta_{n'})_1$ be not supposed to vanish, the foregoing equations are equivalent to

$$\lambda_1 = 0, \quad \left(\frac{d\lambda}{dx}\right)_1 = 0, \quad \left(\frac{d^2\lambda}{dx^2}\right)_1 = 0, \text{ \&c.,} \quad \left(\frac{d^{n'-n-1}\lambda}{dx^{n'-n-1}}\right)_1 = 0. \quad \text{(D)}$$

Now if we proceed in a similar manner to consider the equations derived from the coefficients of

$$\left(\frac{d^{m'-1}\delta z}{dx^{m'-1}}\right)_1, \quad \left(\frac{d^{m'-2}\delta z}{dx^{m'-2}}\right)_1 \cdots\cdots \left(\frac{d^{m'-(m'-m-1)}\delta z}{dx^{m'-(m'-m-1)}}\right)_1;$$

supposing, as before, that

$$\beta'_{m'} \text{ or } \frac{dL}{d \cdot \dfrac{d^{m'}z}{dx^{m'}}}$$

does not vanish at the superior limit, we shall find

$$\lambda_1 = 0, \quad \left(\frac{d\lambda}{dx}\right)_1 = 0 \cdots\cdots \left(\frac{d^{m'-m-1}\lambda}{dx^{m'-m-1}}\right)_1 = 0. \qquad \text{(E)}$$

If, then, we suppose that $n' - n > m' - m$, it is plain that these equations are identical with the first $m' - m$ equations of group (D).

Thus the number of the ancillary equations is diminished by $m' - m$; and if the same supposition be applied to the lower limit, a similar diminution will take place. In this case, therefore, the number of the equations for the determination of the arbitrary constants will be only $2(m' + n') - 2(m' - m) = 2(m + n')$. There will be, therefore, $2(m' - m)$ constants remaining still undetermined.

(2.) Suppose $\lambda_1 >$ or < 0. In this case the first of equations (C) becomes

$$(\beta_{n'})_1 = 0.$$

This supposition, together with $\lambda_1 >$ or < 0, reduces the second of these equations to

$$(\beta_{n'-1})_1 - \left(\frac{d\beta_{n'}}{dx}\right)_1 = 0;$$

and the third becomes

$$\lambda_1\left(\beta_{n'-2} - \frac{d\beta_{n'-1}}{dx} + \frac{d^2\beta_{n'}}{dx^2}\right)_1 + (\beta_{n'-1})_1\left(\frac{d\lambda}{dx}\right)_1 = 0.$$

Without pursuing this discussion any further, it is sufficient to remark that the reduction in the number of the ancillary equations which was found to take place in the preceding case, does not occur here, as it is evident that the equations just found, as given by the coefficients of

$$\delta y_1, \left(\frac{d\delta y}{dx}\right)_1, \text{\&c.},$$

are different from those derived from the coefficients of

$$\delta z_1, \left(\frac{d\delta z}{dx}\right)_1, \text{\&c.}$$

(3.) Again, the first two of the equations (C) may be satisfied by making

$$\lambda_1 = 0, \quad (\beta_{n'})_1 = 0.$$

This supposition reduces the third equation to

$$\left(\frac{d\lambda}{dx}\right)_1\left(\beta_{n'-1} - \frac{2d\beta_{n'}}{dx}\right)_1 = 0,$$

which may be satisfied either by making

$$\left(\frac{d\lambda}{dx}\right)_1 = 0,$$

or by making

$$\left(\beta_{n'-1} - \frac{2d\beta_{n'}}{dx}\right)_1 = 0.$$

If we adopt the former of these suppositions, the first three equations (C) will be equivalent to

$$\lambda_1 = 0, \quad \left(\frac{d\lambda}{dx}\right)_1 = 0, \quad (\beta_{n'})_1 = 0;$$

and if the latter, we shall have

$$\lambda_1 = 0, \quad (\beta_{n'})_1 = 0, \quad \left(\beta_{n'-1} - \frac{2d\beta_{n'}}{dx}\right)_1 = 0.$$

Upon referring to the equations derived from the coefficients of

$$\delta z_1, \left(\frac{d\delta z}{dx}\right)_1, \ \&c.,$$

it is plain that the former of these suppositions annuls the first two, those, namely, which are derived from the coefficients of the variations,

$$\left(\frac{d^{m'-1}\delta z}{dx^{m'-1}}\right)_1, \quad \left(\frac{d^{m'-2}\delta z}{dx^{m'-2}}\right)_1;$$

thus diminishing the number of the ancillary equations by two. The latter will only diminish the same number by one.

Without following this reasoning any further, it sufficiently appears from what has been said,—1. That if either of the quantities, $(\beta_{n'})_1$, $(\beta'_{m'})_1$, be taken to be finite, the number of the ancillary equations will be diminished by $m' - m$; and that if a similar supposition be made with respect to either of the quantities $(\beta_{n'})_0$, $(\beta'_{m'})_0$, a similar diminution will be effected. 2. That if the first equation of each group, those, namely, which are derived from the coefficients of the variations,

$$\left(\frac{d^{n'-1}\delta y}{dx^{n'-1}}\right)_1, \quad \left(\frac{d^{m'-1}\delta z}{dx^{m'-1}}\right)_1,$$

be satisfied by making

$$(\beta_{n'})_1 = 0, \quad (\beta'_{m'})_1 = 0,$$

the number of the ancillary equations may be diminished by any number from 0 to $2(m' - m)$, the particular number depending upon the selection which we make among the many methods of satisfying the remaining equations.

The result of the entire discussion may be stated as follows:

(1.) If $m > m'$, and $n > n'$, the order of the final differential equation will be the greater of the two quantities,

$$2(m + n'), \quad 2(m' + n),$$

and there will be a sufficient number of ancillary equations to determine the arbitrary constants which enter into its solution.

(2.) The same conclusion holds for the case in which $m > m'$ and $n < n'$.

(3.) If $m' > m$ and $n' > n$, the order of the final equation will be, in general,

$$2(m' + n');$$

and its solution may contain any number of indeterminate constants not exceeding the lesser of the two quantities,

$$2(m' - m), \quad 2(n' - n).$$

60. M. Poisson, in his valuable memoir upon the Calculus of Variations, states, that the solution of a question such as that which has been discussed above will, *in general*, contain indeterminate constants, inasmuch as the relation between y and z is given by a *differential* equation: "Toutefois, une partie de ces constantes sera, en général, surabondante, et restera indéterminée —ce qui provient de ce que l'un des inconnues y et z n'est déterminée implicitement au moyen de l'autre que par l'équation différentielle $L = 0$."* M. Poisson has not given the reasoning by which he arrives at this conclusion, which is wholly at variance with that which I have endeavoured to establish. I can, therefore, merely conjecture that it was something of the following kind:

* Mem. de l'Instit. tom. xii. p. 258.

Suppose that the equation

$$L = 0$$

were integrable, so as to give the value of z in terms of y, x, and a number of arbitrary constants. Let this value be

$$z = f\left(x, y, \frac{dy}{dx} \ldots . c, c', \ldots\right),$$

and let the values of

$$\frac{dz}{dx}, \frac{d^2z}{dx^2}, \text{ \&c.}$$

be derived from it, and substituted in the given quantity V. This process will reduce V to the form

$$V = F\left(x, y, \frac{dy}{dx} \ldots . c, c', \ldots .\right);$$

and if we then apply the Calculus of Variations to discover when $\int V dx$ is a maximum or minimum, it is evident that the solution will, in general, contain all the constants which are to be found in V, that is to say, all the constants, c, c', &c., which have been introduced in the integration of the equation

$$L = 0.$$

As, therefore, the method of Lagrange is but a different mode of effecting the same object, it may be expected that the final result will contain a similar number of undetermined constants. To this I reply, that the questions solved by these two methods are essentially different. In the first, in which the quantities c, c', &c. are not supposed to vary, we seek to determine, among all the systems of values of y and z which satisfy the particular equation,

$$z = f\left(x, y, \frac{dy}{dx} \ldots c, c' \ldots\right),$$

that system which will render the given integral a maximum or minimum. But in the method of Lagrange the required system is to be sought, not among those only which satisfy this particular equation, but among all those which satisfy the general equation

$$L = 0.$$

If, therefore, that method be perfect, it ought to furnish means for determining the required system from among all those which satisfy any one of the infinite number of equations which may be deduced from

$$z = f\left(x,\ y,\ \frac{dy}{dx} \ldots c,\ c' \ldots\right),$$

by assigning different values to the constants c, c', &c. It ought, therefore, to furnish a sufficient number of equations for the determination of these constants. A single example will sufficiently elucidate this reasoning.

Suppose that it were required to determine the functions y and z of such a form as to render

$$\int \sqrt{\left(1 + \frac{dy^2}{dx^2} + \frac{dz^2}{dx^2}\right)}\, dx$$

a minimum, y and z being connected by the differential equation

$$x + y\frac{dy}{dx} + z\frac{dz}{dx} = 0. \qquad \text{(A)}$$

As this equation is integrable, the problem may be treated by either of the methods just alluded to, that is to say, we may either integrate the equation (A), and, by means of its integral,

$$x^2 + y^2 + z^2 = a^2,$$

eliminate $\frac{dz}{dx}$ from V previous to the application of the Calculus of Variations; or we may apply that science immediately to the expression

$$\sqrt{\left(1 + \frac{dy^2}{dx^2} + \frac{dz^2}{dx^2}\right)}\, dx,$$

using, according to the method of Lagrange, the equation (A) as an equation of condition. But the problems solved by these two methods are not identical. This will appear most clearly if we consider their respective geometrical significations.

In the first method, where equation (A) is integrated previous to the application of the Calculus of Variations, the object of the problem is to draw a line of minimum length upon a *given* sphere whose radius is a. The surface upon which the line is to be drawn is therefore absolutely fixed, both the centre and radius of the sphere being given. But in treating the problem by the method of Lagrange, we are at liberty to select, as the surface upon which the line is to be drawn, *any* sphere whose centre is at the origin of co-ordinates. And if the solution be complete, this method ought to furnish not only a rule for drawing the line upon any particular sphere so selected, but also an equation for determining the radius of the sphere upon which the line so drawn is less than that similarly drawn upon any of the others; and if we actually apply this method we shall find that it does give the equation

$$a = \infty.$$

To prove this, let it be supposed that the extreme values of x are given, or, in other words, let it be required to determine the functions y and z of such a form as to render

$$\int_{x_0}^{x_1} \sqrt{\left(1 + \frac{dy^2}{dx^2} + \frac{dz^2}{dx^2}\right)} . dx$$

a minimum, the limiting values, x_0, x_1, being given, and the functions y, z being connected by the equation

$$x + y\frac{dy}{dx} + z\frac{dz}{dx} = 0.$$

Here we have

$$V = \sqrt{\left(1 + \frac{dy^2}{dx^2} + \frac{dz^2}{dx^2}\right)},$$

and, therefore,

$$P_1 = \frac{\frac{dy}{dx}}{\sqrt{\left(1 + \frac{dy^2}{dx^2} + \frac{dz^2}{dx^2}\right)}}, \quad P'_1 = \frac{\frac{dz}{dx}}{\sqrt{\left(1 + \frac{dy^2}{dx^2} + \frac{dz^2}{dx^2}\right)}},$$

$$N = 0, \quad N' = 0.$$

We have also, from equation (A),

$$\alpha = \frac{dy}{dx}, \quad \beta = y, \quad \alpha' = \frac{dz}{dx}, \quad \beta' = z.$$

The equations

$$N + \lambda\alpha - \frac{d(P_1 + \lambda\beta)}{dx} + \&c. = 0,$$

$$N' + \lambda\alpha' - \frac{d(P'_1 + \lambda\beta')}{dx} + \&c. = 0,$$

become, therefore,

$$\left.\begin{aligned} y\frac{d\lambda}{dx} + \frac{d}{dx}\left(\frac{\frac{dy}{dx}}{\sqrt{\left(1 + \frac{dy^2}{dx^2} + \frac{dz^2}{dx^2}\right)}}\right) = 0, \\ z\frac{d\lambda}{dx} + \frac{d}{dx}\left(\frac{\frac{dz}{dx}}{\sqrt{\left(1 + \frac{dy^2}{dx^2} + \frac{dz^2}{dx^2}\right)}}\right) = 0, \end{aligned}\right\} \quad \text{(B)}$$

the coefficient of λ in each equation vanishing of itself.

Eliminating $\frac{d\lambda}{dx}$ between these equations, we have

$$z\frac{d}{dx}\left(\frac{\frac{dy}{dx}}{\sqrt{\left(1 + \frac{dy^2}{dx^2} + \frac{dz^2}{dx^2}\right)}}\right) - y\frac{d}{dx}\left(\frac{\frac{dz}{dx}}{\sqrt{\left(1 + \frac{dy^2}{dx^2} + \frac{dz^2}{dx^2}\right)}}\right) = 0.$$

Integrating this equation (which is evidently possible), we find

$$\frac{z\frac{dy}{dx} - y\frac{dz}{dx}}{\sqrt{\left(1 + \frac{dy^2}{dx^2} + \frac{dz^2}{dx^2}\right)}} = c, \quad \text{(B')}$$

c being an arbitrary constant.

To determine this constant we must have recourse to the terms free from the sign of integration. Now since x_1, x_0 are given, and y_1, y_0, z_1, z_0, indeterminate, it is evident that the coefficients of δy_1, δy_0, δz_1, δz_0, must vanish of themselves. Hence we have the equations

$$(P_1 + \lambda\beta)_1 = 0, \quad (P'_1 + \lambda\beta')_1 = 0,$$
$$(P_1 + \lambda\beta)_0 = 0, \quad (P'_1 + \lambda\beta')_0 = 0,$$

or

$$\left.\begin{aligned}&\left(\frac{\frac{dy}{dx}}{\sqrt{\left(1+\frac{dy^2}{dx^2}+\frac{dz^2}{dx^2}\right)}}\right)_1 + \lambda_1 y_1 = 0,\\ &\left(\frac{\frac{dz}{dx}}{\sqrt{\left(1+\frac{dy^2}{dx^2}+\frac{dz^2}{dx^2}\right)}}\right)_1 + \lambda_1 z_1 = 0,\end{aligned}\right\} \qquad \text{(C)}$$

for the superior limit, with two similar equations for the inferior.

Eliminating λ_1, we find

$$\left(\frac{z\frac{dy}{dx} - y\frac{dz}{dx}}{\sqrt{\left(1+\frac{dy^2}{dx^2}+\frac{dz^2}{dx^2}\right)}}\right)_1 = 0.$$

Hence $c = 0$, and equation (B′) is reduced to

$$z\frac{dy}{dx} - y\frac{dz}{dx} = 0;$$

or, by integration,

$$y = mz.$$

The required functions are, therefore, given by the system of equations,

$$y = mz,$$

$$x^2 + y^2 + z^2 = a^2.$$

If now we recur to the first of equations (B), and eliminate y and z by means of the equations just found, we shall have

$$\sqrt{(a^2 - x^2)} \cdot \frac{d\lambda}{dx} - \frac{1}{a} = 0;$$

or, by integration,

$$\lambda - \frac{1}{a}\sin^{-1}\frac{x}{a} = b.$$

Hence, at the superior limit,

$$\lambda_1 - \frac{1}{a}\sin^{-1}\frac{x_1}{a} = b. \qquad \text{(D)}$$

Again, if in the first of equations (C) we substitute for

$$y_1, \left(\frac{dy}{dx}\right)_1, \left(\frac{dz}{dx}\right)_1,$$

their values in terms of x, we find

$$\lambda_1 - \frac{x_1}{a\sqrt{(a^2 - x_1^2)}} = 0. \qquad \text{(E)}$$

Eliminating λ_1 between (D) and (E), we have

$$\frac{1}{a}\left(\frac{x_1}{\sqrt{(a^2 - x_1^2)}} - \sin^{-1}\frac{x_1}{a}\right) = b; \qquad \text{(F)}$$

and in the same way for the other limit,

$$\frac{1}{a}\left(\frac{x_0}{\sqrt{(a^2 - x_0^2)}} - \sin^{-1}\frac{x_0}{a}\right) = b. \qquad \text{(G)}$$

These equations are satisfied by

$$a = \infty, \; b = 0.$$

If a be not infinite, we have, by subtracting these equations,

$$\frac{x_1}{\sqrt{(a^2 - x_1^2)}} - \sin^{-1}\frac{x_1}{a} = \frac{x_0}{\sqrt{(a^2 - x_0^2)}} - \sin^{-1}\frac{x_0}{a},$$

an impossible equation, since

$$x_1 > x_0,$$

and, therefore, as is easily seen, the left hand member is always greater than the other. In fact the differential coefficient of

$$\frac{x}{\sqrt{(a^2 - x^2)}} - \sin^{-1}\frac{x}{a}$$

is always positive while $\sqrt{(a^2 - x^2)}$ is so.

Hence it is evident that the *only* value of a which satisfies the equations (F) and (G) is

$$a = \infty.$$

This example shows conclusively that by the method of Lagrange we are furnished with an equation for the determination of the arbitrary constant introduced in the integration of the equation of condition; and that, therefore, the solution arrived at by that method does not, in general, contain arbitrary constants. With regard to the arbitrary constant m, which this solution does contain, it is easily seen that the existence of this constant is an accidental peculiarity of the individual question, and in no way connected with the fact that the equation of condition is a *differential* equation. In fact, it will readily appear that if we take, as equation of condition, the integral equation,

$$x^2 + y^2 + z^2 = a^2,$$

and eliminate z from

$$\int_{x_0}^{x_1} \sqrt{\left(1 + \frac{dy^2}{dx^2} + \frac{dz^2}{dx^2}\right)}\, dx,$$

previous to the application of the Calculus of Variations, the solution will still contain the arbitrary constant m. Geometrically speaking, the existence of this constant denotes that if it be required to draw between two parallel circles upon a sphere a line of minimum length, the problem is solved by *any* common secondary.

This theory may be extended to the case of three or more dependent variables, and the method to be adopted is precisely similar, both in the deduction of the general differential equations, and in the determination of the arbitrary constants which enter into the solution. Examples of the application of this method will be found in Chapter IV.

Prop. VIII.

61. To determine the form of the function y which will render $\int V dx$ a maximum or minimum, and at the same time satisfy the condition

$$\int V' dx = c,$$

V and V' being functions of

$$x,\ y,\ \frac{dy}{dx},\ \&c.$$

The condition that $\int V dx$ shall be a maximum or minimum gives, as before,

$$D \int V dx = 0;$$

and the condition

$$\int V' dx = c,$$

gives

$$D.\int V' dx = 0.$$

Hence, according to the principle of Lagrange previously laid down, the solution of this question is to be found by multiplying the second of these equations by an indeterminate coefficient λ, and adding it to the former. This will give

$$V_1 dx_1 - V_0 dx_0 + \delta.\int V dx + \lambda(V'_1 dx_1 - V'_0 dx_0 + \delta.\int V' dx) = 0.$$

Now it is plain, from the form of this equation, that it can be satisfied only by making

$$\lambda = \text{const.};$$

for if we substitute for $\delta.\int V dx$ and $\delta.\int V' dx$ their values, and proceed as before, the equation furnished by the terms free from the sign of integration will contain only λ and constants, and will therefore give for λ a constant value. It is plain, therefore, that the given problem will be solved by that value of the function y which renders the integral

$$\int (V + mV') dx$$

an absolute maximum or minimum. This integral, therefore, is to be treated precisely in the same way as that employed in discussing the maximum or minimum values of $\int V dx$ in Prop. II. The value of y, as found by that method, will, of course, contain the arbitrary constant m, inasmuch as the equations furnished by the method of Prop. II. are only sufficient to determine the constants introduced in the integration of the differential equation. This constant is to be determined by calculating the value of $\int V' dx$ from that of y, and equating it to the given constant c. Thus, for example, if it were required to determine, among all the forms of the function y which render

$$\int \sqrt{\left(1 + \frac{dy^2}{dx^2}\right)}\, dx = c,$$

that one which will make $\int y dx$ a maximum, it would be necessary, in the first place, to determine y, such as to render

$$\int\left\{y + m\sqrt{\left(1 + \frac{dy^2}{dx^2}\right)}\right\}dx$$

an absolute maximum, and from the value so found, which will contain the arbitrary constant m, to determine the value of the integral

$$\int\sqrt{\left(1 + \frac{dy^2}{dx^2}\right)}\,dx.$$

Equating then this value to the given constant c, we shall find the value of m. The solution is thus complete.

This theory may be extended with equal facility to the case in which the values of two or more integrals are given. For it will readily appear, by reasoning similar to that of the present Prop., that if

$$\int V'dx, \int V''dx, \text{ \&c.},$$

be the integrals whose values are given, the expression which is to be rendered an absolute maximum is

$$\int(V + mV' + m'V'' + \text{\&c.})\,dx.$$

Problems of this nature were formerly denominated *isoperimetrical*, from a remarkable class which they contain, that, namely, in which it is proposed to determine, among all curves of given *length*, that one which renders a given integral a maximum or minimum. As this name, however, is of too limited a signification to include the entire class, it will be more correct to give them the name of *relative* maxima and minima.

Examples of this class will be found in Chapter IV. It is evidently unnecessary to dwell further upon the theory of such cases, which, after the first step, differs in no respect from that of absolute maxima and minima.*

* Vid. note upon p. 136.

CHAPTER IV.

APPLICATION OF THE CALCULUS OF VARIATIONS TO GEOMETRY.

I.—*Theory of Curves.*

62. The geometrical applications of the Calculus of Variations are confined exclusively to the theory of maxima and minima; and, if that science be defined with especial reference to this theory, may be considered as co-extensive with the science itself, at least as far as it is concerned with integrals of the first two orders. For as, in the Differential Calculus, it is possible to give a geometrical statement of any problem of maxima and minima which does not involve more than two independent variables, so in the Calculus of Variations it is possible to give a geometrical statement of any problem of maxima or minima not involving integrals of an order higher than the second. In the Differential Calculus it is known that any such problem may be stated as follows:—"To find a point in a given curve or surface at which the ordinate of that curve or surface is a maximum or minimum." Here the curve or surface is given. But in the Calculus of Variations the principal question is to determine, not a point on a curve or surface, but the curve or surface itself. Geometrically stated, the problem is this:—"To determine a curve or surface which may have the property of rendering the value of a given integral a maximum or minimum." In fact, if the words "nature of the curve or surface" be substituted in any of the preceding chapters for "form of the function," we shall have the geometrical statement of the problem.* This statement is evidently confined to integrals of the first and second orders, for geometry does not admit of more than two independent variables. But in devoting a chapter to the "Application of the Calculus of Variations to Geometry," we shall give to the words a some-

* This was the form in which the problem of maxima and minima was first stated by John and James Bernouilli.—Vid. Acta Erud. 1696–7.

what more restricted signification, considering those problems only* in which the definite integral represents a purely geometrical quantity, viz., either a length, an area, or a solid, and reserving to a future chapter those problems in which, although the thing to be determined is a curve or surface, and therefore geometrical, the quantity expressed by the definite integral is one which properly belongs to Mechanics, such as a time, a force, or a velocity. We shall commence with the following very general problem, which will be found useful both in the geometrical and in the mechanical applications of our science.

Prop. I.

63. Let ds be the element of a plane curve, and μ a given function of its co-ordinates, and let it be required to determine the nature of the curve, such as to render $\int \mu ds$ a maximum or minimum.

Adopting the method of Lagrange as being the more symmetrical, we shall take s as the independent variable, treating the co-ordinates x and y as functions of s connected by the equation

$$\frac{dx^2}{ds^2} + \frac{dy^2}{ds^2} = 1.$$

The equations (A), p. 117, will become in this case

$$\left.\begin{aligned} \frac{d\mu}{dx} - \frac{d.\lambda\dfrac{dx}{ds}}{ds} &= 0, \\ \frac{d\mu}{dy} - \frac{d.\lambda\dfrac{dy}{ds}}{ds} &= 0. \end{aligned}\right\} \quad (A)$$

Multiplying these equations by $\frac{dx}{ds}$ and $\frac{dy}{ds}$ respectively, and adding them, it is easy to see that we shall have

$$\frac{d\lambda}{ds} = \frac{d\mu}{ds},$$

whence

$$\lambda = \mu + a.$$

* General propositions, which are useful in both kinds of application, are, as will be seen, excepted from this restriction.

The existence of the arbitrary constant a is an ambiguity necessarily introduced by the selection of s for the independent variable. For it is easy to see that the equations arrived at by the method here employed will be the same whether the original integral be $\int \mu ds$ or $\int (\mu + c) ds$. Geometrically speaking, therefore, we shall, by this method, arrive at the same conclusion, whether the *length* of the curve be or be not given. It is easy to see, however, that the question which we are now considering is to be distinguished from the isoperimetrical problem by making $a = 0$. For if we suppose the independent variable to be changed from s to θ, it is plain that the equation

$$\int \left\{ \delta\mu + \lambda \left(\frac{dx}{ds} \cdot \frac{d\delta x}{ds} + \frac{dy}{ds} \cdot \frac{d\delta y}{ds} \right) \right\} ds = 0$$

may be written, putting $\mu + a$ for λ,

$$\int \left(\delta\mu \frac{ds}{d\theta} + (\mu + a) \frac{\left(\frac{dx}{d\theta} \cdot \frac{d\delta x}{d\theta} + \frac{dy}{d\theta} \cdot \frac{d\delta y}{d\theta} \right)}{\frac{ds}{d\theta}} \right) d\theta = 0;$$

or, since

$$\frac{dx}{d\theta} \cdot \frac{d\delta x}{d\theta} + \frac{dy}{d\theta} \cdot \frac{d\delta y}{d\theta} = \frac{ds}{d\theta} \cdot \frac{d\delta s}{d\theta}$$

$$\int \left(\delta\mu \frac{ds}{d\theta} + \mu \frac{d\delta s}{d\theta} + a \frac{d\delta s}{d\theta} \right) d\theta = 0,$$

which is evidently identical with the equation

$$\delta . \int \mu \frac{ds}{d\theta} d\theta + a\delta . \int \frac{ds}{d\theta} d\theta = 0.$$

Unless, then, $a = 0$, it is plain that the problem solved by this method is of the isoperimetrical class, namely, to find among all curves of given length that one which renders $\int \mu \frac{ds}{d\theta} d\theta$ or $\int \mu ds$ a maximum or minimum.

Putting then $a = 0$, we have $\lambda = \mu$, and substituting this value in each of equations (A), we find

$$\frac{d\mu}{dx} - \frac{dx}{ds} \cdot \frac{d\mu}{ds} = \mu \frac{d^2x}{ds^2},$$

$$\frac{d\mu}{dy} - \frac{dy}{ds} \cdot \frac{d\mu}{ds} = \mu \frac{d^2y}{ds^2};$$

or, putting for $\frac{d\mu}{ds}$ its value, $\frac{d\mu}{dx} \cdot \frac{dx}{ds} + \frac{d\mu}{dy} \cdot \frac{dy}{ds}$,

$$\left.\begin{aligned} \frac{dy}{ds}\left(\frac{d\mu}{dx} \cdot \frac{dy}{ds} - \frac{d\mu}{dy} \cdot \frac{dx}{ds}\right) &= \mu \frac{d^2x}{ds^2}, \\ \frac{dx}{ds}\left(\frac{d\mu}{dy} \cdot \frac{dx}{ds} - \frac{d\mu}{dx} \cdot \frac{dy}{ds}\right) &= \mu \frac{d^2y}{ds^2}. \end{aligned}\right\} \quad \text{(B)}$$

Let ρ be the radius of curvature. Multiply the first of these equations by $\frac{dy}{ds}$, and the second by $\frac{dx}{ds}$, and subtract them. Then

$$\mu\left(\frac{dy}{ds} \cdot \frac{d^2x}{ds^2} - \frac{dx}{ds} \cdot \frac{d^2y}{ds^2}\right) = \frac{\mu}{\rho} = \frac{d\mu}{dx} \cdot \frac{dy}{ds} - \frac{d\mu}{dy} \cdot \frac{dx}{ds};$$

or, if α, β be the *acute* angles which the normal makes with the axes of co-ordinates,

$$\frac{1}{\rho} = -\frac{1}{\mu}\left(\frac{d\mu}{dx} \cos\alpha + \frac{d\mu}{dy} \cos\beta\right). \quad \text{(C)}$$

This equation gives the geometrical definition of the curve. It is evidently impossible to proceed further with the integration without previously fixing the form of μ, but the following property may be deduced from the general equation:

If the given integral had been $\int \frac{ds}{\mu}$, it is evident that the equation arrived at would have differed from (C) merely in the sign of its right hand member. It would, in fact, have been

$$\frac{1}{\rho} = +\frac{1}{\mu}\left(\frac{d\mu}{dx} \cos\alpha + \frac{d\mu}{dy} \cos\beta\right).$$

Now both these solutions are included in

$$\frac{1}{\rho^2} = \frac{1}{\mu^2}\left(\frac{d\mu}{dx} \cos\alpha + \frac{d\mu}{dy} \cos\beta\right)^2.$$

Hence it appears that the two curves contained in the equation

$$\rho^2 = f\left(x, y, \frac{dy}{dx}\right)$$

are connected by this relation, that if the one renders $\int \mu ds$ a maximum or minimum, the other will have the same property with regard to $\int \frac{ds}{\mu}$.

64. If the limiting values of x and y be given, the arbitrary constants contained in the solution of (C) will be determined by the substitution of these values. But if the limiting points of the curve be determined by its intersection with two given curves, it may be shown generally that one class of curves included in this problem will intersect their bounding curves at right angles. For the terms outside the sign of integration will furnish two equations:

$$\left.\begin{aligned} \mu_1 \left\{ \left(\frac{dx}{ds}\right)_1 \delta x_1 + \left(\frac{dy}{ds}\right)_1 \delta y_1 \right\} = 0, \\ \mu_0 \left\{ \left(\frac{dx}{ds}\right)_0 \delta x_0 + \left(\frac{dy}{ds}\right)_0 \delta y_0 \right\} = 0. \end{aligned}\right\} \quad \text{(D)}$$

These equations are satisfied either by making

$$\mu_1 = 0, \quad \mu_0 = 0,$$

or by making

$$\left.\begin{aligned} \left(\frac{dx}{ds}\right)_1 \delta x_1 + \left(\frac{dy}{ds}\right)_1 \delta y_1 = 0, \\ \left(\frac{dx}{ds}\right)_0 \delta x_0 + \left(\frac{dy}{ds}\right)_0 \delta y_0 = 0; \end{aligned}\right\} \quad \text{(E)}$$

or by a combination of the first and fourth, or of the second and third, of these suppositions.

Neglecting the suppositions (which are seldom admissible)

$$\mu_1 = 0, \quad \mu_0 = 0,$$

let it be supposed that the equations of the limiting curves are

$$dy_1 = m_1 dx_1, \quad dy_0 = m_0 dx_0.$$

As the conditions of the problem require that the curve, however varied, should terminate in these bounding curves, we have

$$\delta y_1 = m_1 \delta x_1, \quad \delta y_0 = m_0 \delta x_0.$$

Equations (E) become, therefore,

$$\left(\frac{dx}{ds}\right)_1 + m_1 \left(\frac{dy}{ds}\right)_1 = 0,$$
$$\left(\frac{dx}{ds}\right)_0 + m_0 \left(\frac{dy}{ds}\right)_0 = 0, \qquad \text{(F)}$$

which evidently contain the theorem in question.

65. To determine whether the solution at which we have arrived gives a maximum or a minimum, it will be most convenient to put the given integral under the form

$$\int \mu \sqrt{\left(1 + \frac{dy^2}{dx^2}\right)} dx.$$

This will give

$$\frac{d^2 V}{\left(d . \frac{dy}{dx}\right)^2} = \frac{d^2 . \mu \sqrt{\left(1 + \frac{dy^2}{dx^2}\right)}}{\left(d . \frac{dy}{dx}\right)^2} = \frac{\mu}{\left(1 + \frac{dy^2}{dx^2}\right)^{\frac{3}{2}}}. \qquad \text{(G)}$$

The result, therefore, will be (p. 80) a maximum or minimum, according as this quantity is negative or positive, i. e. according as μ and $\sqrt{\left(1 + \frac{dy^2}{dx^2}\right)}$ have opposite or similar signs. But as the former supposition would render the sign of the entire integral negative, it is plain that in that case a *maximum* value would denote the value furthest removed from $-\infty$, and therefore nearest to zero of all those immediately adjoining it. The value arrived at is, therefore, in all cases, so far as its magnitude alone is concerned, a minimum.

It has been shown (p. 88) that it is essential to the existence of a maximum or minimum value, that the coefficient

$$\frac{d^2 V}{\left(d . \frac{dy}{dx}\right)^2}$$

should not change its sign within the limits of integration. Now when the length of a curve is expressed by the integral

$$\int_{x_0}^{x_1} \sqrt{\left(1 + \frac{dy^2}{dx^2}\right)} dx,$$

it is, of course, supposed that the radical is taken positively throughout. Hence, and from equation (G), it appears that the condition just alluded to requires that μ should preserve the same sign for all values of x between x_0 and x_1. With respect to the other conditions of Art. 46, it is plain that these must be considered separately in each individual case. For it is, as has been there shown, essential to this investigation that the complete integral of the differential equation furnished by the Calculus of Variations be known. But we have already seen that this integration cannot be effected while the form of the function μ remains undetermined.

Example 1.

66. To draw the shortest line between two given points or between two given curves.

In this case

$$\mu = 1, \quad \frac{d\mu}{dx} = 0, \quad \frac{d\mu}{dy} = 0.$$

Equation (C) gives, therefore,

$$\frac{1}{\rho} = 0,$$

or

$$\frac{d^2x}{ds^2} = 0, \quad \frac{d^2y}{ds^2} = 0.$$

These equations, being integrated, give

$$\frac{dx}{ds} = a, \quad \frac{dy}{ds} = b, \tag{1}$$

and, therefore,

$$b\frac{dx}{ds} - a\frac{dy}{ds} = 0.$$

A second integration gives

$$bx - ay = c, \tag{2}$$

the equation of a right line.

To determine the constants we have, if the extreme points be given, the equations

$$bx_1 - ay_1 = c, \quad bx_0 - ay_0 = c,$$

and

$$a^2 + b^2 = 1.$$

The last of these equations is obtained by squaring, and adding the equations (1). If the line is to be drawn between two given curves, whose equations are

$$y = f(x), \quad y = F(x),$$

the condition that the required line must cut these curves at right angles gives the two equations,

$$\frac{a}{b} = -f'(x_1) = -F'(x_0).$$

These, with the five equations,

$$y_1 = f(x_1), \quad y_0 = F(x_0),$$

$$bx_1 - ay_1 = c, \quad bx_0 - ay_0 = c,$$

$$a^2 + b^2 = 1,$$

are sufficient to determine the seven quantities,

$$x_1, y_1, \quad x_0, y_0, \quad a, b, c.$$

If the given curves,

$$y = f(x), \quad y = F(x),$$

be so related that every line which is perpendicular to one is also perpendicular to the other, these equations are not independent, and therefore one of the constants remains indeterminate. In this case it is obvious that the curves have the same evolute, and that the intercepted portion of the line is the same for every point.

Example 2.

67. To find a curve such that the surface generated by its revolution about a given line may be a minimum.

Taking the given line as axis of x, the surface generated will be represented by $2\pi\int y ds$. The integral which is to be made a minimum is therefore $\int y ds$, giving

$$\mu = y, \quad \frac{d\mu}{dx} = 0, \quad \frac{d\mu}{dy} = 1.$$

Equation (C) becomes, therefore,

$$\frac{1}{\rho} = -\frac{1}{y}\cos\beta,$$

or

$$\rho = -y\sec\beta = -y\,\mathrm{cosec}\,\alpha. \tag{a}$$

But y cosec α expresses evidently the intercept of the normal between the curve and the axis of x. The required curve is therefore such that the radius of curvature is equal to the normal, and in an opposite direction. This is a well-known property of the catenary, which is therefore the curve required. Its equation may readily be deduced as follows:

The equation

$$\frac{1}{\rho} = -\frac{1}{y}\cos\beta,$$

expressed in rectangular co-ordinates, gives

$$\left(\frac{d^2x}{ds^2}\right)^2 + \left(\frac{d^2y}{ds^2}\right)^2 = \frac{1}{y^2}\cdot\frac{dx^2}{ds^2}. \tag{b}$$

Multiplying this equation by $\frac{dx^2}{ds^2}$, and substituting for $\left(\frac{dx}{ds}\cdot\frac{d^2x}{ds^2}\right)^2$ its value, $\left(\frac{dy}{ds}\cdot\frac{d^2y}{ds^2}\right)^2$, we have

$$\left(\frac{d^2y}{ds^2}\right)^2 = \frac{1}{y^2}\cdot\frac{dx^4}{ds^4};$$

hence

$$\frac{d^2y}{ds^2} = \pm\frac{1}{y}\cdot\frac{dx^2}{ds^2} = \pm\frac{1}{y}\left(1 - \frac{dy^2}{ds^2}\right).$$

As the radius of curvature and the normal are opposed in direction, it is plain that the curve is convex towards the axis of x, and therefore that $\frac{d^2y}{ds^2}$ is positive; and as y is taken positively throughout, the upper sign is the one to be used, giving

$$y\frac{d^2y}{ds^2} = 1 - \frac{dy^2}{ds^2}, \text{ or } y\frac{d^2y}{ds^2} + \frac{dy^2}{ds^2} = 1.$$

Integrating this, and adding an arbitrary constant, we have

$$y\frac{dy}{ds} = s + a;$$

and hence

$$y^2 = (s+a)^2 + b^2.$$

Squaring the former of these equations, and substituting for $\frac{dy^2}{ds^2}$, $1 - \frac{dx^2}{ds^2}$, we find

$$y^2\frac{dx^2}{ds^2} = y^2 - (s+a)^2 = b^2;$$

hence

$$dx = \frac{bds}{\sqrt{\{b^2 + (s+a)^2\}}}.$$

Integrating this, we have

$$x + c = bl\,[s + a + \sqrt{\{b^2 + (s+a)^2\}}];$$

or, eliminating $s + a$,

$$x + c = bl\{y + \sqrt{(y^2 - b^2)}\};$$

or, which is a more convenient form,

$$x = bl.\frac{1}{m}\left\{\frac{y}{b} + \sqrt{\left(\frac{y^2}{b^2} - 1\right)}\right\}. \qquad \text{(c)}$$

This solution contains, as will be seen, but two arbitrary constants. Two others, however, enter into the complete values of x, y, and λ in terms of s. One of these was excluded, by the necessity of distinguishing between the cases in which the length of the curve is given and those in which it is not given. The second disappears in the elimination of s. This latter result might have been foreseen, for as s itself does not enter either into μ or into the equation of condition,

$$\frac{dx^2}{ds^2} + \frac{dy^2}{ds^2} = 1,$$

it is plain that the data of the question, and therefore the result, will not be altered by the substitution of s + const. for s. Hence it might have been expected that one of the arbitrary constants in the final result should enter merely in connexion with s, and should therefore disappear in the elimination of that quantity. Taking into consideration the two constants which have been removed by the particular methods just alluded to, the complete solution will contain altogether four, thus agreeing with the general principle stated in p. 122.

To determine the arbitrary constants in this solution, let it be supposed, in the first place, that the extreme points are given.

Let it be supposed, for the sake of simplicity, that the extreme values of y are equal, and let the axis of y pass through the middle point of the line joining the extreme points of the curve. We shall have then

$$y_0 = y_1, \quad x_0 = -x_1.$$

But as the equation of the curve may readily be put under the form

$$y = \tfrac{1}{2} b \left(m e^{\frac{x}{b}} + \frac{1}{m} e^{-\frac{x}{b}} \right),$$

we have, for the determination of the constants m and b, the equations

$$y_1 = \tfrac{1}{2} b \left(m e^{\frac{x_1}{b}} + \frac{1}{m} e^{-\frac{x_1}{b}} \right),$$

$$y_1 = \tfrac{1}{2} b \left(m e^{-\frac{x_1}{b}} + \frac{1}{m} e^{\frac{x_1}{b}} \right).$$

These equations give, in the first place, $m = \pm 1$. This reduces either of the equations to

$$y_1 = \tfrac{1}{2} b \left(e^{\frac{x_1}{b}} + e^{-\frac{x_1}{b}} \right),$$

in which b must evidently be taken positively. It will readily appear that this equation is not soluble for all systems of values of y_1 and x_1, e. g. for the system $x_1 = a$, $y_1 = 0$, a being any finite quantity, we have the impossible equation,

$$e^{\frac{2a}{b}} + 1 = 0.$$

To find the limits within which the problem is possible, suppose the value of y_1 to be given, i. e. let it be required to join two given equal circles, whose planes are perpendicular to the line joining their centres, by a surface of revolution whose superficies shall be a minimum. We shall now proceed to show that if the distance between these circles exceed a certain determinable quantity, the problem does not admit of a solution. In other words, we shall prove that, for a given value of y_1, x_1 admits of a maximum.

We have from equation (c), making $x = x_1$, $y = y_1$, $m = 1$,

$$x_1 = bl\left\{\frac{y_1}{b} + \sqrt{\left(\frac{y_1^2}{b^2} - 1\right)}\right\}.$$

Putting $x_1 = my_1$, and $y_1 = nb$, this becomes

$$mn = l\{n + \sqrt{(n^2 - 1)}\},$$

in which m is to be a maximum. Differentiating, and making $dm = 0$, we find

$$\frac{n}{\sqrt{(n^2-1)}} = l\{n + \sqrt{(n^2-1)}\}, \quad m = \frac{1}{\sqrt{(n^2-1)}}.$$

The first of these gives $n = 1{\cdot}8$ nearly. Hence $\frac{y_1}{x_1} = \frac{1}{m} = \sqrt{(n^2-1)}$ $= 1{\cdot}4968$ nearly. If, therefore, two lines be drawn through the origin, making with the axis of x angles whose tangent is $1{\cdot}4968$, i. e. angles of $72° \, 20'$ nearly, the problem does not admit of a real solution when the extreme points of the curve lie between these lines and the axis of x.

68. We shall next proceed to consider whether the solution which we have found satisfies the other conditions necessary to the existence of a real minimum.

The general solution being of the form

$$y = \frac{1}{2} c_1 \left(c_2 e^{\frac{x}{c_1}} + \frac{1}{c_2} e^{-\frac{x}{c_1}} \right),$$

we have, putting a, b for c_1, c_2,

$$\frac{dy}{dc_1} = \frac{dy}{da} = \frac{1}{2} \left(be^{\frac{x}{a}} + \frac{1}{b} e^{-\frac{x}{a}} \right) - \frac{x}{2a} \left(be^{\frac{x}{a}} - \frac{1}{b} e^{-\frac{x}{a}} \right),$$

$$\frac{dy}{dc_2} = \frac{dy}{db} = \frac{1}{2} a \left(e^{\frac{x}{a}} - \frac{1}{b^2} e^{-\frac{x}{a}} \right).$$

Hence in p. 86,

$$u = C_1 \left(be^{\frac{x}{a}} + \frac{1}{b} e^{-\frac{x}{a}} \right) + \left(C_2 - C_1 \frac{x}{a} \right) \left(be^{\frac{x}{a}} - \frac{1}{b} e^{-\frac{x}{a}} \right),$$

$$\frac{du}{dx} = \left(\frac{C_2}{a} - \frac{C_1 x}{a^2} \right) \left(be^{\frac{x}{a}} + \frac{1}{b} e^{-\frac{x}{a}} \right);$$

or, putting for b its value as found above,

$$u = C_1 \left(e^{\frac{x}{a}} + e^{-\frac{x}{a}} \right) + \left(C_2 - C_1 \frac{x}{a} \right) \left(e^{\frac{x}{a}} - e^{-\frac{x}{a}} \right),$$

$$\frac{du}{dx} = \frac{1}{a} \left(C_2 - C_1 \frac{x}{a} \right) \left(e^{\frac{x}{a}} + e^{-\frac{x}{a}} \right). \tag{d}$$

Now it is plain that no finite value of x will satisfy the equation,

$$\frac{du}{dx} = \infty;$$

hence (Art. 47) the second variation will remain finite, if it be possible to determine C_1 and C_2 in such a way that u may not vanish within the limits of integration. Putting

$$u = 0,$$

we have

$$\frac{C_1}{C_2} = \frac{e^{\frac{x}{a}} - e^{-\frac{x}{a}}}{\frac{x}{a} \left(e^{\frac{x}{a}} - e^{-\frac{x}{a}} \right) - \left(e^{\frac{x}{a}} + e^{-\frac{x}{a}} \right)}. \tag{e}$$

The existence of a minimum value of the given integral depends, therefore, upon the possibility of assuming a value for $\frac{C_1}{C_2}$ such that no value of x within the limits of integration shall satisfy the foregoing equation. And this will be always possible unless the right-hand member of equation (e) assume successively within the limits of integration all possible values between $\pm\infty$.

Assume

$$X = \frac{e^{\frac{x}{a}} - e^{-\frac{x}{a}}}{\frac{x}{a}\left(e^{\frac{x}{a}} - e^{-\frac{x}{a}}\right) - \left(e^{\frac{x}{a}} + e^{-\frac{x}{a}}\right)}.$$

Then it is plain that when $x = 0$ we have

$$X = 0.$$

For increasing positive values of x, X increases negatively, and when a value x' is reached, which satisfies the equation

$$\frac{x'}{a}\left(e^{\frac{x'}{a}} - e^{-\frac{x'}{a}}\right) - \left(e^{\frac{x'}{a}} + e^{-\frac{x'}{a}}\right) = 0, \qquad \text{(f)}$$

we find

$$X = -\infty.$$

For increasing negative values of x, X increases positively until a value is reached equal and opposite to x', where we find

$$x = +\infty.$$

Hence if there be found, within the limits of integration, a value of x which satisfies equation (f), it is impossible to satisfy the conditions imposed by the theory of Jacobi.

The limits within which a minimum value of the given integral is possible, are therefore found by solving this equation. But, on referring to p. 148, we readily find that the same equation determines the limits within which it is possible to satisfy the several conditions furnished by the equation

$$\delta U = 0,$$

and the given limiting values of y and x.

The reason of this will be seen if we refer to the concluding remark of Art. 52. For, as has been there shown, the limits, within which it is possible to fulfil the conditions imposed by the theory of Jacobi, are found by considering at what point two of the roots of the equation by which the maximum or minimum curve is determined become equal. Now, as in general two roots of an equation become equal in passing from the real to the imaginary state, the cause of the identity of the above-mentioned equations is obvious.

Example 3.

69. To find a curve of given length, such that, if it be made to revolve about a given line, the superficial area of the generated surface may be a minimum.

It is plain, from what has been said p. 139, that the solution of this problem is derived from that of the preceding, simply by the substitution of $y + c$ for y, c being a new arbitrary constant. This gives

$$x = bl\frac{1}{m}\left\{\frac{y+c}{b} + \sqrt{\left(\frac{(y+c)^2}{b^2} - 1\right)}\right\}.$$

The curve is, therefore, still a catenary. With respect to the arbitrary constants, the equations for their determination are those found in the preceding example joined to the condition derived from the given value of the length of the curve.

Example 4.

70. Let μ be a *homogeneous* function of the co-ordinates x and y. Find the curve which will render

$$\int \mu ds$$

a minimum.

The complete integration of equation (C) is of course impossible without a more particular determination of the form of the function μ; but we may deduce from the general equation the following geometrical property belonging to all curves of this class.

Let a curve ACA' (Fig. 4) be described, whose equation is

$$\mu = \text{const.},$$

and let MPM' be the minimum curve. Let O be the origin, and draw OL parallel to the tangent at C, and PN normal to the curve MPM'. Then if m be the degree of the function μ, we shall have

$$PN = -m\rho.$$

Let x, y be the co-ordinates of the point P, x_1, y_1 of C, and x', y' of N. Then

$$\cos\alpha = \frac{x - x'}{PN}, \quad \cos\beta = \frac{y - y'}{PN}.$$

Substituting these values in equation (c), we have

$$\frac{1}{\rho} = -\frac{1}{\mu}\left(\frac{x\dfrac{d\mu}{dx} + y\dfrac{d\mu}{dy}}{PN}\right) - \frac{1}{\mu}\left(\frac{x'\dfrac{d\mu}{dx} + y'\dfrac{d\mu}{dy}}{PN}\right). \qquad \text{(a)}$$

Now since ON is parallel to the tangent at C, if we denote by μ_1 the value of μ after the substitution of x_1, y_1 for x, y, we have

$$x'\frac{d\mu_1}{dx_1} + y'\frac{d\mu_1}{dy_1} = 0. \qquad \text{(b)}$$

But as μ is a homogeneous function, it is plain that

$$\frac{d\mu_1}{dx_1} \cdot \frac{d\mu}{dy} = \frac{d\mu_1}{dy_1} \cdot \frac{d\mu}{dx};$$

and therefore that equation (b) may be written

$$x'\frac{d\mu}{dx} + y'\frac{d\mu}{dy} = 0.$$

Moreover, by the theorem of homogeneous functions,

$$x\frac{d\mu}{dx} + y\frac{d\mu}{dy} = m\mu.$$

Making these substitutions in equation (a), it becomes

$$\frac{1}{\rho} = -\frac{m}{PN}, \text{ or } PN = -m\rho.$$

PROP. II.

71. To find the curve in which

$$\int(\mu ds + \mu' dx)$$

is a maximum or minimum, μ, μ' being given functions of the co-ordinates x, y.

Adopting the same method as in the foregoing Proposition, we shall put the given integral under the form

$$\int\left(\mu + \mu' \frac{dx}{ds}\right) ds.$$

Equations (A), p. 117, become in this case

$$\frac{d}{ds} \cdot \lambda \frac{dx}{ds} + \frac{d\mu'}{ds} = \frac{d\mu}{dx} + \frac{dx}{ds} \cdot \frac{d\mu'}{dx}$$

$$\frac{d}{ds} \cdot \lambda \frac{dy}{ds} = \frac{d\mu}{dy} + \frac{dx}{ds} \cdot \frac{d\mu'}{dy}.$$

Multiplying these equations, as before, by $\frac{dx}{ds}$, $\frac{dy}{ds}$, respectively, and adding them, we find

$$\frac{d\lambda}{ds} + \frac{dx}{ds} \cdot \frac{d\mu'}{ds} = \frac{d\mu}{ds} + \frac{dx}{ds}\left(\frac{d\mu'}{dx} \cdot \frac{dx}{ds} + \frac{d\mu'}{dy} \cdot \frac{dy}{ds}\right);$$

or, since

$$\frac{d\mu'}{ds} = \frac{d\mu'}{dx} \cdot \frac{dx}{ds} + \frac{d\mu'}{dy} \cdot \frac{dy}{ds}$$

$$\frac{d\lambda}{ds} = \frac{d\mu}{ds}, \quad \lambda = \mu,$$

the constant being neglected for the reason stated in Prop. I. Substituting this value, and proceeding as before, we find

$$\frac{1}{\rho} = -\frac{1}{\mu}\left(\cos\alpha \frac{d\mu}{dx} + \cos\beta \frac{d\mu}{dy} + \frac{d\mu'}{dy}\right). \qquad \text{(A)}$$

If the extreme points of the curve be not given, i. e. if it be required to draw it between two given curves, it is easy to see that the equations by which the limiting points are determined are

$$\mu'_1 + \mu_1 \left(\frac{dx}{ds}\right)_1 + \mu_1 m_1 \left(\frac{dy}{ds}\right)_1 = 0,$$

$$\mu'_0 + \mu_0 \left(\frac{dx}{ds}\right)_0 + \mu_0 m_0 \left(\frac{dy}{ds}\right)_0 = 0;$$

the equations of the limiting curves being, as before,

$$dy_1 = m_1\, dx_1, \quad dy_0 = m_0\, dx_0.$$

Let ω_1, ω_0 be the angles at which the curve intersects its two limiting curves, respectively, and θ_1, θ_0 the angles which the tangents to the limiting curves at these points make with the axis of x. Then it is plain that the foregoing equations may be written

$$\left.\begin{aligned} \mu_1 \cos\omega_1 + \mu'_1 \cos\theta_1 = 0, \\ \mu_0 \cos\omega_0 + \mu'_0 \cos\theta_0 = 0. \end{aligned}\right\} \quad \text{(B)}$$

Example 1.

72. To draw between two given points or curves a curve of given length, such that the area included between the arc and its chord may be a maximum.

The area included between the arc and chord is evidently expressed by

$$\int_{x_0}^{x_1} ydx - \tfrac{1}{2}(y_1 + y_0)\ (x_1 - x_0).$$

Hence it is evident, from the general method of solving isoperimetrical problems, that the integral which is to be made a maximum is

$$\int (ydx - ads).^*$$

Here, therefore, we have $\mu = -a$, $\mu' = y$; and the equation found in this Proposition becomes

$$\frac{1}{\rho} = \frac{1}{a},$$

or rad. of curv. = const. The required curve is therefore a circle.

---

* For an explanation of the negative sign, vid. Art. 76.

The complete integral of the equation

$$\rho = a$$

is, as is well known,

$$(x - b)^2 + (y - c)^2 = a^2,$$

an equation containing three arbitrary constants. If the extreme points of the curve be given, these constants are determined by the equations

$$(x_1 - b)^2 + (y_1 - c)^2 = a^2,$$
$$(x_0 - b)^2 + (y_0 - c)^2 = a^2,$$

combined with the equation (derived from the given length of the arc joining these points)

$$(x_1 - x_0)^2 + (y_1 - y_0)^2 = 4a^2 \sin^2 \tfrac{1}{2}\frac{\sigma}{a},$$

σ being the given arc.

If the extreme points be not given, but merely situated on given curves, it is evident that the equations derived from the part of the variation which depends upon a change in the limits will be

$$\left\{ y_1 - a\left(\frac{dx}{ds}\right)_1 - \tfrac{1}{2}(y_1 + y_0) \right\} \delta x_1 - \left\{ a\left(\frac{dy}{ds}\right)_1 + \tfrac{1}{2}(x_1 - x_0) \right\} \delta y_1 = 0,$$

$$\left\{ y_0 - a\left(\frac{dx}{ds}\right)_0 - \tfrac{1}{2}(y_1 + y_0) \right\} \delta x_0 - \left\{ a\left(\frac{dy}{ds}\right)_0 - \tfrac{1}{2}(x_1 - x_0) \right\} \delta y_0 = 0.$$

Let the equations of the limiting curves be, as before,

$$dy_1 = m_1 dx_1, \quad dy_0 = m_0 dx_0;$$

and since

$$\delta y_1 = m_1 \delta x_1, \quad \delta y_0 = m_0 \delta x_0,$$

these equations become

$$\left.\begin{aligned} \tfrac{1}{2}(y_1 - y_0) - \tfrac{1}{2} m_1 (x_1 - x_0) &= a\left\{ \left(\frac{dx}{ds}\right)_1 + m_1 \left(\frac{dy}{ds}\right)_1 \right\} \\ \tfrac{1}{2}(y_0 - y_1) - \tfrac{1}{2} m_0 (x_0 - x_1) &= a\left\{ \left(\frac{dx}{ds}\right)_0 + m_0 \left(\frac{dy}{ds}\right)_0 \right\}. \end{aligned}\right\} \quad \text{(a)}$$

To interpret these equations, let P_1, P_0 (Fig. 5) be the extreme points of the curve, and O the middle point of the line joining them. Let also $P_1 T_1$, $P_0 T_0$, be the tangents to the limiting curves at the points P_1, P_0. Let fall the perpendiculars Op_1, Op_0. We have then

$$Op_1 = \frac{\frac{1}{2}(y_0 - y_1) - \frac{1}{2}m_1(x_0 - x_1)}{\sqrt{(1+m_1^2)}},$$

$$Op_0 = \frac{\frac{1}{2}(y_1 - y_0) - \frac{1}{2}m_0(x_1 - x_0)}{\sqrt{(1+m_0^2)}}.$$

Let C be the centre of the circle, and let fall the perpendiculars, $C\pi_1$, $C\pi_0$. Then it is plain that

$$\sin CP_1\pi_1 = -\frac{\left(\frac{dx}{ds}\right)_1 + m_1\left(\frac{dy}{ds}\right)_1}{\sqrt{(1+m_1^2)}},$$

$$\sin CP_0\pi_0 = -\frac{\left(\frac{dx}{ds}\right)_0 + m_0\left(\frac{dy}{ds}\right)_0}{\sqrt{(1+m_0^2)}}.$$

Equations (a) become, therefore,

$$Op_1 = CP_1 \sin CP_1\pi_1 = C\pi_1,$$
$$Op_0 = CP_0 \sin CP_0\pi_0 = C\pi_0.$$

These equations denote either—1. That the points O and C coincide; or, 2. That the tangents $P_1\,T_1$, $P_0\,T_0$, are parallel to OC. In the first case the segment will evidently be a semicircle. In the second, the tangents $P_1\,T_1$, $P_0\,T_0$ are perpendicular to the line P_1P_0, joining the extreme points; and the length of this line is, therefore, itself a minimum.

If the area which is to be made a maximum be the entire of the space included between the curve, the extreme ordinates, and the axis of x, it is plain that the expression which is to be made a maximum is simply

$$\int_{x_0}^{x_1}(ydx - ads),$$

the term $\frac{1}{2}(y_0 + y_1)(x_1 - x_0)$ being omitted. The general solution is therefore the same as before, with the exception of the conditions to be observed at the limits.

If the extreme points of the curve be given, the arbitrary constants are determined, as in the foregoing case, by the substitution of the given quantities, $x_0\,y_0$, $x_1\,y_1$, in the general solution.

If the extreme points of the curve be situated upon two given curves, the equations (B) of Art. 71 become

$$a \cos \omega_1 = y_1 \cos \theta_1,$$
$$a \cos \omega_0 = y_0 \cos \theta_0.$$

If the limiting values of x only be given, i. e., if the bounding curves become right lines perpendicular to the axis of x, we should have

$$\cos \theta_1 = 0, \quad \cos \theta_0 = 0,$$

and therefore the first of the above equations would give either

$$a \cos \omega_1 = 0, \quad \text{or} \quad y_1 = \infty.$$

Rejecting the latter of these suppositions, which would render the area infinite, we have

$$a \cos \omega_1 = 0,$$

and, similarly,

$$a \cos \omega_0 = 0.$$

But it is easily seen that these equations are impossible. The given area does not therefore admit of a finite maximum. This case has been already noticed as an exception to the general theory of maxima and minima.*

73. As a more general case, let it be required to draw between two given curves a curve of given length, such that the area included between it and a curve of given species passing through its extreme points may be a maximum.

It is evident that the equation of the curve of given species which forms one boundary of the space in question will be of the form

$$y = F(x, x_0, x_1).$$

The area included between it and the curve which it is our object to determine, will therefore be

$$\int \{y - F(x, x_0, x_1)\} dx;$$

and the entire expression which is to be made a maximum,

$$\int [\{y - F(x, x_0, x_1)\} dx - ads];$$

or, denoting the integral $\int F(x, x_0, x_1) dx$ by $F_1(x, x_0, x_1)$, (F being a given function),

$$\int (ydx - ads) - F_1(x_1, x_0, x_1) + F_1(x_0, x_0, x_1).$$

---

* Vid. Chap. III. pp. 48, 63.

Since, then, the quantity under the sign of integration is the same as before, it is plain that the differential equation of the curve is not altered. It is, therefore, still a circle. With regard to the terms which appear outside the sign of integration, we shall have, in the first place,

$$y_1 \delta x_1 - a \left\{ \left(\frac{dy}{ds}\right)_1 \delta y_1 + \left(\frac{dx}{ds}\right)_1 \delta x_1 \right\},$$

$$- y_0 \delta x_0 + \&c.$$

arising from $\int (y dx - a ds)$. It remains, then, to consider those arising from the other part of the expression, namely,

$$- F_1 (x_1, x_0, x_1) + F_1 (x_0, x_0, x_1) = - \int_{x_0}^{x_1} F(x, x_0, x_1).$$

Putting

$$u = \int_{x_0}^{x_1} F(x, x_0, x_1)\, dx,$$

we have

$$\frac{du}{dx_0} = - F(x_0, x_0, x_1) + \int_{x_0}^{x_1} \frac{dF}{dx_0}\, dx,$$

$$\frac{du}{dx_1} = F(x_1, x_0, x_1) + \int_{x_0}^{x_1} \frac{dF}{dx_1}\, dx.$$

But since both the given curve and the required curve pass through the same points, $x_0 y_0$, $x_1 y_1$, it is evident that

$$y_0 = F(x_0, x_0, x_1),$$

$$y_1 = F(x_1, x_0, x_1).$$

Hence

$$\frac{du}{dx_0} = - y_0 + \int_{x_0}^{x_1} \frac{dF}{dx_0}\, dx,$$

$$\frac{du}{dx_1} = y_1 + \int_{x_0}^{x_1} \frac{dF}{dx_1}\, dx.$$

The complete change in u is therefore

$$\frac{du}{dx_0} \delta x_0 + \frac{du}{dx_1} \delta x_1 = \left(- y_0 + \int_{x_0}^{x_1} \frac{dF}{dx_0}\, dx \right) \delta x_0 + \left(y_1 + \int_{x_0}^{x_1} \frac{dF}{dx_1}\, dx \right) \delta x_1;$$

adding this, with its proper sign, to the terms previously found, and equating separately to zero the terms connected with each limit, we have

$$a\left(\frac{dy}{ds}\right)_0 \delta y_0 + \left\{a\left(\frac{dx}{ds}\right)_0 - \int_{x_0}^{x_1} \frac{dF}{dx_0}\,dx\right\}\delta x_0 = 0,$$

$$a\left(\frac{dy}{ds}\right)_1 \delta y_1 + \left\{a\left(\frac{dx}{ds}\right)_1 + \int_{x_0}^{x_1} \frac{dF}{dx_1}\,dx\right\}\delta x_1 = 0.$$

Or, if the equations of the bounding curves be, as before,

$$dy_0 = m_0\,dx_0, \quad dy_1 = m_1\,dx_1,$$

$$a\left\{m_0\left(\frac{dy}{ds}\right)_0 + \left(\frac{dx}{ds}\right)_0\right\} - \int_{x_0}^{x_1} \frac{dF}{dx_0}\,dx = 0,$$

$$a\left\{m_1\left(\frac{dy}{ds}\right)_1 + \left(\frac{dx}{ds}\right)_1\right\} + \int_{x_0}^{x_1} \frac{dF}{dx_1}\,dx = 0.$$

If these equations be transformed, as in the general proposition, by the introduction of the angles $\omega_0\,\omega_1$, $\theta_0\,\theta_1$, they will become

$$a\cos\omega_0 - \cos\theta_0 \int_{x_0}^{x_1} \frac{dF}{dx_0}\,dx = 0,$$

$$a\cos\omega_1 + \cos\theta_1 \int_{x_0}^{x_1} \frac{dF}{dx_1}\,dx = 0.$$

Example 2.

74. Of all isoperimetrical curves described upon the given base AB, to determine a curve ACB (Fig. 6), such that the area of a second curve $AC'B$, which is derived from the first by the condition that each of its ordinates, $C'D$, shall be a given function of the corresponding ordinate, CD, may be a maximum.*

If the given function be denoted by Y, the area which is to be made a maximum will evidently be expressed by

$$\int_{x_0}^{x_1} Y dx.$$

Hence in the general equation of Prop. II. we shall have

$$\mu = -a, \quad \mu' = Y,$$

* This is the first case of the celebrated isoperimetrical problem of James Bernouilli.

and, therefore,

$$\frac{1}{\rho} = \frac{1}{a} \cdot \frac{dY}{dy};$$

or, putting for ρ its value in terms of x and y, and integrating

$$\frac{1}{\sqrt{\left(1 + \frac{dy^2}{dx^2}\right)}} = \frac{Y}{a} + c.$$

Hence we easily find

$$dx = \frac{\left(\frac{Y}{a} + c\right) dy}{\sqrt{\left\{1 - \left(\frac{Y}{a} + c\right)^2\right\}}},$$

and, therefore,

$$x + c' = \int \frac{\frac{Y}{a} + c}{\sqrt{\left\{1 - \left(\frac{Y}{a} + c\right)^2\right\}}} dy,$$

which gives the equation of the curve when the form of the function Y is known.

If $Y = y$, the curves ACB, $AC'B$ become identical, and the problem is the same with that discussed in Example 1 of the present Proposition.

Example 3.

75. To find a curve of given length, such that the volume of the solid generated by its revolution round a given line may be a maximum.

The volume of the generated solid being represented by $\pi \int y^2 dx$, it is evident that the integral which is to be rendered a maximum is

$$\int \left(y^2 \frac{dx}{ds} - a^2\right) ds,$$

a being an arbitrary constant.

Here, therefore,

$$\mu = a^2, \quad \frac{d\mu}{dx} = 0, \quad \frac{d\mu}{dy} = 0, \quad \mu' = y^2, \quad \frac{d\mu'}{dy} = 2y.$$

Equation (A) becomes, therefore,

$$\frac{1}{\rho} = \frac{2y}{a^2}.$$

The curve is, therefore, such that the radius of curvature is inversely proportional to the ordinate. We may obtain a first integral of this equation by substituting for $\frac{1}{\rho}$ its value,

$$\frac{-\frac{dp}{dx}}{(1+p^2)^{\frac{3}{2}}}, \text{ or } \frac{-\frac{pdp}{dy}}{(1+p^2)^{\frac{3}{2}}},$$

$\left(\text{where } p = \frac{dy}{dx}\right)$. This will give

$$-\frac{pdp}{(1+p^2)^{\frac{3}{2}}} = \frac{2ydy}{a^2};$$

or, integrating, and adding an arbitrary constant,

$$\frac{1}{\sqrt{(1+p^2)}} = \frac{y^2+b^2}{a^2}.$$

The further integration of this equation is evidently impossible.

Example 4.

76. To find a curve such that the surface generated by its revolution round a given line may be given, and that the volume of the generated solid may be a maximum.

It is easy to see that the integral which is, in this case, to be rendered an absolute maximum, is

$$\int(y^2dx + ayds),$$

a being an arbitrary constant. We have, therefore,

$$\mu = ay, \quad \frac{d\mu}{dx} = 0, \quad \frac{d\mu}{dy} = a, \quad \mu' = y^2, \quad \frac{d\mu'}{dy} = 2y.$$

Equation (A) becomes, therefore,

$$-\frac{1}{\rho}=\frac{2}{a}+\frac{\cos\beta}{y}.$$

Let n be the normal, and since $n=y\sec\beta$, this equation becomes

$$-\frac{1}{\rho}=\frac{2}{a}+\frac{1}{n},$$

or

$$\frac{1}{\rho}+\frac{1}{n}=-\frac{2}{a}.$$

Now since the principal radii of curvature at each point of the surface of revolution are ρ and n, this equation expresses the property that the sum of the curvatures is the same at each point in the surface. We shall see afterwards that this property belongs to a more general case.

It is easily shown that to render the result a real maximum, a must be negative. For if the original integral be put under the form

$$\int Vdx=\int\{y^2+ay\sqrt{(1+p^2)}\}\,dx,$$

we have

$$\frac{d^2V}{dp^2}=\frac{ay}{(1+p^2)^{\frac{3}{2}}}.$$

Now, in order that the integral may be a maximum, this quantity must be negative, and, therefore, since y and $\sqrt{(1+p^2)}$ are both positive, a must be taken negatively. Putting, then, $-a$ for a, we have

$$\frac{1}{\rho}+\frac{1}{n}=\frac{2}{a}.$$

Substituting for $\frac{1}{\rho}$ and $\frac{1}{n}$ their respective values,

$$\frac{-\frac{pdp}{dy}}{(1+p^2)^{\frac{3}{2}}} \text{ and } \frac{1}{y\sqrt{(1+p)^2}},$$

and multiplying by ydy, this becomes

$$\frac{dy}{\sqrt{(1+p^2)}} - \frac{ypdp}{(1+p^2)^{\frac{3}{2}}} = \frac{2ydy}{a}.$$

Integrating this, and adding an arbitrary constant,

$$\frac{y}{\sqrt{(1+p^2)}} = \frac{y^2+c}{a}.$$

This equation cannot, in general, be integrated further. But suppose the surface to be a *closed* one, i. e., suppose either that the curve by which it is generated terminates at both ends in the axis of revolution, or that it is a *closed* curve, lying entirely at one side of the axis of x, thus giving an *annular* surface.

In the latter case it is evident that p must go through all values between $\pm\infty$ for *positive* values of y. Now if we put θ for the angle which the tangent at any point makes with the axis of x, the equation

$$\frac{y}{\sqrt{(1+p^2)}} = \frac{y^2+c}{a}$$

may be put under the form

$$y^2 - ay\cos\theta + c = 0.$$

Now since *both* values of y, corresponding to the same value of θ, are positive, it is plain that c must be essentially positive. But since the curve is closed, and of finite curvature, the angle θ must pass through all values between 0 and 2π. Hence there must be at least two points on the curve, for which

$$\cos\theta = 0,$$

and, therefore,

$$y^2 + c = 0,$$

an impossible equation, inasmuch as c is essentially positive. The equation

$$\frac{y}{\sqrt{(1+p^2)}} = \frac{y^2+c}{a}$$

cannot, therefore, represent a closed curve, lying entirely at one side of the axis of x. We must, therefore, adopt the former supposition, namely, that the curve terminates at both ends in the axis of revolution. Then it is plain that the solution of this equation must be such as to admit of the value $y = 0$. But since $\sqrt{(1+p^2)}$ can never vanish, this condition becomes possible only when we suppose $c = 0$. This supposition reduces the equation to

$$y\sqrt{(1+p^2)} = a,$$

from which we obtain

$$dx = \frac{ydy}{\sqrt{(a^2-y^2)}},$$

and, by integration,

$$(x+c)^2 + y^2 = a^2,$$

the equation of a circle whose centre is in the axis of revolution. The sphere is therefore the only *closed* surface of revolution which possesses the property of including a maximum solid under a given surface. This reasoning is equally applicable, if it be only supposed that one extremity of the generating curve terminates in the axis of revolution. If then it be required to erect upon a given circle a closed segment of a surface of revolution, such that under a given superficial area it may include a maximum volume, the surface which solves the question is still spherical. In both these cases it is evident that the radius of the sphere is determined from the given value of the superficial area.

Let this area be S, and R the required radius. Then, in the first case, where the surface is completely closed,

$$R^2 = \frac{S}{4\pi}.$$

In the second case let r be the radius of the circle on which the segment is to be erected, and it is readily found that

$$R^2 = \frac{\frac{S}{4\pi}}{1 - \frac{\pi r^2}{S}}.$$

If the circle upon which the segment is to be erected be not given, but merely a section of a given surface of revolution, in other words, if one extremity of the generating curve be situated upon a given curve, the first of equations (B), p. 154, becomes

$$a y_1 \cos \omega_1 = y_1^2 \cos \theta_1,$$

giving either $y_1 = 0$, or $a \cos \omega_1 = y_1 \cos \theta_1$.

The first of these denotes that the generating curve terminates at *both* ends in the axis of revolution, thus rendering the surface a complete sphere, and, as is easily seen, rendering the volume an absolute maximum. To interpret the second, let BA (Fig. 7) be the axis of revolution, P_1A the limiting curve, P_1 the point at which the generating circle intersects it, C the centre of this circle, and P_1N, P_1Y, the normal and ordinate of the limiting curve at the point P_1. Then since $\omega_1 = CP_1N$ and $\theta_1 = YP_1N$, we have

$$\frac{CP_1}{YP_1} = \frac{a}{y} = \frac{\cos \theta_1}{\cos \omega_1} = \frac{\cos YP_1N}{\cos CP_1N}.$$

It is evident, therefore, that the normal P_1N is in this point perpendicular to the axis of revolution. In other words, P_1Y is a maximum or minimum ordinate. If P_1Y be a *minimum*, it is easy to see that the volume of the segment will be an absolute maximum. If P_1Y be a *maximum*, the volume will be a *relative* minimum, i. e. a minimum compared with similarly described *spherical* segments, although not so when compared with those of other surfaces of revolution.

Prop. III.

77. Of all isoperimetrical curves described upon a given base, AB (Fig. 8), to determine ACB, such that the area of the curve, $AC'B$, whose ordinate, $C'D$, is at each point a given function of the arc AC, may be a maximum.*

Let the given function be represented by S. Then it is evident that the area of the curve, $AC'B$, will be represented by the integral

$$\int S \frac{dx}{ds}\, ds.$$

* This is the second case of James Bernouilli's isoperimetrical problem.

Hence, according to the general theory of isoperimetrical problems, the integral which is to be made a maximum is

$$\int\left(S\frac{dx}{ds}+m\right)ds,$$

the functions x, y being, as before, connected by the equation

$$\frac{dx^2}{ds^2}+\frac{dy^2}{ds^2}=1. \qquad \text{(A)}$$

We have, therefore,

$$V=S\frac{dx}{ds}+m, \quad N=0, \quad N'=0, \quad P_1=S, \quad P'_1=0,$$

$$\alpha=0, \quad \alpha'=0, \quad \beta=2\frac{dx}{ds}, \quad \beta'=2\frac{dy}{ds}.$$

Substituting these values in the equations

$$N+\lambda\alpha-\frac{d(P_1+\lambda\beta)}{ds}=0,$$

$$N'+\lambda\alpha'-\frac{d(P'_1+\lambda\beta')}{ds}=0,$$

and integrating, we find

$$\left.\begin{aligned} S+2\lambda\frac{dx}{ds}&=a,\\ 2\lambda\frac{dy}{ds}&=b. \end{aligned}\right\} \qquad \text{(B)}$$

Hence, and from equation (A), we have

$$2\lambda=\sqrt{\{b^2+(a-S)^2\}}.$$

Substituting this value in the equations (B), and integrating, we find

$$x+c=\int\frac{a-S}{\sqrt{\{b^2+(a-S)^2\}}}\,ds,$$

$$y+d=\int\frac{b}{\sqrt{\{b^2+(a-S)^2\}}}\,ds.$$

Eliminating s between these equations, we have the equation of the required curve.

This solution contains, as is evident, four arbitrary constants. Of these two are determined by the condition that the required curve shall pass through the given points A, B, one by the given length of the arc ACB, and the fourth by the condition that the independent variable s shall be reckoned from some determinate point, as, for example, A. The solution of the problem is therefore complete.

The reader will find no difficulty in applying the foregoing formulæ to the case in which

$$S = s,$$

and deducing the equation of the catenary.

Prop. IV.

78. Let ρ be the radius of curvature of a plane curve, and let $\mu = \phi(\rho)$. Required to determine the curve which will render $\int \mu ds$ a maximum or minimum.

Adopting, as before, the method of Lagrange, we shall have

$$\int \left\{ \delta\mu + \lambda \left(\frac{dx}{ds} \cdot \frac{d\delta x}{ds} + \frac{dy}{ds} \cdot \frac{d\delta y}{ds} \right) \right\} ds = 0.$$

But since

$$\mu = \phi(\rho),$$

and

$$\frac{1}{\rho^2} = \left(\frac{d^2x}{ds^2}\right)^2 + \left(\frac{d^2y}{ds^2}\right)^2,$$

$$\delta\mu = \frac{d\mu}{d\rho} \cdot \delta\rho = -\mu'\rho^3 \left(\frac{d^2x}{ds^2} \cdot \frac{d^2\delta x}{ds^2} + \frac{d^2y}{ds^2} \cdot \frac{d^2\delta y}{ds^2} \right),$$

putting

$$\frac{d\mu}{d\rho} = \mu'.$$

Hence it is evident that the equations furnished by the terms under the sign of integration are

$$\frac{d^2 . \mu'\rho^3 \frac{d^2x}{ds^2}}{ds^2} + \frac{d . \lambda \frac{dx}{ds}}{ds} = 0,$$

$$\frac{d^2 . \mu'\rho^3 \frac{d^2y}{ds^2}}{ds^2} + \frac{d . \lambda \frac{dy}{ds}}{ds} = 0. \qquad \text{(A)}$$

Integrating these equations, we find

$$\frac{d . \mu'\rho^3 \frac{d^2x}{ds^2}}{ds} + \lambda \frac{dx}{ds} = a,$$

$$\frac{d . \mu'\rho^3 \frac{d^2y}{ds^2}}{ds} + \lambda \frac{dy}{ds} = b;$$

or

$$\mu'\rho^3 . \frac{d^3x}{ds^3} + \frac{d^2x}{ds^2} . \frac{d . \mu'\rho^3}{ds} + \lambda \frac{dx}{ds} = a,$$

$$\mu'\rho^3 . \frac{d^3y}{ds^3} + \frac{d^2y}{ds^2} . \frac{d . \mu'\rho^3}{ds} + \lambda \frac{dy}{ds} = b. \qquad \text{(B)}$$

Multiplying the first of these equations by $\frac{dy}{ds}$, and the second by $\frac{dx}{ds}$, and subtracting, we have

$$\mu'\rho^3 \left(\frac{dy}{ds} . \frac{d^3x}{ds^3} - \frac{dx}{ds} . \frac{d^3y}{ds^3}\right) + \frac{d . \mu'\rho^3}{ds} \left(\frac{dy}{ds} . \frac{d^2x}{ds^2} - \frac{dx}{ds} . \frac{d^2y}{ds^2}\right) = a \frac{dy}{ds} - b \frac{dx}{ds}.$$

Integrating again, and adding an arbitrary constant, we find

$$\mu'\rho^3 \left(\frac{dy}{ds} . \frac{d^2x}{ds^2} - \frac{dx}{ds} . \frac{d^2y}{ds^2}\right) = ay - bx + c;$$

or since

$$\frac{1}{\rho} = \frac{dy}{ds} . \frac{d^2x}{ds^2} - \frac{dx}{ds} . \frac{d^2y}{ds^2},$$

$$\mu'\rho^2 = ay - bx + c. \qquad \text{(C)}$$

Previously to proceeding further, we shall consider the mode of determining the constants a, b, c.

Let the extreme points of the curve be given, so that

$$\delta x_1 = 0, \quad \delta y_1 = 0, \quad \delta x_0 = 0, \quad \delta y_0 = 0.$$

Then the terms which remain outside the sign of integration are

$$\mu'_1 \rho_1^3 \left(\frac{d^2x}{ds^2} \cdot \frac{d\delta x}{ds} + \frac{d^2y}{ds^2} \cdot \frac{d\delta y}{ds}\right)_1,$$

$$\mu'_0 \rho_0^3 \left(\frac{d^2x}{ds^2} \cdot \frac{d\delta x}{ds} + \frac{d^2y}{ds^2} \cdot \frac{d\delta y}{ds}\right)_0.$$

Eliminating either of the variations, e. g. $\left(\frac{d\delta y}{ds}\right)_1$, from the first of these, by means of the equation

$$\left(\frac{dx}{ds}\right)_1 \cdot \left(\frac{d\delta x}{ds}\right)_1 + \left(\frac{dy}{ds}\right)_1 \cdot \left(\frac{d\delta y}{ds}\right)_1 = 0,$$

it becomes

$$\mu'_1 \rho_1^3 \frac{\left(\frac{dy}{ds} \cdot \frac{d^2x}{ds^2} - \frac{dx}{ds} \cdot \frac{d^2y}{ds^2}\right)_1}{\left(\frac{dy}{ds}\right)_1} \left(\frac{d\delta x}{ds}\right),$$

or

$$\mu'_1 \rho_1^2 \frac{\left(\frac{d\delta x}{ds}\right)_1}{\left(\frac{dy}{ds}\right)_1}.$$

Since, then, $\left(\frac{dy}{ds}\right)_1$ cannot become infinite, this term can only be made to vanish by putting $\mu'_1 \rho_1^2 = 0$. Similarly we shall have $\mu'_0 \rho_0^2 = 0$. This will give

$$\left.\begin{aligned} ay_1 - bx_1 + c &= 0, \\ ay_0 - bx_0 + c &= 0; \end{aligned}\right\} \tag{D}$$

or if the origin be taken at the point $x_0 y_0$, and the line joining the given points be made the axis of x,

$$b = 0, \quad c = 0.$$

The equation (C) becomes, therefore,

$$\mu' \rho^2 = ay. \tag{E}$$

Since μ, and therefore μ', is a function of ρ, this equation gives the value of ρ in terms y. Hence we infer, *The plane curve which renders $\int\phi(\rho)ds$, taken between two fixed points, a maximum or minimum, is such that all points of equal curvature are equally distant from the line which joins these extremities.* And since equations (D) hold equally whether these points be or be not given, it is easily seen that this theorem is equally true if for fixed points we substitute two fixed curves.

To find the equation of the curves in finite terms, suppose (E) to be solved so as to give

$$\frac{1}{\rho} = Y.$$

Substituting for $\frac{1}{\rho}$ its value,

$$\frac{-\frac{pdp}{dy}}{(1+p^2)^{\frac{3}{2}}},$$

and integrating, we have

$$\frac{1}{\sqrt{(1+p^2)}} = \int Ydy = Y_1 + e;$$

and, therefore,

$$dx = \frac{(Y_1+e)dy}{\sqrt{\{1-(Y_1+e)^2\}}}.$$

Hence the equation of the curve is

$$x + f = \int \frac{(Y_1+e)dy}{\sqrt{\{1-(Y_1+e)^2\}}}, \qquad \text{(F)}$$

the constants e and f being determined from the conditions that y shall vanish for $x = 0$ and for $x = x_1$, and the remaining constant, a, depending upon the given length of the curve.

If the curve be bounded not by given points, but by given curves, whose equations are

$$dy_1 = m_1 dx_1, \quad dy_0 = m_0 dx_0,$$

the terms depending upon the variations δx_1, δy_1, which are evidently

$$\left(\frac{d.\mu'\rho^3\frac{d^2x}{ds^2}}{ds}+\lambda\frac{dx}{ds}\right)_1\delta x_1+\left(\frac{d.\mu'\rho^3\frac{d^2y}{ds^2}}{ds}+\lambda\frac{dy}{ds}\right)_1\delta y_1,$$

become

$$\left\{\left(\frac{d.\mu'\rho^3\frac{d^2x}{ds^2}}{ds}+\lambda\frac{dx}{ds}\right)_1+m_1\left(\frac{d.\mu'\rho^3\frac{d^2y}{ds^2}}{ds}+\lambda\frac{dy}{ds}\right)_1\right\}\delta x_1,$$

which are reduced by equations (B) to

$$(a+m_1 b)\,\delta x_1.$$

Hence we have

$$a+m_1 b=0.$$

But if the line joining the extreme points be made the axis of x, we have before seen that $b=0$. Hence, as it is plain that a cannot vanish, we must have $m_1=\infty$, and similarly $m_0=\infty$. The line joining the extremities of the curve is therefore perpendicular to the tangents to the limiting curves at the points where it intersects them. Hence we infer

The plane curve which renders $\int\phi(\rho)ds$, taken between its points of intersection with two given curves, a maximum or minimum, will intersect these curves in two points such that the rectilinear distance between them will be itself a maximum or minimum.

79. Before giving examples of this proposition, as applied to particular cases, it will be necessary to recall an observation made at the commencement of Chap. III., as to the meaning of the words *maximum* and *minimum*. These words, as was there stated, do not signify values which are absolutely the *greatest*, or absolutely the *least*, but values which are greater or less than any others which can be formed by an indefinitely small change in any of the varying elements. In a problem, therefore, the statement of which involves *curvature*, it is to be remembered that the curve which the Calculus of Variations gives for its solution must be compared only with curves which can be deduced from it without an abrupt change in the curvature. If, then, we had found, by that method, a curve without a *cusp* between its extreme points as the solution of such a problem, it would be no exception to the truth of the solution that a curve might be found which rendered the given function greater or less than such a maximum or mini-

mum, if that curve contained between its extreme points one or more cusps. Similarly, if the curve found by the method of Variations contain any number of cusps, it must be compared only with curves having the same number. It is the more necessary to keep this in mind, because an abrupt change in curvature does not necessarily involve an abrupt change in geometrical position, i. e. the several points of the curve may still be indefinitely near those of the curve with which it is compared, and it might, therefore, be thought that such a comparison was legitimate. This we have seen not to be true. An example of this principle will be found in the following problem.

Example.

80. To find a curve of given length, such that the area bounded by the curve itself, its two extreme radii of curvature, and the arc of the evolute between them, may be a minimum.

Denoting, as before, by ρ the radius of curvature, and by ds the element of the curve, it is evident that the element of the area is ρds. The solution of this problem will, therefore, be obtained from the foregoing proposition by making

$$\mu = \rho + m, \quad \mu' = 1.$$

Making these substitutions in equation (E), it becomes

$$\rho^2 = ay. \tag{a}$$

The equation in finite terms may be deduced from the general equation (F) as follows:

The equation

$$\rho^2 = ay$$

gives

$$Y = \frac{1}{\sqrt{ay}},$$

and, therefore,

$$Y_1 = 2\sqrt{\frac{y}{a}}.$$

Equation (F) becomes

$$x + f = \int \frac{\left(2\sqrt{\frac{y}{a}} + e\right) dy}{\sqrt{\left\{1 - \left(2\sqrt{\frac{y}{a}} + e\right)^2\right\}}}.$$

Putting $a = 4b$, $e = \sin\alpha$, and $\sqrt{\frac{y}{b}} + e = \sin\theta$, we have

$$\begin{aligned} x + f &= 2b\int \sin\theta\,(\sin\theta - \sin\alpha)d\theta \\ &= b\,(\theta - \sin\theta\cos\theta + 2\sin\alpha\cos\theta). \end{aligned} \qquad \text{(b)}$$

Now since $y = 0$ when $x = 0$ and when $x = x_1$,

$$f = b\,(\alpha + \sin\alpha\cos\alpha)$$

$$x_1 + f = b\,(\pi - \alpha - \sin\alpha\cos\alpha).$$

Whence

$$b = \frac{x_1}{\pi - 2(\alpha + \sin\alpha\cos\alpha)}.$$

This determines the constants b and f in terms of α. This remaining constant is determined by means of the given length of the curve as follows:

Differentiating equations (b), we have

$$dy = 2b(\sin\theta - \sin\alpha)\cos\theta d\theta,$$

$$dx = 2b(\sin\theta - \sin\alpha)\sin\theta d\theta.$$

Hence we find

$$ds = 2b\,(\sin\theta - \sin\alpha)\,d\theta.$$

Integrating this expression between the limits

$$\theta = \alpha, \quad \theta = \pi - \alpha,$$

and denoting the given length of the curve by $2s_1$, we have

$$s_1 = 2b\{\cos\alpha + (\alpha - \tfrac{1}{2}\pi)\sin\alpha\};$$

or, putting for b its value in terms of α,

$$s_1 = 2x_1\frac{\cos\alpha + (\alpha - \frac{1}{2}\pi)\sin\alpha}{\pi - 2(\alpha + \sin\alpha\cos\alpha)}.$$

Hence the constant α is determined.

The equation of the curve is found by eliminating θ between the equations

$$\frac{y}{b} = (\sin\theta - \sin\alpha)^2,$$

and

$$\frac{x}{b} = \theta - \alpha + 2\sin\alpha\cos\theta - \sin\theta\cos\theta - \sin\alpha\cos\alpha.$$

If the origin be transferred to the middle point of the line joining the extreme points of the curve, which is done by substituting

$$x + \tfrac{1}{2}x_1, \text{ or } x + b\left(\tfrac{1}{2}\pi - \alpha - \sin\alpha\cos\alpha\right)$$

for x, the second equation becomes

$$\frac{x}{b} = \theta + (2\sin\alpha - \sin\theta)\cos\theta - \tfrac{1}{2}\pi,$$

or

$$\frac{x}{b} + \cos^{-1}\left(\sin\alpha + \sqrt{\frac{y}{b}}\right)$$

$$= \left(\sin\alpha - \sqrt{\frac{y}{b}}\right)\sqrt{\left\{1 - \left(\sin\alpha + \sqrt{\frac{y}{b}}\right)^2\right\}}.$$

A remarkable geometrical property belonging to curves of this class may be derived from the first integral of the equation

$$\rho^2 = 4by.$$

If, as before, we substitute for $\frac{1}{\rho}$ its value,

$$\frac{-\frac{pdp}{dy}}{(1+p^2)^{\frac{3}{2}}},$$

and integrate, we shall have

$$\frac{1}{\sqrt{(1+p^2)}} = \sqrt{\frac{y}{b}} + e;$$

or dividing by y, and putting for $\frac{1}{y\sqrt{(1+p^2)}}$ and $\frac{1}{\sqrt{by}}$ their respective values, $\frac{1}{\text{normal}}$ and $\frac{2}{\rho}$,

$$\frac{1}{n} = \frac{2}{\rho} + \frac{e}{y}.$$

81. If the length of the curve between the extreme points be not given, i. e. if it be required to determine, among all curves which can be described between two given points, that curve for which the above-mentioned area is a minimum, the superfluous constant, α, will be determined by expressing the area as a function of that constant, and equating its differential to zero.

Now

$$\rho ds = 2\sqrt{by}\sqrt{(1+p^2)}\,dx = \frac{2\sqrt{by}\sqrt{(1+p^2)}\,dy}{p}$$

$$= 2b\,.\,\frac{\sqrt{\frac{y}{b}}\,dy}{\sqrt{\left\{1-\left(\sqrt{\frac{y}{b}}+e\right)^2\right\}}}.$$

Putting, as before,

$$\sqrt{\frac{y}{b}}+e = \sin\theta, \text{ and } e = \sin\alpha,$$

we have

$$\int \rho ds = 4b^2 \int (\sin\theta - \sin\alpha)^2 d\theta$$

$$= 2b^2\{\theta\,(1+2\sin^2\alpha) - \sin\theta\cos\theta + 4\sin\alpha\cos\theta\}.$$

Now since the maximum value of y corresponds to $\theta = \frac{\pi}{2}$, it is easy to see that the value of the area in question is

$$8b^2.\int_\alpha^{\frac{\pi}{2}} (\sin\theta - \sin\alpha)^2\, d\theta$$

$$= 4b^2\left\{\left(\frac{\pi}{2}-\alpha\right)(1+2\sin^2\alpha) - 3\sin\alpha\cos\alpha\right\};$$

or, putting for b its value in terms of α, and for α, $\frac{\pi}{2}-\beta$,

$$\text{area} = x_1^2.\frac{\beta(1+2\cos^2\beta) - 3\sin\beta\cos\beta}{(\beta - \sin\beta\cos\beta)^2}.$$

Equating to zero the differential of the logarithm of this expression, we have

$$\frac{\sin\beta\,(\sin\beta - \beta\cos\beta)}{\beta + 2\beta\cos^2\beta - 3\sin\beta\cos\beta} = \frac{\sin^2\beta}{\beta - \sin\beta\cos\beta}.$$

Neglecting the value $\beta = 0$ (which would make $b = \infty$, and render the area also infinite), and clearing of fractions, we find

$$\cos\beta\,(\beta\sin\beta\cos\beta + \beta^2 - 2\sin^2\beta) = 0.$$

This may be satisfied either by making

$$\cos\beta = 0, \text{ or } \beta\sin\beta\cos\beta + \beta^2 - 2\sin^2\beta = 0.$$

But it may readily be shown that the latter of these equations is impossible,* the left hand member being always positive. Hence we must have

$$\cos\beta = \sin\alpha = 0.$$

The equation

$$\frac{1}{\sqrt{(1+p^2)}} = \sqrt{\frac{y}{b}} + e$$

becomes, therefore,

$$\frac{1}{\sqrt{(1+p^2)}} = \sqrt{\frac{y}{b}},$$

or

$$dx = \frac{ydy}{\sqrt{(by - y^2)}},$$

the equation of a cycloid. And as at each of the limiting points $\rho = 0$, it is plain that the curve is a complete cycloid.

---

* This may be shown as follows:

Multiply the left hand member of the equation by 4, and put $2\beta = \theta$. This member may then be written

$$u = \theta^2 + \theta\sin\theta + 4\cos\theta - 4.$$

If for $\sin\theta$ and $\cos\theta$ we put

$$\theta - \frac{\theta^3}{1.2.3} + \frac{\theta^5}{1.2.3.4.5} - \&c.$$

and

$$1 - \frac{\theta^2}{1.2} + \frac{\theta^4}{1.2.3.4} - \&c.,$$

it will at once appear that u is positive for small values of θ. As, therefore, u vanishes for $\theta = 0$, if it vanish for any other value, $\theta = \theta_1$, it must attain a maximum for some intermediate value. Hence we have for some intermediate value,

$$\frac{du}{d\theta} = 2\theta + \theta\cos\theta - 3\sin\theta = 0.$$

This also vanishes for $\theta = 0$, and is positive for small values of θ; if, then, it be supposed to vanish for any other value, we must have, by the same reasoning,

$$\frac{d^2u}{d\theta^2} = 2(1 - \cos\theta) - \theta\sin\theta = 0,$$

for some intermediate value. But this would give $\frac{1}{2}\theta - \tan\frac{1}{2}\theta = 0$, which is manifestly impossible for any other value than $\theta = 0$.

Before concluding this example, it may be well to refer to a remark made with respect to the cases of this proposition generally, namely, that although a curve may be found, differing infinitely little in *position* from that which the Calculus of Variations gives as the solution of the problem, and rendering the given integral less than it is found to be in this latter curve, that solution may, nevertheless, be a real minimum. To exemplify this in the present case, let ACB (Fig. 9) be the cycloid described as above stated. The area which is required to be a minimum will then be $\frac{AB^2}{\pi}$. Now if Bb be taken indefinitely small, and cycloids be described upon Ab, bB, the mixed curve so formed will be indefinitely near to ACB in *position*, i. e., its several points will be indefinitely near to the corresponding points of the other. But the corresponding *area* for this latter curve is $\frac{Ab^2 + Bb^2}{\pi}$, which is evidently less than $\frac{AB^2}{\pi}$. This latter is nevertheless a real minimum in the proper sense of that term, inasmuch as the mixed curve, AbB, has a *cusp* at b, and therefore cannot be deduced from ACB by a legitimate variation, requiring at the point b a change of curvature *not* indefinitely small. In fact the variations, δy and $\delta \rho$, are, at this point, of different orders of magnitude.

Prop. V.

82. To find the nature of the curve which will render $\int \mu ds$ a maximum or minimum, ds being the element of a curve traced on a given surface, and μ a given function of the co-ordinates of any of its points.

Adopting, as before, the method of Lagrange, we shall consider x, y, z as functions of s, connected by the equations

$$\left.\begin{aligned} \frac{dx^2}{ds^2} + \frac{dy^2}{ds^2} + \frac{dz^2}{ds^2} &= 1, \\ u &= 0, \end{aligned}\right\} \quad (A)$$

(the equation of the given surface). Proceeding as before, let λ

be the factor for the first of these equations, and λ' that for the second. Then it is plain that the complete variation will be

$$\int\left\{\left(\frac{d\mu}{dx}+\lambda'\frac{du}{dx}\right)\delta x+\left(\frac{d\mu}{dy}+\lambda'\frac{du}{dy}\right)\delta y+\left(\frac{d\mu}{dz}+\lambda'\frac{du}{dz}\right)\delta z\right.$$

$$\left.+\lambda\left(\frac{dx}{ds}\cdot\frac{d\delta x}{ds}+\frac{dy}{ds}\cdot\frac{d\delta y}{ds}+\frac{dz}{ds}\cdot\frac{d\delta z}{ds}\right)\right\}ds.$$

Integrating the last three terms by parts, and equating to zero the coefficients of δx, δy, δz, we have the three equations,

$$\frac{d\mu}{dx}+\lambda'\frac{du}{dx}-\frac{d}{ds}.\lambda\frac{dx}{ds}=0,$$

$$\frac{d\mu}{dy}+\lambda'\frac{du}{dy}-\frac{d}{ds}.\lambda\frac{dy}{ds}=0, \qquad \text{(B)}$$

$$\frac{d\mu}{dz}+\lambda'\frac{du}{dz}-\frac{d}{ds}.\lambda\frac{dz}{ds}=0.$$

Multiplying the first of these equations by $\frac{dx}{ds}$, the second by $\frac{dy}{ds}$, and the third by $\frac{dz}{ds}$, adding them, and observing the conditions,

$$\frac{d\mu}{dx}\cdot\frac{dx}{ds}+\frac{d\mu}{dy}\cdot\frac{dy}{ds}+\frac{d\mu}{dz}\cdot\frac{dz}{ds}=\frac{d\mu}{ds},$$

$$\frac{dx}{ds}\cdot\frac{d^2x}{ds^2}+\frac{dy}{ds}\cdot\frac{d^2y}{ds^2}+\frac{dz}{ds}\cdot\frac{d^2z}{ds^2}=0,$$

we find

$$\frac{d\mu}{ds}-\frac{d\lambda}{ds}=0, \quad \lambda=\mu+a.$$

If the length of the curve be not given, it will appear, from considerations similar to those of Prop. I., that we must take $a=0$, and therefore $\lambda=\mu$. The three equations become, therefore,

$$\frac{d\mu}{dx}+\lambda'\frac{du}{dx}-\mu\frac{d^2x}{ds^2}-\frac{dx}{ds}\cdot\frac{d\mu}{ds}=0.$$

$$\frac{d\mu}{dy}+\lambda'\frac{du}{dy}-\mu\frac{d^2y}{ds^2}-\frac{dy}{ds}\cdot\frac{d\mu}{ds}=0, \qquad \text{(C)}$$

$$\frac{d\mu}{dz}+\lambda'\frac{du}{dz}-\mu\frac{d^2z}{ds^2}-\frac{dz}{ds}\cdot\frac{d\mu}{ds}=0.$$

Let α, β, γ be the angles made with the co-ordinate planes by the plane of the normal section which touches the curve at any point. Then since this plane is perpendicular to the surface,

$$\cos\alpha\,\frac{du}{dx} + \cos\beta\,\frac{du}{dy} + \cos\gamma\,\frac{du}{dz} = 0;$$

and since it passes through the tangent to the curve,

$$\cos\alpha\,\frac{dx}{ds} + \cos\beta\,\frac{dy}{ds} + \cos\gamma\,\frac{dz}{ds} = 0.$$

Hence if we multiply the equations (C) by $\cos\alpha$, $\cos\beta$, $\cos\gamma$, respectively, and add them, we shall evidently have

$$\cos\alpha\,\frac{d\mu}{dx} + \cos\beta\,\frac{d\mu}{dy} + \cos\gamma\,\frac{d\mu}{dz}$$

$$-\mu\left(\cos\alpha\,\frac{d^2x}{ds^2} + \cos\beta\,\frac{d^2y}{ds^2} + \cos\gamma\,\frac{d^2z}{ds^2}\right) = 0. \qquad \text{(D)}$$

But if α', β', γ' be the angles which ρ, the radius of absolute curvature of the curve, makes with the axes of co-ordinates, it is known that

$$\cos\alpha' = \rho\,\frac{d^2x}{ds^2},$$

$$\cos\beta' = \rho\,\frac{d^2y}{ds^2},$$

$$\cos\gamma' = \rho\,\frac{d^2z}{ds^2}.$$

Equation (D) may therefore be written,

$$\cos\alpha\,\frac{d\mu}{dx} + \cos\beta\,\frac{d\mu}{dy} + \cos\gamma\,\frac{d\mu}{dz}$$

$$= \frac{\mu}{\rho}\left(\cos\alpha\cos\alpha' + \cos\beta\cos\beta' + \cos\gamma\cos\gamma'\right). \qquad \text{(E)}$$

Let ω be the angle between the osculating plane to the curve and the plane of the normal section. We have then

$$\cos\alpha\cos\alpha' + \cos\beta\cos\beta' + \cos\gamma\cos\gamma' = \sin\omega.$$

We have also, by Meunier's theorem,

$$\sin^2\omega = 1 - \frac{\rho'^2}{\rho^2},$$

ρ' being the radius of curvature of the normal section. Hence equation (E) may be written in any one of the three forms,

$$\frac{1}{\rho^2} - \frac{1}{\rho'^2} = \frac{1}{\mu^2}\left(\cos\alpha\frac{d\mu}{dx} + \cos\beta\frac{d\mu}{dy} + \cos\gamma\frac{d\mu}{dz}\right)^2,$$

$$\frac{\sin\omega}{\rho} = \frac{1}{\mu}\left(\cos\alpha\frac{d\mu}{dx} + \cos\beta\frac{d\mu}{dy} + \cos\gamma\frac{d\mu}{dz}\right), \qquad \text{(F)}$$

$$\frac{\tan\omega}{\rho'} = \frac{1}{\mu}\left(\cos\alpha\frac{d\mu}{dx} + \cos\beta\frac{d\mu}{dy} + \cos\gamma\frac{d\mu}{dz}\right).$$

If the given integral had been $\int\frac{ds}{\mu}$, it is evident (as in the case of a plane curve) that the first of these equations would remain wholly unaltered. Hence we infer, as before,

If the curve (traced on a given surface), whose equation is

$$\rho = f\left(x, y, z, \frac{dy}{dx}\right),$$

possess the property of rendering $\int\mu ds$ a maximum or minimum, the curve whose equation is

$$\rho = -f\left(x, y, z, \frac{dy}{dx}\right)$$

will have the property of rendering $\int\frac{ds}{\mu}$ a maximum or minimum. A remarkable example of this theorem will occur in the applications to Mechanics. With respect to the terms without the sign of integration, it is easy to see that they furnish the equations

$$\mu_1\left\{\left(\frac{dx}{ds}\right)_1\delta x_1 + \left(\frac{dy}{ds}\right)_1\delta y_1 + \left(\frac{dz}{ds}\right)_1\delta z_1\right\} = 0,$$

$$\mu_0\left\{\left(\frac{dx}{ds}\right)_0\delta x_0 + \left(\frac{dy}{ds}\right)_0\delta y_0 + \left(\frac{dz}{ds}\right)_0\delta z_0\right\} = 0. \qquad \text{(G)}$$

(1.) If the extreme points of the curve be fixed, these equations are satisfied of themselves, inasmuch as

$$\delta x_1 = 0,\ \delta y_1 = 0,\ \delta z_1 = 0,\ \delta x_0 = 0,\ \delta y_0 = 0,\ \delta z_0 = 0.$$

(2.) If the limiting points be not given, but merely restricted to the given curves (drawn upon the given surface), whose equations are

$$dx_1 = m_1 dz_1, \qquad dx_0 = m_0 dz_0,$$
$$dy_1 = n_1 dz_1, \qquad dy_0 = n_0 dz_0,$$

it is evident that we must have

$$\delta x_1 = m_1 \delta z_1, \qquad \delta x_0 = m_0 \delta z_0,$$
$$\delta y_1 = n_1 \delta z_1, \qquad \delta y_0 = n_0 \delta z_0,$$

and that, therefore, the two equations (G) become

$$\mu_1 \left\{ m_1 \left(\frac{dx}{ds}\right)_1 + n_1 \left(\frac{dy}{ds}\right)_1 + \left(\frac{dz}{ds}\right)_1 \right\} = 0,$$

$$\mu_0 \left\{ m_0 \left(\frac{dx}{ds}\right)_0 + n_0 \left(\frac{dy}{ds}\right)_0 + \left(\frac{dz}{ds}\right)_0 \right\} = 0.$$

Hence, if we neglect the suppositions, $\mu_1 = 0$, $\mu_0 = 0$, we infer, *If a curve be traced upon a given surface such as to render $\int \mu ds$ a maximum or minimum, it will cut its bounding curves at right angles.*

83. The equations of the curve, which render $\int \mu ds$ a maximum or minimum, may also be expressed as follows:

Let the equation of the given surface be

$$dz = p dx + q dy.$$

Then, since the cosines of the angles made by the osculating plane with the co-ordinate planes are

$$\rho \left(\frac{dz}{ds} \cdot \frac{d^2 y}{ds^2} - \frac{dy}{ds} \cdot \frac{d^2 z}{ds^2} \right),$$

$$\rho \left(\frac{dx}{ds} \cdot \frac{d^2 z}{ds^2} - \frac{dz}{ds} \cdot \frac{d^2 x}{ds^2} \right),$$

$$\rho \left(\frac{dy}{ds} \cdot \frac{d^2 x}{ds^2} - \frac{dx}{ds} \cdot \frac{d^2 y}{ds^2} \right),$$

it is evident that

$$\sin\omega =$$

$$\frac{\rho\left\{\frac{dy}{ds}\cdot\frac{d^2x}{ds^2}-\frac{dx}{ds}\cdot\frac{d^2y}{ds^2}-p\left(\frac{dz}{ds}\cdot\frac{d^2y}{ds^2}-\frac{dy}{ds}\cdot\frac{d^2z}{ds^2}\right)-q\left(\frac{dx}{ds}\cdot\frac{d^2z}{ds^2}-\frac{dz}{ds}\cdot\frac{d^2x}{ds^2}\right)\right\}}{\sqrt{(1+p^2+q^2)}}.$$

Substituting this in the second of equations (F), and eliminating either $\frac{d^2x}{ds^2}$ or $\frac{d^2y}{ds^2}$ by means of the equation

$$\frac{dx}{ds}\cdot\frac{d^2x}{ds^2}+\frac{dy}{ds}\cdot\frac{d^2y}{ds^2}+\frac{dz}{ds}\cdot\frac{d^2z}{ds^2}=0,$$

the required curve may be represented by either of the equations

$$\left.\begin{aligned}\frac{d^2x}{ds^2}+p\frac{d^2z}{ds^2}&=\frac{\sqrt{(1+p^2+q^2)}}{\mu}\left(\cos\alpha\frac{d\mu}{dx}+\cos\beta\frac{d\mu}{dy}+\cos\gamma\frac{d\mu}{dz}\right)\frac{dy}{ds},\\ \frac{d^2y}{ds^2}+q\frac{d^2z}{ds^2}&=-\frac{\sqrt{(1+p^2+q^2)}}{\mu}\left(\cos\alpha\frac{d\mu}{dx}+\cos\beta\frac{d\mu}{dy}+\cos\gamma\frac{d\mu}{dz}\right)\frac{dx}{ds},\end{aligned}\right\}\quad(H)$$

combined with the equation of the given surface.

84. If from the same point, O (Fig. 10), on a surface, there be drawn to two indefinitely near points, T, T', of a given curve, two curves, OT, OT', each possessing the property of rendering $\int\mu ds$ a maximum or minimum; and if we denote by $d\sigma$ the arc TT', and by θ the angle $OT'T$, the difference between the values of the integral for these curves is ultimately

$$\mu_1\cos\theta d\sigma,$$

μ_1 being the value of μ for either of the indefinitely near points, T, T'.

For since each of these curves renders $\int\mu ds$ a maximum, and thus causes the variation under the sign of integration to vanish, and since, moreover, one of the limiting points is fixed, it is evident that the complete variation which the integral receives in passing from OT to OT' is

$$\lambda_1\left\{\left(\frac{dx}{ds}\right)_1\delta x_1+\left(\frac{dy}{ds}\right)_1\delta y_1+\left(\frac{dz}{ds}\right)_1\delta z_1\right\}.$$

But we have before seen that $\lambda = \mu$, and since the values of x_1, y_1, z_1 are varied by passing from one point of the curve AB to the consecutive point, it is plain that

$$\delta x_1 = \left(\frac{dx}{d\sigma}\right)_1 d\sigma, \quad \delta y_1 = \left(\frac{dy}{d\sigma}\right)_1 d\sigma, \quad \delta z_1 = \left(\frac{dz}{d\sigma}\right)_1 d\sigma.$$

Making these substitutions, the complete variation becomes

$$\mu_1 \left\{\left(\frac{dx}{ds}\right)_1 \left(\frac{dx}{d\sigma}\right)_1 + \left(\frac{dy}{ds}\right)_1 \left(\frac{dy}{d\sigma}\right)_1 + \left(\frac{dz}{ds}\right)_1 \left(\frac{dz}{d\sigma}\right)_1\right\} d\sigma;$$

and since

$$\cos\theta = \left(\frac{dx}{ds}\right)_1 \left(\frac{dx}{d\sigma}\right)_1 + \left(\frac{dy}{ds}\right)_1 \left(\frac{dy}{d\sigma}\right)_1 + \left(\frac{dz}{ds}\right)_1 \left(\frac{dz}{d\sigma}\right)_1,$$

the truth of the proposition is evident.

85. If μ be a homogeneous function of x, y, z, we may deduce a theorem analogous to that established (p. 152) for plane curves.

Let the intersection of the normal plane to the curve with the tangent plane to the surface be called the normal to the curve. This line is evidently perpendicular to the plane of the normal section before alluded to, and therefore makes with the axes the angles α, β, γ. Let a surface be described whose equation is

$$\mu = c,$$

and let a plane be drawn through the origin *conjugate** (with regard to this surface) to the line drawn from the origin to the point on the curve. Produce the normal to the curve till it meet this plane, and let the line so produced be called n. We have then evidently

$$\cos\alpha = \frac{x - x'}{n}, \quad \cos\beta = \frac{y - y'}{n}, \quad \cos\gamma = \frac{z - z'}{n},$$

$x'y'z'$ being the point in which n cuts the conjugate plane.

* That is, parallel to the tangent plane to the surface

$$\mu = c,$$

at the point in which the line to which it is conjugate cuts that surface. It is easily seen that, as μ is a homogeneous function, this plane is unique.

Equation (F) may therefore be written

$$\frac{n \sin \omega}{\rho} = \frac{1}{\mu}\left\{\left(x\frac{d\mu}{dx} + y\frac{d\mu}{dy} + z\frac{d\mu}{dz}\right) - \left(x'\frac{d\mu}{dx} + y'\frac{d\mu}{dy} + z'\frac{d\mu}{dz}\right)\right\}.$$

But since $x'y'z'$ is a point in the conjugate plane, we have, as in p. 152,

$$x'\frac{d\mu}{dx} + y'\frac{d\mu}{dy} + z'\frac{d\mu}{dz} = 0;$$

and since μ is a homogeneous function,

$$x\frac{d\mu}{dx} + y\frac{d\mu}{dy} + z\frac{d\mu}{dz} = m\mu,$$

m denoting the degree of the function. Hence

$$n \sin \omega = m\rho.$$

Hence, preserving the foregoing definitions, we have the following theorem:

If μ be a homogeneous function of the co-ordinates of a point upon a given surface, the curve which renders $\int \mu ds$ a maximum or minimum is such, that if the normal at any point be projected upon the osculating plane, its projection is equal to m times the radius of curvature.

A similar substitution will reduce the first of the foregoing equations (F) to

$$\frac{1}{\rho^2} - \frac{1}{\rho'^2} = \frac{m^2}{n^2}.$$

Either of these theorems will furnish means for obtaining the osculating plane and the radius of curvature at any point, if the direction of the tangent be known.

Example.

86. To find the shortest line which can be drawn between two points upon a given surface.

This is the simplest case of the preceding proposition, from which it may be deduced by making $\mu = 1$. This gives

$$\frac{d\mu}{dx} = 0, \quad \frac{d\mu}{dy} = 0, \quad \frac{d\mu}{dz} = 0,$$

and therefore we have, from equations (F),

$$\rho = \rho', \quad \omega = 0.$$

Hence the radius of curvature is the same as that of the normal section which touches it; or, in other words, the osculating plane is at every point perpendicular to the surface.

As a complete discussion of the properties of shortest or, as they are generally termed, *geodetic* lines, would be too extensive to find a place in a treatise like the present, we shall content ourselves with giving one or two of the most important.

(1.) The equations of a geodetic line, the arc being the independent variable, are readily deduced from the general equations (H), by making

$$\mu = 1, \quad \frac{d\mu}{dx} = 0, \quad \frac{d\mu}{dy} = 0, \quad \frac{d\mu}{dz} = 0.$$

We have, in this way,

$$\frac{du}{dz} \cdot \frac{d^2x}{ds^2} - \frac{du}{dx} \cdot \frac{d^2z}{ds^2} = 0,$$

$$\frac{du}{dz} \cdot \frac{d^2y}{ds^2} - \frac{du}{dy} \cdot \frac{d^2z}{ds^2} = 0;$$

or, if the equation of the given surface be

$$dz = pdx + qdy,$$

$$\frac{d^2x}{ds^2} + p\frac{d^2z}{ds^2} = 0, \qquad \text{(a)}$$

$$\frac{d^2y}{ds^2} + q\frac{d^2z}{ds^2} = 0.$$

(2.) The equations of a geodetic line referred to x as the independent variable, may readily be deduced from those just given; but it is, perhaps, simpler to derive them at once from the consideration that the osculating plane is at all points perpendicular to the surface.

If the equation of the surface be

$$dz = pdx + qdy,$$

this consideration gives

$$p\,(dyd^2z - dzd^2y) + q\,(dzd^2x - dxd^2z) = dxd^2y - dyd^2x.$$

Putting $d^2x = 0$, and substituting for dz and d^2z their values given by the equations

$$dz = pdx + qdy,$$
$$d^2z = rdx^2 + 2sdxdy + tdy^2 + qd^2y,$$

we find the equation

$$(1 + p^2 + q^2)\frac{d^2y}{dx^2} + \left(q - p\frac{dy}{dx}\right)\left(r + 2s\frac{dy}{dx} + t\frac{dy^2}{dx^2}\right) = 0, \quad \text{(b)}$$

which, with the equation

$$dz - pdx - qdy = 0,$$

represents the geodetic line.

(3.) If from the same point, O, on a surface, geodetic lines be drawn to two indefinitely near points, T, T', of a curve, and if we assume $d\sigma$ = the indefinitely small arc TT', and $\theta = TT'O$, we shall have for the ultimate value of the increment $OT' - OT$,

$$OT' - OT = d\sigma \,.\, \cos\theta.$$

This is evidently a particular case of the general proposition stated in p. 182. And it appears, from the conclusion there arrived at, that this property, which is well known to belong to right lines, is a property of geodetic lines generally, and peculiar to them. In the application, therefore, of the method of infinitesimals to curves drawn upon a given surface, it is easy to see that geodetic lines may, in cases in which their *length* only is concerned, be, in general, treated as right lines.

(4.) It will immediately appear, from the general principles laid down in pp. 90, 91, that a geodetic line is not necessarily a line of minimum length between any two of its points. For we have there seen that the curve, whose differential equation is

$$\beta = 0,$$

will not, in general, possess the property of rendering

$$\int_{x_0}^{x_1} f\left(x, y, \frac{dy}{dx}\right) dx$$

a maximum or minimum, if the limits of integration be such as to render it possible to draw a second curve indefinitely near to the first, satisfying the same differential equation, and passing through the extreme points A, B, or through any two points lying between A and B. Hence we have the following rule for determining at what point a *geodetic* line ceases to be a *shortest* line.

Let A (Fig. 2) be one extremity of the geodetic line. Draw through A a second geodetic line, making with the first an indefinitely small angle. This curve will, in general, intersect the first at some other point, C. If, then, we commence to measure the length of the geodetic line from A, it will, in general, cease to be a shortest line when we pass C.

(5.) To find the equation of a geodetic line upon the surface of an ellipsoid.

Let the equation of the ellipsoid be

$$\frac{x^2}{a^2} + \frac{y^2}{b^2} + \frac{z^2}{c^2} = 1.$$

The general equations (a), p. 185, become

$$\left.\begin{aligned} \frac{d^2x}{ds^2} &= \frac{c^2x}{a^2z} \cdot \frac{d^2z}{ds^2}, \\ \frac{d^2y}{ds^2} &= \frac{c^2y}{b^2z} \cdot \frac{d^2z}{ds^2}. \end{aligned}\right\} \qquad (c)$$

Assume

$$u = \frac{x^2}{a^4} + \frac{y^2}{b^4} + \frac{z^2}{c^4},$$

$$v = \frac{1}{a^2} \cdot \frac{dx^2}{ds^2} + \frac{1}{b^2} \cdot \frac{dy^2}{ds^2} + \frac{1}{c^2} \cdot \frac{dz^2}{ds^2}.$$

Then if we differentiate this latter equation, and substitute for $\frac{d^2x}{ds^2}, \frac{d^2y}{ds^2}$ their values, derived from (c), we find

$$\frac{dv}{ds} = \frac{2c^2}{z}\left(\frac{x}{a^4} \cdot \frac{dx}{ds} + \frac{y}{b^4} \cdot \frac{dy}{ds} + \frac{z}{c^4} \cdot \frac{dz}{ds}\right)\frac{d^2z}{ds^2} = \frac{c^2}{z} \cdot \frac{du}{ds} \cdot \frac{d^2z}{ds^2}.$$

Again, differentiating the equation of the surface twice, and substituting for $\frac{d^2x}{ds^2}, \frac{d^2y}{ds^2}$, as before,

$$v = -\frac{c^2}{z}u\frac{d^2z}{ds^2}.$$

Dividing these equations one by the other, we find

$$\frac{1}{v}\frac{dv}{ds} = -\frac{1}{u}\frac{du}{ds};$$

or, by integration,

$$uv = \text{const.}$$

Replacing u and v by their values, we find, for the equation of a geodetic line,

$$\left(\frac{x^2}{a^4}+\frac{y^2}{b^4}+\frac{z^2}{c^4}\right)\left(\frac{1}{a^2}\cdot\frac{dx^2}{ds^2}+\frac{1}{b^2}\cdot\frac{dy^2}{ds^2}+\frac{1}{c^2}\cdot\frac{dz^2}{ds^2}\right) = \text{const.} \qquad \text{(d)}$$

Let p be the perpendicular from the centre upon the tangent plane, and d the semidiameter drawn parallel to the tangent to the geodetic line. Then since

$$\frac{1}{p^2} = \frac{x^2}{a^4}+\frac{y^2}{b^4}+\frac{z^2}{c^4},$$

$$\frac{1}{d^2} = \frac{1}{a^2}\cdot\frac{dx^2}{ds^2}+\frac{1}{b^2}\cdot\frac{dy^2}{ds^2}+\frac{1}{c^2}\cdot\frac{dz^2}{ds^2},$$

the equation of the geodetic line is equivalent to

$$pd = \text{const.} = m^2.$$

If ρ be the radius of curvature of the geodetic line at any point, it is known that

$$\rho = \frac{d^2}{p}.\text{*}$$

* This may readily be proved as follows:

We have, in general,

$$\frac{1}{\rho^2} = \left(\frac{d^2x}{ds^2}\right)^2+\left(\frac{d^2y}{ds^2}\right)^2+\left(\frac{d^2z}{ds^2}\right)^2;$$

or, substituting for $\frac{d^2x}{ds^2}$ and $\frac{d^2y}{ds^2}$ their values derived from the equations of the geodetic line,

$$\frac{1}{\rho^2} = \left(\frac{x^2}{a^4}+\frac{y^2}{b^4}+\frac{z^2}{c^4}\right)\left(\frac{d^2z}{ds^2}\right)^2\cdot\frac{c^4}{z^2}:$$

Substituting for d its value $\frac{m^2}{p}$, we have

$$\rho = \frac{m^4}{p^3}.$$

Hence *along the same geodetic line the radius of curvature varies inversely as the cube of the perpendicular on the tangent plane.*

These two theorems are due to M. Joachimstal.

Prop. VI.

87. To draw, between two points or curves situated upon a given surface, a curve such that the definite integral

$$\int(\mu ds + \mu' dx)$$

may be a maximum or minimum, μ, μ' being functions of x, y, z.

Taking, as before, the arc as the independent variable, and putting, therefore, the given integral under the form

$$\int\left(\mu + \mu'\frac{dx}{ds}\right)ds,$$

it is easy to see that the method of Lagrange will give the equations

$$\lambda\frac{d^2x}{ds^2} + \frac{dx}{ds}\cdot\frac{d\lambda}{ds} - \lambda'\frac{du}{dx} - \frac{d\mu}{dx} - \frac{dx}{ds}\cdot\frac{d\mu'}{dx} + \frac{d\mu'}{ds} = 0,$$

$$\lambda\frac{d^2y}{ds^2} + \frac{dy}{ds}\cdot\frac{d\lambda}{ds} - \lambda'\frac{du}{dy} - \frac{d\mu}{dy} - \frac{dx}{ds}\cdot\frac{d\mu'}{dy} = 0, \qquad \text{(A)}$$

$$\lambda\frac{d^2z}{ds^2} + \frac{dz}{ds}\cdot\frac{d\lambda}{ds} - \lambda'\frac{du}{dz} - \frac{d\mu}{dz} - \frac{dx}{ds}\cdot\frac{d\mu'}{dz} = 0.$$

---

but we have seen above that

$$v = -\frac{c^2}{z}u\frac{d^2z}{ds^2};$$

hence

$$\frac{c^4}{z^2}\left(\frac{d^2z}{ds^2}\right)^2 = \frac{v^2}{u^2} = \frac{p^4}{d^4}.$$

Substituting this in the value of ρ, we have

$$\frac{1}{\rho^2} = \frac{p^2}{d^4}, \text{ or } \rho = \frac{d^2}{p}.$$

Multiplying these equations respectively by $\frac{dx}{ds}, \frac{dy}{ds}, \frac{dz}{ds}$, adding them, and recollecting that

$$\frac{d\mu'}{ds} = \frac{d\mu'}{dx} \cdot \frac{dx}{ds} + \frac{d\mu'}{dy} \cdot \frac{dy}{ds} + \frac{d\mu'}{dz} \cdot \frac{dz}{ds},$$

it readily appears that we have, as in Prop. V.,

$$\frac{d\lambda}{ds} - \frac{d\mu}{ds} = 0.$$

Hence, as before, if the length of the curve be given, $\lambda = \mu + a$, and if it be not given, $\lambda = \mu$. Taking the former of these two cases, multiply, as in Prop. V., the three equations (A) by $\cos a$, $\cos \beta$, $\cos \gamma$, respectively, and add them. We have then, putting $\mu + a$ for λ, as in that proposition,

$$(\mu + a) \left(\cos a \frac{d^2x}{ds^2} + \cos \beta \frac{d^2y}{ds^2} + \cos \gamma \frac{d^2z}{ds^2} \right)$$

$$= \cos a \frac{d\mu}{dx} + \cos \beta \frac{d\mu}{dy} + \cos \gamma \frac{d\mu}{dz}$$

$$- \cos a \frac{d\mu'}{ds} + \frac{dx}{ds} \left(\cos a \frac{d\mu'}{dx} + \cos \beta \frac{d\mu'}{dy} + \cos \gamma \frac{d\mu'}{dz} \right).$$

Substituting for

$$\cos a \frac{d^2x}{ds^2} + \cos \beta \frac{d^2y}{ds^2} + \cos \gamma \frac{d^2z}{ds^2}$$

its value, $\frac{\sin \omega}{\rho}$, and for $\frac{d\mu'}{ds}$,

$$\frac{d\mu'}{dx} \cdot \frac{dx}{ds} + \frac{d\mu'}{dy} \cdot \frac{dy}{ds} + \frac{d\mu'}{dz} \cdot \frac{dz}{ds},$$

this equation becomes

$$(\mu + a) \frac{\sin \omega}{\rho} = \cos a \frac{d\mu}{dx} + \cos \beta \frac{d\mu}{dy} + \cos \gamma \frac{d\mu}{dz}$$

$$+ \frac{d\mu'}{dy} \left(\cos \beta \frac{dx}{ds} - \cos a \frac{dy}{ds} \right) + \frac{d\mu'}{dz} \left(\cos \gamma \frac{dx}{ds} - \cos a \frac{dz}{ds} \right).$$

But since α, β, γ are the angles which the perpendicular to the plane of the normal section makes with the axes, and $\frac{dx}{ds}, \frac{dy}{ds}, \frac{dz}{ds}$ the cosines of the angles which the tangent to the curve makes with the same axes, it is known that, if we denote by α', β', γ', the angles made by the perpendicular to the plane of these lines (which is evidently the normal to the surface), we shall have

$$\cos\alpha' = \cos\beta \frac{dz}{ds} - \cos\gamma \frac{dy}{ds},$$

$$\cos\beta' = \cos\gamma \frac{dx}{ds} - \cos\alpha \frac{dz}{ds},$$

$$\cos\gamma' = \cos\alpha \frac{dy}{ds} - \cos\beta \frac{dx}{ds}.$$

The equation, at which we have arrived, may therefore, as before, be written in any one of the three forms,

$$\frac{1}{\rho^2} - \frac{1}{\rho'^2} =$$

$$\frac{1}{(\mu+a)^2}\left(\cos\alpha \frac{d\mu}{dx} + \cos\beta \frac{d\mu}{dy} + \cos\gamma \frac{d\mu}{dz} + \cos\gamma' \frac{d\mu'}{dy} - \cos\beta' \frac{d\mu'}{dz}\right)^2,$$

$$\frac{\sin\omega}{\rho} = \frac{1}{\mu+a}\left(\cos\alpha \frac{d\mu}{dx} + \cos\beta \frac{d\mu}{dy} + \cos\gamma \frac{d\mu}{dz} + \cos\gamma' \frac{d\mu'}{dy} - \cos\beta' \frac{d\mu'}{dz}\right), \quad \text{(B)}$$

$$\frac{\tan\omega}{\rho'} = \frac{1}{\mu+a}\left(\cos\alpha \frac{d\mu}{dx} + \cos\beta \frac{d\mu}{dy} + \cos\gamma \frac{d\mu}{dz} + \cos\gamma' \frac{d\mu'}{dy} - \cos\beta' \frac{d\mu'}{dz}\right).$$

The solution of the case in which the length of the curve is not given is immediately deduced from this by simply putting $a = 0$. If the limiting points of the curve be given, it is evident that the terms free from the sign of integration disappear as before. If the limiting points be not given, but merely restricted to the curves whose equations are

$$dy_0 = m_0 dx_0, \quad dy_1 = m_1 dx_1,$$
$$dz_0 = n_0 dx_0, \quad dz_1 = n_1 dx_1,$$

we shall have, by reasoning similar to that employed in Prop. II.,

$$(\mu_0 + a)\left(\frac{dx}{ds} + m_0\frac{dy}{ds} + n_0\frac{dz}{ds}\right)_0 + \mu'_0 = 0,$$

$$(\mu_1 + a)\left(\frac{dx}{ds} + m_1\frac{dy}{ds} + n_1\frac{dz}{ds}\right)_1 + \mu'_1 = 0;$$

or if $\theta_0\,\theta_1$ be the angles at which it cuts its bounding curves, and $\phi_0\,\phi_1$ the angles which the tangents to these curves at the points of intersection make with the axis of x,

$$\left.\begin{aligned}(\mu_0 + a)\cos\theta_0 + \mu'_0\cos\phi_0 = 0,\\ (\mu_1 + a)\cos\theta_1 + \mu'_1\cos\phi_1 = 0.\end{aligned}\right\}\quad \text{(C)}$$

All the conclusions arrived at in this proposition are adapted to the case in which the length of the curve is not given, by simply omitting a.

Example.

88. Given two points on a surface, and any curve connecting them, to draw between these points a curve of given length, which shall include, with the given curve, the greatest possible area.*

Since the superficial area of any surface is represented by

$$\iint\sqrt{(1 + p^2 + q^2)}\,dxdy,$$

it is plain that, if the equation of the surface be given, it may always be put under the form $\int\mu' dx$, in which

$$\mu' \left(= \int\sqrt{(1 + p^2 + q^2)}\,dy\right)$$

is a given function of x and y. The solution of this problem is, therefore, derived from the general equations (B), by making

$$\mu = 0,\quad \mu' = \int\sqrt{(1 + p^2 + q^2)}\,dy,$$

$$\frac{d\mu'}{dy} = \sqrt{(1 + p^2 + q^2)} = \sec\gamma',\quad \frac{d\mu'}{dz} = 0.$$

* The investigation of this curve is (as far as I am aware) due to M. Delaunay.—Liouville, Journal de Math., tom. viii. p. 241.

These equations become, therefore,

$$\frac{1}{\rho^2} - \frac{1}{\rho'^2} = \frac{1}{a^2},$$

$$\rho = a \sin \omega, \qquad \text{(a)}$$

$$\rho' = a \tan \omega.$$

Hence if the direction of the tangent at any point be known, the osculating plane and radius of curvature may be found by the following construction:

Let P (Fig. 12) be the point on the curve, and let the plane of the paper be the normal plane to the curve. Let C' be the centre of curvature of the normal section, and draw $C'A$ perpendicular to PC' and $= a$. Join PA, and draw $C'C$, PC, respectively parallel, and perpendicular to PA. Then C is the centre of curvature of the required curve, and a plane through PC perpendicular to the plane of the paper is the osculating plane. The reason of this construction is evident from the equations (a).*

If it be required to draw the curve between two given curves in such a way that the area included between the curve so drawn, and a curve of given species passing through its extremities, may be a maximum, it will appear, by reasoning precisely similar to that of p. 157, that the equation of the curve is still the same as before. With regard to the terms outside the sign of integration, the mode of treating them is also perfectly analogous to that which is there employed. For it is evident that the area included between the two curves may be expressed by

$$\int_{x_0}^{x_1} \mu' dx - \int_{x_0}^{x_1} F(x, x_0, x_1)\, dx,$$

in which μ' has the same signification as in the first part of this

* If the given surface be an ellipsoid, the direction of the osculating plane may be found from that of the tangent by the following construction:

Let C (Fig. 13) be the centre of the ellipsoid, CP the perpendicular on the tangent plane, and CD the semi-diameter parallel to the tangent to the curve. Draw PO perpendicular to the plane PCD and equal to $\frac{CD^2}{a}$. The plane DCO is parallel to the osculating plane required. For

$$\tan PCO = \frac{OP}{CP} = \frac{d^2}{ap} = \frac{\rho'}{a} = \tan \omega.$$

example, and $F(x, x_0, x_1)$ is a given function, namely, the value of the integral $\int \surd(1+p^2+q^2)\,dy$ for the given curve which passes through the two extreme points. Hence, by proceeding exactly as in the case of the plane curve, we shall arrive at the equations,

$$a \cos\omega_0 + \cos\theta_0 \int_{x_0}^{x_1} \frac{dF}{dx_0}\,dx = 0,$$

$$a \cos\omega_1 - \cos\theta_1 \int_{x_0}^{x_1} \frac{dF}{dx_1}\,dx = 0.$$

Prop. VII.

89. To draw between two given points a curve whose curvature shall be constant, and such that the length of the arc between the two points may be a minimum.

Adopting Lagrange's method, we shall take the arc of the curve as the independent variable, which will give

$$V = 1,$$

the dependent variables, x, y, z, being connected by the equations

$$\frac{dx^2}{ds^2} + \frac{dy^2}{ds^2} + \frac{dz^2}{ds^2} = 1,$$

and

$$\left(\frac{d^2x}{ds^2}\right)^2 + \left(\frac{d^2y}{ds^2}\right)^2 + \left(\frac{d^2z}{ds^2}\right)^2 = \frac{1}{m^2}.$$

Since, then, none of the variables enter into V, it is evident that the equations furnished by the coefficients of δx, δy, δz are

$$\frac{d^2.\lambda'\dfrac{d^2x}{ds^2}}{ds^2} - \frac{d.\lambda\dfrac{dx}{ds}}{ds} = 0,$$

$$\frac{d^2.\lambda'\dfrac{d^2y}{ds^2}}{ds^2} - \frac{d.\lambda\dfrac{dy}{ds}}{ds} = 0, \qquad \text{(A)}$$

$$\frac{d^2.\lambda'\dfrac{d^2z}{ds}}{ds^2} - \frac{d.\lambda\dfrac{dz}{ds}}{ds} = 0.$$

These equations, being immediately integrable, give

$$\frac{d\,.\,\lambda'\dfrac{d^2x}{ds^2}}{ds} - \lambda\,\frac{dx}{ds} = a,$$

$$\frac{d\,.\,\lambda'\dfrac{d^2y}{ds^2}}{ds} - \lambda\,\frac{dy}{ds} = b,$$

$$\frac{d\,.\,\lambda'\dfrac{d^2z}{ds^2}}{ds} - \lambda\,\frac{dz}{ds} = c;$$

or

$$\left.\begin{aligned}\frac{d^2x}{ds^2}\cdot\frac{d\lambda'}{ds} + \lambda'\frac{d^3x}{ds^3} - \lambda\frac{dx}{ds} &= a,\\ \frac{d^2y}{ds^2}\cdot\frac{d\lambda'}{ds} + \lambda'\frac{d^3y}{ds^3} - \lambda\frac{dy}{ds} &= b,\\ \frac{d^2z}{ds^2}\cdot\frac{d\lambda'}{ds} + \lambda'\frac{d^3z}{ds^3} - \lambda\frac{dz}{ds} &= c.\end{aligned}\right\} \quad (B)$$

Eliminating λ between the first two of these equations, we have

$$\left(\frac{dy}{ds}\cdot\frac{d^2x}{ds^2} - \frac{dx}{ds}\cdot\frac{d^2y}{ds^2}\right)\frac{d\lambda'}{ds} + \left(\frac{dy}{ds}\cdot\frac{d^3x}{ds^3} - \frac{dx}{ds}\cdot\frac{d^3y}{ds^3}\right)\lambda' = a\frac{dy}{ds} - b\,\frac{dx}{ds}.$$

This equation, which is also integrable, gives

$$\lambda'\left(\frac{dy}{ds}\cdot\frac{d^2x}{ds^2} - \frac{dx}{ds}\cdot\frac{d^2y}{ds^2}\right) = ay - bx + f.$$

Similarly, by the elimination of λ between the first and third, and between the second and third, we have

$$\left.\begin{aligned}\lambda'\left(\frac{dx}{ds}\cdot\frac{d^2z}{ds^2} - \frac{dz}{ds}\cdot\frac{d^2x}{ds^2}\right) &= cx - az + f',\\ \lambda'\left(\frac{dz}{ds}\cdot\frac{d^2y}{ds^2} - \frac{dy}{ds}\cdot\frac{d^2z}{ds^2}\right) &= bz - cy + f''.\end{aligned}\right\} \quad (C)$$

Previous to proceeding further, we may determine the values of the constants introduced in these integrations. For this purpose we shall suppose the origin of co-ordinates to have been taken at one of the given points, which will render

$$x_0 = 0, \quad y_0 = 0, \quad z_0 = 0;$$

and, since both limits are fixed, we shall have

$$\delta x_0 = 0, \quad \delta y_0 = 0, \quad \delta z_0 = 0, \quad \delta x_1 = 0, \quad \delta y_1 = 0, \quad \delta z_1 = 0.$$

It is easy to see, then, that the equations furnished by the terms which are without the sign of integration are

$$\left.\begin{aligned} \lambda'_1 \left(\frac{d^2x}{ds^2} \cdot \frac{d\delta x}{ds} + \frac{d^2y}{ds^2} \cdot \frac{d\delta y}{ds} + \frac{d^2z}{ds^2} \cdot \frac{d\delta z}{ds}\right)_1 = 0, \\ \lambda'_0 \left(\frac{d^2x}{ds^2} \cdot \frac{d\delta x}{ds} + \frac{d^2y}{ds^2} \cdot \frac{d\delta y}{ds} + \frac{d^2z}{ds^2} \cdot \frac{d\delta z}{ds}\right)_0 = 0. \end{aligned}\right\} \quad (D)$$

These equations are evidently satisfied either by making

$$\lambda'_1 = 0, \quad \lambda'_0 = 0,$$

or by making

$$\left(\frac{d^2x}{ds^2} \cdot \frac{d\delta x}{ds} + \frac{d^2y}{ds^2} \cdot \frac{d\delta y}{ds} + \frac{d^2z}{ds^2} \cdot \frac{d\delta z}{ds}\right)_1 = 0,$$

$$\left(\frac{d^2x}{ds^2} \cdot \frac{d\delta x}{ds} + \frac{d^2y}{ds^2} \cdot \frac{d\delta y}{ds} + \frac{d^2z}{ds^2} \cdot \frac{d\delta z}{ds}\right)_0 = 0.$$

The conditions,

$$\lambda'_0 = 0, \quad \lambda'_1 = 0,$$

will evidently give, in equations (C),

$$f = 0, \quad f' = 0, \quad f'' = 0,$$

and

$$\frac{a}{x_1} = \frac{b}{y_1} = \frac{c}{z_1}.$$

With regard to the other factors,

$$\left(\frac{d^2x}{ds^2} \cdot \frac{d\delta x}{ds} + \frac{d^2y}{ds^2} \cdot \frac{d\delta y}{ds} + \frac{d^2z}{ds^2} \cdot \frac{d\delta z}{ds}\right)_0,$$

$$\left(\frac{d^2x}{ds^2} \cdot \frac{d\delta x}{ds} + \&c. \ldots\ldots\ldots\ldots\right)_1,$$

it is easy to see that it is impossible to make them vanish, preserving the independence of two of the variations,

$$\left(\frac{d\delta x}{ds}\right)_0, \text{ \&c.}$$

For, since these variations are only connected by the equations

$$\left(\frac{dx}{ds}\cdot\frac{d\delta x}{ds}+\frac{dy}{ds}\cdot\frac{d\delta y}{ds}+\frac{dz}{ds}\cdot\frac{d\delta z}{ds}\right)_0=0,$$

$$\left(\frac{dx}{ds}\cdot\frac{d\delta x}{ds}+\frac{dy}{ds}\cdot\frac{d\delta y}{ds}+\frac{dz}{ds}\cdot\frac{d\delta z}{ds}\right)_1=0,$$

it is evident that in order to make these factors vanish, without restricting their generality, we should have

$$\left(\frac{d^2x}{ds^2}\right)_0=\nu_0\left(\frac{dx}{ds}\right)_0, \qquad \left(\frac{d^2x}{ds^2}\right)_1=\nu_1\left(\frac{dx}{ds}\right)_1,$$

$$\left(\frac{d^2y}{ds^2}\right)_0=\nu_0\left(\frac{dy}{ds}\right)_0, \qquad \left(\frac{d^2y}{ds^2}\right)_1=\nu_1\left(\frac{dy}{ds}\right)_1,$$

$$\left(\frac{d^2z}{ds^2}\right)_0=\nu_0\left(\frac{dz}{ds}\right)_0, \qquad \left(\frac{d^2z}{ds^2}\right)_1=\nu_1\left(\frac{dz}{ds}\right)_1.$$

Squaring the first system, and adding, we have

$$\frac{1}{m^2}=\left(\frac{d^2x}{ds^2}\right)_0^2+\left(\frac{d^2y}{ds^2}\right)_0^2+\left(\frac{d^2z}{ds^2}\right)_0^2=\nu_0^2.$$

Again, multiplying these equations respectively by

$$\left(\frac{dx}{ds}\right)_0, \quad \left(\frac{dy}{ds}\right)_0, \quad \left(\frac{dz}{ds}\right)_0,$$

$$\left(\frac{dx}{ds}\right)_0\left(\frac{d^2x}{ds^2}\right)_0+\left(\frac{dy}{ds}\right)_0\left(\frac{d^2z}{ds^2}\right)_0+\left(\frac{dz}{ds}\right)_0\left(\frac{d^2z}{ds^2}\right)_0=\nu_0.$$

But it is plain that the first side of this equation is identically zero, hence we should have

$$\frac{1}{m}=\nu_0=0,$$

an impossible equation, as m is a given quantity. The same reasoning obviously applies to the second system. These equations, then, leading to impossible results, it is evident that we must have

$$\lambda'_1 = 0, \quad \lambda'_0 = 0,$$

and, therefore,

$$f = 0, \quad f' = 0, \quad f'' = 0.$$

Hence, if we neglect f, f', f'', and multiply equations (C) by

$$\frac{dz}{ds}, \ \frac{dy}{ds}, \ \frac{dx}{ds},$$

respectively, we shall have

$$(ay - bx)\,dz + (cx - az)\,dy + (bz - cy)\,dx = 0,$$

or

$$a\,(ydz - zdy) + b\,(zdx - xdz) + c\,(xdy - ydx) = 0.$$

This equation, which is easily integrated by making

$$y = ux, \quad z = vx,$$

gives

$$ay - bx = m\,(az - cx), \tag{E}$$

the equation of a plane passing through the origin. The required curve is therefore, in this case, a plane curve; and since its radius of curvature is constant, it is evidently a circle.

It appears, therefore, that the curve of constant curvature, whose length between two given points is a minimum, will be a circle if the position of either of the extreme tangents be undetermined. But, on the other hand, it is evident that the circle cannot be the general solution of the case in which the position of the extreme tangents is given. For this position might be such as to render it impossible to describe a circle which should fulfil the requisite conditions.*

If the position of the tangents at the extreme points be given, such that the equation

$$cf + bf' + af'' = 0, \tag{F}$$

holds, the curve is still plane. For if the equations (C) be multiplied by

* For example, the given tangents might not be in the same plane.

$$\frac{dz}{ds},\quad \frac{dy}{ds},\quad \frac{dx}{ds},$$

respectively, and added, we find

$$(ay - bx + f)\,dz + (cx - az + f')\,dy + (bz - cy + f'')\,dx = 0. \qquad \text{(G)}$$

Now if the origin be transferred to any point in the line whose equations are

$$ay - bx + f = 0, \quad cx - az + f' = 0,$$

it is easy to see that, in consequence of the relation

$$cf + bf' + af'' = 0,$$

the three absolute terms will disappear, and the equation be reduced to

$$(ay - bx)\,dz + (cx - az)\,dy + (bz - cy)\,dx = 0,$$

the same form as before. In this case, therefore, the curve is still a circle.

If the limiting points be not given, but merely restricted to the two surfaces,

$$u_0 = 0, \quad u_1 = 0,$$

it is plain that the terms which involve

$$\delta x_0,\ \delta y_0,\ \delta z_0, \quad \delta x_1,\ \delta y_1,\ \delta z_1,$$

will be

$$\left(\frac{d.\lambda'\frac{d^2x}{ds^2}}{ds} - \lambda\frac{dx}{ds}\right)_0 \delta x_0 + \left(\frac{d.\lambda'\frac{d^2y}{ds^2}}{ds} - \lambda\frac{dy}{ds}\right)_0 \delta y_0 + \left(\frac{d.\lambda'\frac{d^2z}{ds^2}}{ds} - \lambda\frac{dz}{ds}\right)_0 \delta z_0,$$

$$\left(\frac{d.\lambda'\frac{d^2x}{ds^2}}{ds} - \&c.\right)_1 \delta x_1 + \&c.$$

These terms are reduced by equations (B) to

$$a\delta x_0 + b\delta y_0 + c\delta z_0,$$
$$a\delta x_1 + b\delta y_1 + c\delta z_1;$$

and as the variations δx_0, &c., δx_1, &c., are connected by the equations

$$\frac{du_0}{dx_0}\delta x_0 + \frac{du_0}{dy_0}\delta y_0 + \frac{du_0}{dz_0}\delta z_0 = 0,$$

$$\frac{du_1}{dx_1}\delta x_1 + \frac{du_1}{dy_1}\delta y_1 + \frac{du_1}{dz_1}\delta z_1 = 0,$$

the terms involving these variations give the equations

$$\frac{1}{a}\cdot\frac{du_0}{dx_0} = \frac{1}{b}\cdot\frac{du_0}{dy_0} = \frac{1}{c}\cdot\frac{du_0}{dz_0},$$

$$\frac{1}{a}\cdot\frac{du_1}{dx_1} = \frac{1}{b}\cdot\frac{du_1}{dy_1} = \frac{1}{c}\cdot\frac{du_1}{dz_1}.$$

But since the equations

$$\lambda'_0 = 0, \quad \lambda'_1 = 0,$$

give, in general,

$$ay_0 - bx_0 + f = 0, \quad cx_0 - az_0 + f' = 0, \quad bz_0 - cy_0 + f'' = 0,$$
$$ay_1 - bx_1 + f = 0, \quad cx_1 - az_1 + f' = 0, \quad bz_1 - cy_1 + f'' = 0,$$

we find by subtracting

$$\frac{x_1 - x_0}{a} = \frac{y_1 - y_0}{b} = \frac{z_1 - z_0}{c}.$$

From these, combined with the preceding equations, we have

$$\left.\begin{aligned}\frac{1}{x_1 - x_0}\cdot\frac{du_0}{dx_0} = \frac{1}{y_1 - y_0}\cdot\frac{du_0}{dy_0} = \frac{1}{z_1 - z_0}\cdot\frac{du_0}{dz_0},\\ \frac{1}{x_1 - x_0}\cdot\frac{du_1}{dx_1} = \frac{1}{y_1 - y_0}\cdot\frac{du_1}{dy_1} = \frac{1}{z_1 - z_0}\cdot\frac{du_1}{dz_1}.\end{aligned}\right\} \quad (H)$$

Hence it is evident that the line joining the extreme points is perpendicular at each of its extremities to the bounding surface. The length of this line is consequently a minimum.

If the minimum curve be required to touch its bounding surfaces, the preceding discussion becomes of course inapplicable, and the curve is no longer, in general, a circle. We may, however, deduce from the terms without the sign of integration, an interesting geometrical theorem as to the position of the osculating planes at the extreme points. For the sake of simplicity, we shall suppose the extreme points to be given. The problem will then be stated as follows:—To draw between two points, each situated on a given surface, a curve of constant curvature, which shall touch the given surface at its extreme points, and whose length shall be a minimum.

The extreme points being fixed, we shall have, as before,

$$\delta x_0 = 0,\ \delta y_0 = 0,\ \delta z_0 = 0,\ \delta x_1 = 0,\ \delta y_1 = 0,\ \delta z_1 = 0,$$

and the equations furnished by the terms free from the sign of integration will still be

$$\left.\begin{aligned} \lambda'_0\left(\frac{d^2x}{ds^2}\cdot\frac{d\delta x}{ds}+\frac{d^2y}{ds^2}\cdot\frac{d\delta y}{ds}+\frac{d^2z}{ds^2}\cdot\frac{d\delta z}{ds}\right)_0 &= 0,\\ \lambda'_1\left(\frac{d^2x}{ds^2}\cdot\frac{d\delta x}{ds}+\frac{d^2y}{ds^2}\cdot\frac{d\delta y}{ds}+\frac{d^2z}{ds^2}\cdot\frac{d\delta z}{ds}\right)_1 &= 0. \end{aligned}\right\}\quad (D)$$

But since the tangents to the curve at each extremity are, by hypothesis, in the tangent planes to the limiting surfaces, we have

$$\frac{du_0}{dx_0}\left(\frac{dx}{ds}\right)_0+\frac{du_0}{dy_0}\left(\frac{dy}{ds}\right)_0+\frac{du_0}{dz_0}\left(\frac{dz}{ds}\right)_0 = 0,$$

$$\frac{du_1}{dx_1}\left(\frac{dx}{ds}\right)_1+\frac{du_1}{dy_1}\left(\frac{dy}{ds}\right)_1+\frac{du_1}{dz_1}\left(\frac{dz}{ds}\right)_1 = 0.$$

Hence it is easy to see that the variations

$$\left(\frac{d\delta x}{ds}\right)_0,\quad \left(\frac{d\delta y}{ds}\right)_0,\quad \left(\frac{d\delta z}{ds}\right)_0$$

are connected by the equations

$$\frac{du_0}{dx_0}\left(\frac{d\delta x}{ds}\right)_0+\frac{du_0}{dy_0}\left(\frac{d\delta y}{ds}\right)_0+\frac{du_0}{dz_0}\left(\frac{d\delta z}{ds}\right)_0 = 0,$$

$$\left(\frac{dx}{ds}\right)_0\left(\frac{d\delta x}{ds}\right)_0+\left(\frac{dy}{ds}\right)_0\left(\frac{d\delta y}{ds}\right)_0+\left(\frac{dz}{ds}\right)_0\left(\frac{d\delta z}{ds}\right)_0 = 0.$$

Eliminating by means of these equations two of the variations,

$$\left(\frac{d\delta x}{ds}\right)_0,\quad \left(\frac{d\delta y}{ds}\right)_0,\quad \left(\frac{d\delta z}{ds}\right)_0,$$

from the first of equations (D), we have

$$\left(\frac{dx}{ds}\cdot\frac{d^2y}{ds^2}-\frac{dy}{ds}\cdot\frac{d^2x}{ds^2}\right)_0\frac{du_0}{dz_0}+\left(\frac{dz}{ds}\cdot\frac{d^2x}{ds^2}-\frac{dx}{ds}\cdot\frac{d^2z}{ds^2}\right)_0\frac{du_0}{dy_0}$$
$$+\left(\frac{dy}{ds}\cdot\frac{d^2z}{ds^2}-\frac{dz}{ds}\cdot\frac{d^2y}{ds^2}\right)_0\frac{du_0}{dx_0}=0.$$

A similar equation holds, of course, for the other extremity of the curve. Hence we infer, that if a curve be drawn as required by the conditions of this problem, the osculating plane at each extremity will be normal to the limiting surface.*

Before quitting the subject of this proposition we may observe, that the solution which we have found for the case in which the position of the extreme tangents is undetermined, appears at first sight an exception to the general theory of p. 127. For it might be supposed, in accordance with that theory, that as the equations furnished by the terms free from the sign of integration have been satisfied by making

$$\lambda'_0=0,\quad \lambda'_1=0,$$

the solution ought to contain indeterminate constants. The exception is, however, only apparent, and results from the fact, that in consequence of the peculiar way in which λ' enters into the equations (C), each of the equations

$$\lambda'_0=0,\quad \lambda'_1=0,$$

is really equivalent to three. This will be shown most clearly by squaring and adding the equations (C). For we shall then have (as is easily seen)

$$\lambda'^2=m^2\left\{(ay-bx+f)^2+(cx-az+f')^2+(bz-cy+f'')^2\right\}.$$

Hence if λ' vanish for any given system of values of x, y, z, it is plain that this system must satisfy the *three* equations,

$$ay-bx+f=0,\quad cx-az+f'=0,\quad bz-cy+f''=0.$$

---

* The equations (D) may also be satisfied by making $\lambda'_0=0$, $\lambda'_1=0$. But this supposition would, as we have seen, render the curve a circle, and would, therefore, lead to an impossible result. In fact it would be necessary, if we adopted this hypothesis, to describe a circle with a given radius passing through two given points, and touching two given planes.

CHAPTER V.

ON MULTIPLE INTEGRALS IN GENERAL.

90. THE difficulty which attends the application of the Calculus of Variations to the solution of problems which involve integrals of any degree higher than the first, arises mainly from the difference between the nature of the limits employed in simple integration and that of the limits which it is necessary to adopt in almost every question connected with multiple integrals.

Hitherto no part of the process of integration itself has been affected by the peculiar nature of the limits employed, inasmuch as the process is supposed to be completed before the consideration of limits is entertained. We have seen, therefore, that in the investigation of the variation of a simple integral, the terms which depend upon the variation of the limits are, as might have been anticipated, wholly free from the sign of integration, and may, therefore, be treated as ordinary algebraic quantities. Such terms add but little to the difficulty of the problem.

But it is otherwise with multiple integrals. Here each successive integration is supposed to be definite in itself, i. e. it is supposed that, after each integration, and previous to the next, the variable, with regard to which that integration has been performed, is taken within certain assigned limits,—limits which are in general functions of all the remaining variables. It is plain, then, that the nature of the limits employed in each integration will affect every subsequent integration, inasmuch as it will affect the form of the function to be integrated. It may, therefore, be expected, as it will be found to be the case, that in the variation of a multiple integral the terms arising from the variation of the limits will not be free from the sign of integration. This adds materially to the difficulty of the question.

91. As a clear conception of the meaning of the symbol of multiple integration, $\iiint \ldots\ldots$, is essential to our present purpose, we shall proceed to enumerate the several operations which are

denoted by that symbol, firstly, in its most general acceptation, and, secondly, in the more limited sense in which it becomes the subject of the Calculus of Variations.

To render the subject as definite as possible, we shall consider the triple integral,

$$u = \int_{x_0}^{x_1}\int_{y_0}^{y_1}\int_{z_0}^{z_1} . \, V dx\, dy\, dz,$$

in which V is a function of x, y, z. This symbol denotes the following operations:—1. The operation of definite integration with regard to one variable, z for example, the limits being certain functions of x and y. 2. The operation of definite integration with regard to y, the limits being certain functions of x. 3. The operation of definite integration with regard to x, the limits being certain constants. If these functions and constants are independent of each other, the symbol is understood in its most general sense. But with this general sense we have at present no concern, inasmuch, as in the existing state of the Calculus of Variations, and the problems to which it is applied, questions which involve such integrals never occur. We shall, therefore, proceed to consider the process of multiple integration in the more limited sense in which it occurs in practice.

92. In the problems to which the Calculus of Variations is applied, and generally in most problems connected with multiple integrals, the several limits of integration are supposed to be assigned in such a way that the final result may express the limit of the sum of the elements $V dx\, dy\, dz$ for all systems of values of x, y, z, which do not cause a given function, $\phi(x, y, z)$, to change sign, i. e. for all systems of values consistent with the supposition

$$\phi(x, y, z) > 0 \text{ or } < 0.*$$

It is, then, in the first place, essential to consider how this may be effected. Now it is known that a definite integral, such as

$$\int_{z_0}^{z_1} V dz,$$

denotes the limit of the sum of the elements $V dz$ for all values of z lying between z_0 and z_1; if therefore we wish that this integral

* The case of two limiting functions is easily reduced to this. *Vide infra*, p. 216.

should represent the limit of that sum for *all* values of z which are consistent with a certain supposition, it is evident that we must take, as the limits of integration, the *extreme* values of z which are consistent with that supposition. To apply this to the case in point. As functions do not change sign except in passing through zero or infinity, it is manifest that the extreme, and therefore limiting values of the variable with which we commence the integration will be found by equating $\phi(x, y, z)$ to zero or infinity. Neglecting the latter supposition, which seldom occurs in practice, and supposing that the integration has been commenced with regard to z, let the values of that variable furnished by the equation

$$\phi(x, y, z) = 0$$

be z_0, z_1.* These quantities, which are, in general, functions of y and x, are evidently the limits to be employed in the first integration. Let

$$V_1 = \int_{z_0}^{z_1} V dz,$$

and let values of any of the variables which are consistent with the conditions of the question be denominated *possible* values. Then since, in the integration with regard to z, x and y are considered

---

* If the equation $\phi = 0$, when solved for z, give more than two values of that quantity, let the several values of z which it does give be

$$z', z'', \ldots . z^{(n)}.$$

Then if n be even, and if ϕ change sign for each of these values, we shall evidently have, as the expression of the integral when taken between the given limits,

$$\int V dz = \int_{z'}^{z''} V dz + \int_{z'''}^{z''''} V dz + \&c. + \int_{z^{(n-1)}}^{z^{(n)}} V dz.$$

Each of these integrals is then to be treated as

$$\int_{z_0}^{z} V dz$$

in the text.

If n be not even, the last integral in the foregoing series will be

$$\int_{z^{(n)}}^{\infty} V dz.$$

It is further evident that any value of z which, although satisfying the equation

$$\phi(x, y, z) = 0,$$

does not cause $\phi(x, y, z)$ to change sign, is to be neglected in the process here described.

constants, it is evident that V_1 denotes the limit of the sum of the elements Vdz, for all possible values of z corresponding to fixed values of x and y. If now this quantity be multiplied by dy, and integrated between the extreme values of y, the result,

$$\int_{y_0}^{y_1} V_1 dy,$$

will, for the same reason, denote the limit of the sum of the elements $V_1 dy$, for all possible values of y corresponding to a given value of x, and therefore of the elements $Vdydz$ for all possible *systems* of values of y and z corresponding to a given value of x. Now the extreme, i. e. the maximum and minimum values of y corresponding to a given value of x, are evidently found by differentiating the equation

$$\phi(x, y, z) = 0,$$

under the supposition that x is constant, and putting $dy = 0$. Performing this operation, we find

$$\frac{d\phi}{dy} dy + \frac{d\phi}{dz} dz = 0 ;$$

or, since $dy = 0$,

$$\frac{d\phi}{dz} = 0.$$

Hence the limits with regard to y are found by eliminating z between the equations

$$\phi = 0, \quad \frac{d\phi}{dz} = 0.$$

Let the result of this elimination be

$$\phi_1(x, y) = 0 ;$$

and it appears, by precisely similar reasoning, that the limiting values of x are found by eliminating y between the equations

$$\phi_1 = 0 \text{ and } \frac{d\phi_1}{dy} = 0.$$

But since $\phi_1 = 0$ is the result of the elimination of z between $\phi = 0$ and another equation, we have

$$\frac{d\phi_1}{dy} = \frac{d\phi}{dy} + \frac{d\phi}{dz} \cdot \frac{dz}{dy} = 0;$$

or, since

$$\frac{d\phi}{dz} = 0, \quad \frac{d\phi}{dy} = 0.$$

Hence the limiting values of x are found by eliminating y and z between the equations

$$\phi = 0, \quad \frac{d\phi}{dz} = 0, \quad \frac{d\phi}{dy} = 0.$$

We have, therefore, the following rule:

1. Integrate with regard to z, taking as limits the values of z given by the equation

$$\phi(x, y, z) = 0.$$

2. Integrate with regard to y, taking as limits the values of y derived from the equations

$$\phi = 0, \quad \frac{d\phi}{dy} = 0,$$

by the elimination of z. 3. Integrate with regard to x, taking as limits the values of x derived from the equations

$$\phi = 0, \quad \frac{d\phi}{dz} = 0, \quad \frac{d\phi}{dy} = 0,$$

by the elimination of y and z.

93. To exemplify this process, let it be required to find the value of

$$u = \iiint dx dy dz,$$

in which the integral is to be extended to all systems of values of x, y, z which render the function

$$\frac{x^2}{a^2} + \frac{y^2}{b^2} + \frac{z^2}{c^2} - 1$$

negative.

(1.) Integrating with regard to z between the limits

$$z = \pm c \sqrt{\left(1 - \frac{x^2}{a^2} - \frac{y^2}{b^2}\right)},$$

we find

$$u = 2c \iint \sqrt{\left(1 - \frac{x^2}{a^2} - \frac{y^2}{b^2}\right)} dx\, dy.$$

(2.) Integrating with regard to y, which is easily effected by putting $y = b\sin\theta\sqrt{\left(1-\frac{x^2}{a^2}\right)}$, we find, as the value of the indeterminate integral, the expression

$$\int\sqrt{\left(1-\frac{x^2}{a^2}-\frac{y^2}{b^2}\right)}dy =$$

$$\frac{1}{2}\left\{b\left(1-\frac{x^2}{a^2}\right)\sin^{-1}.\frac{y}{b\sqrt{\left(1-\frac{x^2}{a^2}\right)}}+y\sqrt{\left(1-\frac{x^2}{a^2}-\frac{y^2}{b^2}\right)}\right\}.$$

The limiting values of y are found, as above stated, by eliminating z between $\phi = 0$ and $\frac{d\phi}{dz} = 0$, i. e. between

$$\frac{x^2}{a^2}+\frac{y^2}{b^2}+\frac{z^2}{c^2}-1=0, \text{ and } \frac{z}{c}=0;$$

they are, therefore,

$$y = \pm b\sqrt{\left(1-\frac{x^2}{a^2}\right)}.$$

Hence

$$u = \pi bc\int\left(1-\frac{x^2}{a^2}\right)dx.$$

(3.) Integrating with regard to x between the limits found by eliminating y and z from the equations

$$\phi = 0, \quad \frac{d\phi}{dz} = 0, \quad \frac{d\phi}{dy} = 0,$$

or

$$\frac{x^2}{a^2}+\frac{y^2}{b^2}+\frac{z^2}{c^2}-1=0, \quad \frac{z}{c}=0, \quad \frac{y}{b}=0,$$

i. e. between the limits $x = \pm a$, we find

$$u = \frac{4\pi}{3}.abc.$$

Again, let it be required to find

$$u = \iiint(x^2+y^2)\,dxdydz,$$

the form of the function ϕ being the same as before.

Here, as in the preceding example,

$$u = 2c\iint (x^2 + y^2) \sqrt{\left(1 - \frac{x^2}{a^2} - \frac{y^2}{b^2}\right)} dxdy.$$

Assume, as before,

$$y = b \sin\theta \sqrt{\left(1 - \frac{x^2}{a^2}\right)}.$$

Then

$$(x^2 + y^2) \sqrt{\left(1 - \frac{x^2}{a^2} - \frac{y^2}{b^2}\right)} dy =$$

$$\left\{ x^2 + b^2\left(1 - \frac{x^2}{a^2}\right)\sin^2\theta \right\} b\left(1 - \frac{x^2}{a^2}\right) \cos^2\theta d\theta.$$

Hence, as the limiting values of θ are evidently $\pm \frac{\pi}{2}$,

$$\int (x^2 + y^2) \sqrt{\left(1 - \frac{x^2}{a^2} - \frac{y^2}{b^2}\right)} dy =$$

$$bx^2 \left(1 - \frac{x^2}{a^2}\right) . \int_{-\frac{\pi}{2}}^{\frac{\pi}{2}} \cos^2\theta d\theta + b^3\left(1 - \frac{x^2}{a^2}\right)^2 . \int_{-\frac{\pi}{2}}^{\frac{\pi}{2}} \sin^2\theta \cos^2\theta d\theta.$$

Hence

$$u = \pi bc . \int_{-a}^{+a} x^2 \left(1 - \frac{x^2}{a^2}\right) dx + \tfrac{1}{4}\pi b^3 c . \int_{-a}^{+a} \left(1 - \frac{x^2}{a^2}\right)^2 dx = \frac{4\pi}{15} abc(a^2 + b^2).$$

94. It has been thought necessary to describe multiple integration as a process purely analytical, partly to facilitate the extension of the method to integrals of all orders, and partly to prevent the supposition that this process *necessarily* involves geometrical considerations. Such considerations will, however, greatly aid the conceptions of the student in the discussion of double and triple integrals. Thus, for example, if $\phi(x, y, z) = 0$ be the equation of a surface enclosing a finite solid, V_1 will denote the limit of the sum of the elements, Vdz, for all points of that portion of the solid which is bounded by four planes, two parallel to the plane of xz, and two parallel to the plane of yz, at the distances y, $y + dy$, x, $x + dx$, respectively. V_2 denotes the sum of the elements $Vdxdy$ for all points of the solid which lie between

two planes parallel to yz, and at the distances x, $x+dx$; and V_3, or $\iiint Vdxdydz$, denotes the sum of the elements $Vdxdydz$, for all points of the solid.

Thus in the first of the preceding examples

$$\int_{z_0}^{z_1} dxdydz$$

denotes the solid contents of a prism whose length is finite, and parallel to the axis of z, and whose other dimensions are evanescent;

$$\int_{y_0}^{y_1}\int_{z_0}^{z_1} dxdydz,$$

a section whose length and breadth are finite, and whose thickness is evanescent; and

$$\int_{x_0}^{x_1}\int_{y_0}^{y_1}\int_{z_0}^{z_1} dxdydz$$

the entire solid.

This view of the subject being given at full length in Lacroix's Traitè de Calc. Int. tom. ii. p. 196, it is unnecessary here to pursue it further; but it may be remarked, as the geometrical explanation of the limits, that the limiting values of z correspond to the two points on the surface for which the values of x and y are the same; that the limiting values of y correspond to the two points for which x has the same value, and for which the tangent plane is perpendicular to xy; and that the limiting values of x correspond to the two points for which the tangent plane is perpendicular to the axis of x.

It is obvious that in the process of multiple integration, as here described, the order of the integration is indifferent. For with whatever variable we commence, the final result will always be the same, viz., the limit of the sum of the elements $Vdxdydz$ for all values of xyz consistent with the supposition $\phi >$ or < 0. This, which is not true of definite integrals in general, is necessary to render the rules of the Calculus of Variations applicable to such quantities.

ON THE CHANGE OF THE INDEPENDENT VARIABLES IN MULTIPLE INTEGRALS IN GENERAL.

95. (1.) If it be required to change the independent variables previously to any integration, it is known* that in the case of a double integral, if x, y be changed into x', y', and if

$$dx = Pdx' + Qdy',$$
$$dy = P'dx' + Q'dy',$$

the quantity to be substituted for $dxdy$ is $(PQ' - P'Q)dx'dy'$; and, in the case of a triple integral, if

$$dx = Pdx' + Qdy' + Rdz',$$
$$dy = P'dx' + Q'dy' + R'dz',$$
$$dz = P''dx' + Q''dy' + R''dz',$$

the quantity to be substituted for $dxdydz$ is

$$\{P(Q'R'' - Q''R') + P'(Q''R - QR'') + P''(QR' - Q'R)\}\, dx'dy'dz';$$

and, in general, if the n independent variables, $x_1, x_2, x_3, \ldots . x_n$, be changed into n other variables, $x'_1, x'_2, x'_3, \ldots . x'_n$, and if

$$dx_1 = X_1dx'_1 + X_2dx'_2 + \&c. + X_ndx'_n,$$
$$dx_2 = X'_1dx'_1 + X'_2dx'_2 + \&c.$$
$$dx_3 = \&c.$$

the quantity to be substituted for $dx_1dx_2dx_3 \ldots . dx_n$ is

$$Rdx'_1dx'_2dx'_3 \ldots . dx'_n,$$

where R is the *resultant* of the n letters, $X_1, X_2 \ldots . X_n$.

It is unnecessary to enter further into this case, which has been fully discussed in the article already quoted.

(2.) Let it be required to change the independent variables after the performance of the first integration. Let

$$\iint(P_1 - P_0)\,dydz, \quad \iint(Q_1 - Q_0)\,dxdz, \quad \iint(R_1 - R_0)\,dxdy,$$

be three double integrals derived by the process of definite integration from the triple integrals,

* Lacroix, tom. ii. pp. 205, 208.

$$\iiint \frac{dP}{dx} dxdydz, \quad \iiint \frac{dQ}{dy} dxdydz, \quad \iiint \frac{dR}{dy} dxdydz,$$

in which the limits are determined by the equation

$$\phi(x, y, z) = 0,$$

and let it be required to reduce them by changing the independent variables to three other integrals, in which these variables shall be the same.

Since ϕ has a positive value previously to vanishing for $x = x_0$, and a negative value previously to vanishing for $x = x_1$, it is evident that $\frac{d\phi}{dx}$ is negative for $x = x_0$, and positive for $x = x_1$; hence if $\sqrt{\left(\frac{d\phi^2}{dx^2}\right)}$ be taken positively, we have

$$P_1 = \left(\frac{P\frac{d\phi}{dx}}{\sqrt{\left(\frac{d\phi^2}{dx^2}\right)}}\right)_1, \quad -P_0 = \left(\frac{P\frac{d\phi}{dx}}{\sqrt{\left(\frac{d\phi^2}{dx^2}\right)}}\right)_0.$$

Hence

$$\iint (P_1 - P_0)\, dydz = \iint \left(\frac{P\frac{d\phi}{dx}}{\sqrt{\left(\frac{d\phi^2}{dx^2}\right)}}\right)_1 dydz + \iint \left(\frac{P\frac{d\phi}{dx}}{\sqrt{\left(\frac{d\phi^2}{dx^2}\right)}}\right)_0 dydz.$$

Now it is evident that these terms may be consolidated into one, *sc.*

$$\iint \frac{P\frac{d\phi}{dx}}{\sqrt{\left(\frac{d\phi^2}{dx^2}\right)}} dydx,$$

provided it be understood that this integral is extended only to such systems of values of xyz as satisfy the equation

$$\phi(x, y, z) = 0.$$

We have then, in the first place, the equations

$$\iint (P_1 - P_0)\, dydz = \iint \frac{P\frac{d\phi}{dx}}{\sqrt{\left(\frac{d\phi^2}{dx^2}\right)}} dydz;$$

and, similarly,

$$\iint (Q_1 - Q_0)\, dzdx = \iint \frac{Q\,\dfrac{d\phi}{dy}}{\sqrt{\left(\dfrac{d\phi^2}{dy^2}\right)}}\, dzdx, \qquad \text{(A)}$$

$$\iint (R_1 - R_0)\, dxdy = \iint \frac{R\,\dfrac{d\phi}{dz}}{\sqrt{\left(\dfrac{d\phi^2}{dz^2}\right)}}\, dxdy.$$

To reduce these, we may either take two new independent variables, a, b, or reduce the first two integrals with regard to x and y. We shall adopt the latter, as being (although not symmetrical) more convenient in practice.

Suppose, then, that it were required to transform

$$\iint \frac{P\,\dfrac{d\phi}{dx}}{\sqrt{\left(\dfrac{d\phi^2}{dx^2}\right)}}\, dydz$$

into an integral of the form

$$\iint M dxdy.$$

Here we have to change but one variable, namely, z into x. Now since, in the integration of

$$\iint \frac{P\,\dfrac{d\phi}{dx}}{\sqrt{\left(\dfrac{d\phi^2}{dx^2}\right)}}\, dydz,$$

with regard to z, and of

$$\iint M dydx$$

with regard to x, y is considered constant, it is evident that we have the equation

$$\frac{d\phi}{dx}\, dx + \frac{d\phi}{dz}\, dz = 0.$$

Hence it appears that $\dfrac{d\phi}{dx}\, dx$ and $\dfrac{d\phi}{dz}\, dz$ are equal in magnitude.

But as the increments dx, dz are always, in the process of definite integration, taken positively, it is evident that the quantity to be substituted for dz is

$$\frac{\sqrt{\left(\frac{d\phi^2}{dx^2}\right)}}{\sqrt{\left(\frac{d\phi^2}{dz^2}\right)}}\,dx.$$

Hence

$$\iint \frac{P\frac{d\phi}{dx}}{\sqrt{\left(\frac{d\phi^2}{dx^2}\right)}}\,dydz = \iint \frac{P\frac{d\phi}{dx}}{\sqrt{\left(\frac{d\phi^2}{dz^2}\right)}}\,dxdy.$$

Similarly,

$$\iint \frac{Q\frac{d\phi}{dy}}{\sqrt{\left(\frac{d\phi^2}{dy^2}\right)}}\,dxdz = \iint \frac{Q\frac{d\phi}{dy}}{\sqrt{\left(\frac{d\phi^2}{dz^2}\right)}}\,dxdy, \qquad \text{(B)}$$

$$\iint \frac{R\frac{d\phi}{dz}}{\sqrt{\left(\frac{d\phi^2}{dz^2}\right)}}\,dxdy = \iint \frac{R\frac{d\phi}{dz}}{\sqrt{\left(\frac{d\phi^2}{dz^2}\right)}}\,dxdy.$$

Differentiate the equation

$$\phi(x, y, z) = 0,$$

and let the result be denoted by

$$dz = pdx + qdy.$$

We have then

$$\frac{d\phi}{dx} = -p\frac{d\phi}{dz}, \quad \frac{d\phi}{dy} = -q\frac{d\phi}{dz}.$$

Substituting these values in the equations (B), we find

$$\iint (P_1 - P_0)\,dydz = \iint \frac{P\frac{d\phi}{dx}}{\sqrt{\left(\frac{d\phi^2}{dx^2}\right)}}\,dydz = -\iint \frac{Pp\frac{d\phi}{dz}}{\sqrt{\left(\frac{d\phi^2}{dz^2}\right)}}\,dxdy,$$

$$\iint (Q_1 - Q_0)\, dxdz = \iint \frac{Q\frac{d\phi}{dy}}{\sqrt{\left(\frac{d\phi^2}{dy^2}\right)}}\, dxdz = -\iint \frac{Qq\frac{d\phi}{dz}}{\sqrt{\left(\frac{d\phi^2}{dz^2}\right)}}\, dxdy, \quad \text{(C)}$$

$$\iint (R_1 - R_0)\, dxdy = \ldots\ldots\ldots\ldots\ldots\ldots \iint \frac{R\frac{d\phi}{dz}}{\sqrt{\left(\frac{d\phi^2}{dz^2}\right)}}\, dxdy.$$

Symmetrical expressions, which are sometimes useful, may be deduced from equations (A) by assuming

$$ds^2 = dx^2dy^2 + dx^2dz^2 + dy^2dz^2.$$

It is easy to see, then, that these equations may be written

$$\iint (P_1 - P_0)\, dydz = \iint P\Omega \frac{d\phi}{dx}\, ds,$$

$$\iint (Q_1 - Q_0)\, dxdz = \iint Q\Omega \frac{d\phi}{dy}\, ds, \qquad \text{(D)}$$

$$\iint (R_1 - R_0)\, dxdy = \iint R\Omega \frac{d\phi}{dz}\, ds,$$

where

$$\Omega = \left(\frac{d\phi^2}{dx^2} + \frac{d\phi^2}{dy^2} + \frac{d\phi^2}{dz^2}\right)^{-\frac{1}{2}}.$$

These formulæ may also be deduced geometrically. For if

$$\phi(x, y, z) = 0$$

be the equation of a surface, ds the element of its superficial area, and α, β, γ the angles which the normal makes with the axes of co-ordinates, it is manifest that, through the whole of the integral $\iint P_1 dydz$, we have

$$dydz = \cos\alpha\, ds;$$

and through the whole of the integral $\iint P_0 dydz$,

$$dydz = -\cos\alpha\, ds.$$

Hence

$$\iint (P_1 - P_0)\, dydz = \iint P\cos\alpha\, ds;$$

and, similarly,

$$\iint (Q_1 - Q_0)\, dxdz = \iint Q \cos\beta ds,$$
$$\iint (R_1 - R_0)\, dxdy = \iint R \cos\gamma ds,$$

which are identical with formulæ (D), inasmuch as

$$\cos a = \frac{\dfrac{d\phi}{dx}}{\sqrt{\left(\dfrac{d\phi^2}{dx^2} + \dfrac{d\phi^2}{dy^2} + \dfrac{d\phi^2}{dz^2}\right)}},$$
$$\cos\beta = \&c.$$
$$\cos\gamma = \&c.$$

In these formulæ the radical is taken positively.

96. It is frequently necessary, in problems connected with multiple integrals, to determine the limits in such a way that the result may represent the limit of the sum of the elements

$$Vdxdydz \ldots$$

for all systems of values of the independent variables which render the signs of *two* given functions,

$$\phi(x, y, z, \ldots) \quad \psi(x, y, z, \ldots)$$

different. This case may be reduced to the foregoing by a very simple consideration.

Let it be supposed that the given integral is to be extended to all systems of values of $x, y, z, \ldots$ which satisfy the conditions

$$\phi(x, y, z, \ldots) < 0, \quad \psi(x, y, z, \ldots) > 0.$$

Now it is in general supposed that every system of values which satisfies the condition

$$\phi(x, y, z, \ldots) > 0,$$

will also satisfy the condition

$$\psi(x, y, z, \ldots) > 0;$$

and, similarly, every system which satisfies the condition

$$\psi(x, y, z, \ldots) < 0$$

is supposed also to satisfy the condition

$$\phi(x, y, z, \ldots) < 0.$$

Hence we see evidently that

$$\iiint \ldots V dx dy dz \ldots = (\iiint \ldots V dx dy dz \ldots) - [\iiint \ldots V dx dy dz \ldots],$$

in which the integral ( ) is extended to all systems of values which render ϕ negative, and the integral [ ] to all systems which render ψ negative. The given integral is thus reduced to two others, each similar in its nature to that already considered, and to be treated according to the same rules. The analytical conception of this case is also much facilitated by considering its geometrical signification, when the given integral is of the second or third order. In the case of a single limiting function, we have seen that the integral

$$\iint V dx dy \text{ or } \iiint V dx dy dz$$

is extended to every point lying *within* the curve or surface

$$\phi = 0.$$

In the present case it is easy to see that the portion of space with which we are concerned is that which lies *between* the two curves or surfaces,

$$\phi = 0, \quad \psi = 0;$$

and that the hypothesis, that every system of values which satisfies either of the conditions,

$$\phi > 0 \text{ or } \psi < 0,$$

also satisfies the corresponding condition,

$$\psi > 0 \text{ or } \phi < 0,$$

is equivalent to the geometrical supposition that the curve or surface whose equation is

$$\psi = 0$$

is entirely included within the curve or surface whose equation is

$$\phi = 0.$$

97. After what has been said, no difficulty will be found in applying to any multiple integrals whatever, the rules which have been given with special reference to the case of triple integrals.

For this purpose it will only be necessary to substitute for the double integrals,

$$\iint (P_1 - P_0)\, dydz, \quad \iint (Q_1 - Q_0)\, dxdz, \text{ \&c.}$$

the multiple integrals,

$$\iiint \ldots (P_1 - P_0)\, dydzdt \ldots$$
$$\iiint \ldots (Q_1 - Q_0)\, dxdzdt \ldots$$
$$\text{\&c. \&c.}$$

and to make

$$ds^2 = dx^2dy^2dt^2 \ldots + dx^2dz^2dt^2 \ldots + dy^2dz^2dt^2 \ldots + dx^2dy^2dz^2 \ldots + \text{\&c.}$$

It is not necessary to dwell further upon this extension, which is comparatively useless for our present purpose.

Formulæ (C) and (D) are readily adapted to the case of a double integral, by removing the independent variable y, and putting y for z in the result. It is easy to see then that the first two of equations (C) and (D) become

$$\left.\begin{aligned} \int (P_1 - P_0)\, dy &= -\int P \frac{dy}{dx} \frac{\dfrac{d\phi}{dy}}{\sqrt{\left(\dfrac{d\phi^2}{dy^2}\right)}}\, dx = \int \frac{P \dfrac{d\phi}{dx}}{\sqrt{\left(\dfrac{d\phi^2}{dx^2} + \dfrac{d\phi^2}{dy^2}\right)}}\, ds, \\ \int (Q_1 - Q_0)\, dx &= \int \frac{Q \dfrac{d\phi}{dy}}{\sqrt{\left(\dfrac{d\phi^2}{dy^2}\right)}}\, dx = \int \frac{Q \dfrac{d\phi}{dy}}{\sqrt{\left(\dfrac{d\phi^2}{dx^2} + \dfrac{d\phi^2}{dy^2}\right)}}\, ds; \end{aligned}\right\} \quad \text{(E)}$$

in which the value of the coefficient $\dfrac{dy}{dx}$ is derived from the differentiation of the equation

$$\phi(x, y) = 0,$$

and

$$ds = \sqrt{(dx^2 + dy^2)}.$$

CHAPTER VI.

FUNCTIONS OF TWO OR MORE INDEPENDENT VARIABLES.

Prop. I.

98. Let u be an indeterminate function of any number of independent variables, x, y, z, &c., and let it be required to find the complete increment of the differential coefficient,

$$\frac{d^{m+n+p+\&c.}u}{dx^m dy^n dz^p \ldots}.$$

Adopting, as before, the principles of Art. 6, Chap. I., we shall investigate successively the several increments which this function receives from a change in each of its varying elements. Now it is plain that the quantity which we are now considering can be varied only in one of the following ways:

(1.) By a change in some one of the independent variables, x, y, z, &c.

(2.) By a change in the form of the function u.

We shall consider these increments separately.

(1.) Let x receive the increment dx, the remaining variables, y, z, &c., as well as the form of the function u, remaining unchanged. Then it is plain that the corresponding increment of the given differential coefficient will be

$$\frac{d^{m+1+n+p+\&c.}u}{dx^{m+1} dy^n dz^p \ldots} dx.$$

Similarly, if y receive the increment dy, while x, z, &c., and the form of the function, remain unchanged, the corresponding increment will be

$$\frac{d^{m+n+1+p+\&c.}u}{dx^m dy^{n+1} dz^p \ldots} dy,$$

and so on for the remaining variables.

(2.) Let the form of the function u vary, the independent variables, x, y, z, &c., remaining unchanged. Then since the symbol of derivation,

$$\frac{d^{m+n+p+\&c.}}{dx^m dy^n dz^p \ldots}$$

satisfies the condition

$$F.\phi + F.\psi = F(\phi + \psi)$$

it appears from the principle of Art. 5 that

$$\delta . \frac{d^{m+n+p+\&c.} u}{dx^m dy^n dz^p \ldots} = \frac{d^{m+n+p+\&c.} \delta u}{dx^m dy^n dz^p \ldots}.$$

If then we add, according to the principle of Art. 6, the several terms found above, we shall have, for the complete increment,

$$D . \frac{d^{m+n+p+\&c.} u}{dx^m dy^n dz^p \ldots} = \frac{d^{m+1+n+p+\&c.} u}{dx^{m+1} dy^n dz^p \ldots} dx + \frac{d^{m+n+1+p+\&c.} u}{dx^m dy^{n+1} dz^p \ldots} dy + \&c. \quad \text{(A)}$$

$$+ \frac{d^{m+n+p+\&c.} \delta u}{dx^m dy^n dz^p \ldots}.$$

PROP. II.

99. To find the complete increment of

$$V = f\left(x, y, z, \ldots u, \frac{du}{dx}, \frac{du}{dy}, \frac{du}{dz}, \ldots \frac{d^2u}{dx^2}, \frac{d^2u}{dxdy}, \ldots\right),$$

a determinate function of the several quantities contained within the parentheses, namely: 1. Any number of independent variables, $x, y, z, \ldots$ 2. One dependent variable, u, which is supposed to be an indeterminate function of $x, y, z, \ldots$ 3. The several partial differential coefficients of u with regard to these variables.

The number of terms which result from the several differential coefficients of V with regard to the quantities

$$\frac{du}{dx}, \frac{du}{dy}, \&c. \quad \frac{d^2u}{dx^2}, \frac{d^2u}{dxdy}, \&c.$$

being necessarily very great, it will be advisable, before proceeding further, to establish a distinct notation to express these coefficients. We shall, then, use the symbol

$$V_{x^m y^n z^p} \ldots$$

to denote the coefficient

$$\frac{dV}{d \cdot \dfrac{d^{m+n+p+\&c.}\, u}{dx^m dy^n dz^p \ldots}},*$$

and the symbols

$$X,\ Y,\ Z, \ldots\ U,$$

to denote the coefficients

$$\frac{dV}{dx},\ \frac{dV}{dy},\ \frac{dV}{dz},\ \ldots \frac{dV}{du},$$

respectively.

This notation being established, we shall have

$$dV = Xdx + Ydy + Zdz + \&c. + Udu$$

$$+ V_x d.\frac{du}{dx} + V_y d.\frac{du}{dy} + V_z d.\frac{du}{dz} + \&c. + V_{x^2} d.\frac{d^2u}{dx^2} + V_{xy} d.\frac{d^2u}{dxdy} + \&c.$$

Now let x receive the increment dx, the remaining independent variables, y, z, &c., as well as the form of the function u, remaining unchanged. Then it is plain that the corresponding increment which V receives is

$$\left(X + U\frac{du}{dx} + V_x \frac{d^2u}{dx^2} + V_y \frac{d^2u}{dxdy} + V_z \frac{d^2u}{dxdz} + \&c. \right.$$
$$\left. + V_{x^2} \frac{d^3u}{dx^3} + V_{xy} \frac{d^3u}{dx^2dy} + \&c. \right) dx.$$

This may be represented by

$$\left(\frac{dV}{dx}\right) dx,$$

denoting by that symbol the complete differential coefficient of V with respect to x.

* This notation was suggested to me by the Rev. William Roberts.

Similarly the increment which results from a change in y will be

$$\left(\frac{dV}{dy}\right) dy,$$

and so on for the other independent variables.

Now let the form of the function u vary, while the independent variables, x, y, z, &c., remain constant. Then since V is a determinate function of x, y, z, u, $\frac{du}{dx}$, &c., it is evident, as in Art. 7 (5), that

$$\delta V = U\delta u + V_x\delta . \frac{du}{dx} + V_y\delta . \frac{du}{dy} + V_z\delta . \frac{du}{dz} + V_{x^2}\delta . \frac{d^2u}{dx^2}$$
$$+ V_{xy}\delta . \frac{d^2u}{dxdy} + \&c.$$

Substituting for

$$\delta . \frac{du}{dx}, \quad \delta . \frac{du}{dy}, \quad \&c.$$

their values as found in Prop. I., and adding the terms which arise from a change in the independent variables, we find, as the expression of the complete increment,

$$\begin{aligned} DV = \left(\frac{dV}{dx}\right) dx + \left(\frac{dV}{dy}\right) dy + \left(\frac{dV}{dz}\right) dz + \&c. \\ + U\delta u + V_x \frac{d\delta u}{dx} + V_y \frac{d\delta u}{dy} + V_z \frac{d\delta u}{dz} + \&c. \\ + V_{x^2} \frac{d^2\delta u}{dx^2} + V_{xy} \frac{d^2\delta u}{dxdy} + \&c. \end{aligned} \qquad \text{(A)}$$

It is evident that the increment which V receives in consequence solely of a change in the form of u will be given by the equation

$$\delta V = U\delta u + V_x \frac{d\delta u}{dx} + V_y \frac{d\delta u}{dy} + V_z \frac{d\delta u}{dz} + \&c.$$
$$+ V_{x^2} \frac{d^2\delta u}{dx^2} + V_{xy} \frac{d^2\delta u}{dxdy} + \&c.$$

100. If the integral under consideration be of the second or-

der, it is usual to take z as the dependent variable. In this case we shall have, putting

$$Z = \frac{dV}{dz},$$

$$\delta V = Z\delta z + V_x \frac{d\delta z}{dx} + V_y \frac{d\delta z}{dy} + V_{x^2} \frac{d^2\delta z}{dx^2} + V_{xy} \frac{d^2\delta z}{dxdy} + V_{y^2} \frac{d^2\delta z}{dy^2} + \&c.$$

In investigating the variation of a multiple integral, to which we shall next proceed, we shall, for the sake of simplicity, confine our attention to the cases of integrals of the second and third orders. The principles on which these cases are discussed are indeed, as will be seen, common to integrals of all orders. But the number of terms in the result increases so rapidly with each new independent variable, that by proceeding further we should merely complicate the formulæ without any practical utility.

For a similar reason we shall suppose that V contains no differential coefficients of an order higher than the second.

It readily appears, from the discussions in the preceding chapter, that the value of a multiple integral, in the limited sense in which we have agreed to employ that term, depends upon two different elements, namely, 1. The form of the function ϕ, which determines the limits. 2. The form of the function to be integrated, V. A change in either of these will produce a corresponding change in the integral, and the total variation will be, according to the principle laid down in Chap. I., the sum of these partial variations. To find this complete variation, then, it will be necessary to investigate separately the variations produced by changes in the forms of the above-mentioned functions. We shall commence with that which results from a change in the form of the limiting function ϕ, and for the convenience of reference, we shall consider successively the cases of double and triple integrals.

Prop. III.

101. Let V be a function of two independent variables, x, y, and let it be proposed to find the variation in the definite integral, $\iint Vdxdy$, caused by an indefinitely small change in the form of the limiting function, $\phi(x, y)$.

Suppose, as before, that the equation

$$\phi(x, y) = 0,$$

when solved for y, gives but two values, y_0, y_1. Suppose also that, in consequence of a change in the form of the function ϕ, these roots become respectively $y_0 + \delta y_0$ and $y_1 + \delta y_1$. It is plain, then, that the variation which the definite integral will undergo at the *first* integration will be

$$\int V_1 \delta y_1 dx - \int V_0 \delta y_0 dx.$$

To find the variation which arises at the *second* integration, assume

$$V' = \int_{y_0}^{y_1} V dy,$$

and let the variations in the limiting values of x be δx_0, δx_1. Then it appears, in the same way, that the variation which arises at the second integration is

$$V'_1 \delta x_1 - V'_0 \delta x_0,$$

where V'_0, V'_1 denote what V' becomes, when for x we substitute successively x_0, x_1. But since x_1, x_0 are found by eliminating y between the equations

$$\phi = 0, \quad \frac{d\phi}{dy} = 0,$$

it is manifest that when $x = x_0$ or $x = x_1$, $y_0 = y_1$. For it is known that if $\phi = 0$ be considered as an equation (having two roots) in which y is the unknown quantity, the roots of that equation will be equal if $\frac{d\phi}{dy} = 0$. But the values $x = x_0$, $x = x_1$, satisfy the equation $\frac{d\phi}{dy} = 0$, and therefore render equal the roots of the equation $\phi = 0$. Since, then, the limits of integration in V'_0, V'_1 become equal, it is plain that $V'_0 = 0$, $V'_1 = 0$, and therefore this part of the variation vanishes of itself.

This reasoning may be extended to integrals of all degrees. Thus, for example, in the case of the triple integral,

$$\iiint V dx dy dz,$$

if we assume

$$V' = \int_{z_0}^{z_1} Vdz,$$

it will appear, by precisely the same reasoning, that the variation which arises at the integration with regard to y will vanish of itself; and if we assume

$$V'' = \int_{y_0}^{y_1}\int_{z_0}^{z} Vdydz,$$

we shall arrive at a similar conclusion with regard to the variation which arises at the integration with regard to x. Hence, with regard to that part of the variation which arises from a change in the form of the limiting function, we may conclude generally as follows:

If V be a function of any number of independent variables, x, y, z, &c., the variation in the definite integral,

$$\iiint \ldots Vdxdydz \ldots$$

arising from a change in the form of the limiting function ϕ, will be

$$\iint \ldots V_1\delta z_1 dxdy \ldots \quad - \iint \ldots V_0\delta z_0 dxdy \ldots$$

in which V_1, V_0 denote the values which V receives by the successive substitution of z_1 and z_0 for z.

It is, perhaps, unnecessary to remind the reader that the reduction of this part of the variation to two terms does not hold for *all* definite multiple integrals, the reasoning by which it was arrived at being only applicable to the class which we have been considering.

Prop. IV.

102. Let V be a determinate function of

$$x,\ y,\ z,\ \frac{dz}{dx},\ \frac{dz}{dy},\ \frac{d^2z}{dx^2},\ \frac{d^2z}{dxdy},\ \frac{d^2z}{dy^2},$$

where z is an indeterminate function of the independent variables x, y; and let it be required to find the variation produced in the definite integral, $\iint Vdxdy$, by an indefinitely small change in the form of the function z.

Since the limiting function ϕ is here supposed to remain invariable, it is evident that the operation denoted by the symbol $\iint$ satisfies the criterion in Art. 5; hence

$$\delta . \iint V dx dy = \iint \delta V dx dy.$$

Substituting for δV its value, as given in Prop. II., we find

$$\delta \iint V dx dy = \tag{A}$$

$$\iint \left(Z\delta z + V_x \frac{d\delta z}{dx} + V_y \frac{d\delta z}{dy} + V_{x^2} \frac{d^2\delta z}{dx^2} + V_{xy} \frac{d^2\delta z}{dxdy} + V_{y^2} \frac{d^2\delta z}{dy^2} \right) dxdy.$$

But since, in the peculiar kind of definite integration with which we are at present concerned, the *order* of the integrations is indifferent, we find, by the method of integration by parts,

$$\iint V_x \frac{d\delta z}{dx} dxdy = \int V_x \delta z dy - \iint \frac{dV_x}{dx} \delta z dxdy,$$

$$\iint V_y \frac{d\delta z}{dy} dxdy = \int V_y \delta z dx - \iint \frac{dV_y}{dy} \delta z dxdy,$$

$$\iint V_{x^2} \frac{d^2\delta z}{dx^2} dxdy = \int V_{x^2} \frac{d\delta z}{dx} dy - \iint \frac{dV_{x^2}}{dx} \frac{d\delta z}{dx} dxdy$$

$$= \int V_{x^2} \frac{d\delta z}{dx} dy - \int \frac{dV_{x^2}}{dx} \delta z dy + \iint \frac{d^2 V_{x^2}}{dx^2} \delta z dxdy,$$

$$\iint V_{xy} \frac{d^2\delta z}{dxdy} dxdy = \int V_{xy} \frac{d\delta z}{dy} dy - \int \frac{dV_{xy}}{dx} \delta z dx + \iint \frac{d^2 V_{xy}}{dxdy} \delta z dxdy,$$

(or if we commence by integrating with regard to y),

$$\iint V_{xy} \frac{d^2\delta z}{dxdy} dxdy = \int V_{xy} \frac{d\delta z}{dx} dx - \int \frac{dV_{xy}}{dy} \delta z dy + \iint \frac{d^2 V_{xy}}{dxdy} \delta z dxdy,$$

$$\iint V_{y^2} \frac{d^2\delta z}{dy^2} dxdy = \int V_{y^2} \frac{d\delta z}{dy} dx - \int \frac{dV_{y^2}}{dy} \delta z dx + \iint \frac{d^2 V_{y^2}}{dy^2} \delta z dxdy.$$

In these expressions the terms which appear with but a single sign of integration have, for the sake of brevity, been only written once, but as in every such term the operation of definite integration is supposed to have been previously performed, it will readily be understood that if we would write these terms at full length, we must substitute for each the difference between two others, one of which refers to the first, and the other to the second limiting value of z.

Thus for $\int V_x \delta z dy$ we should write,

$$(\int V_x \delta z dy)_1 - (V_x \delta z dy)_0,$$

and similarly for the other terms.

Collecting the several terms given above, and using, for the sake of symmetry, the semi-sum of the two values of

$$\iint V_{xy} \frac{d^2 \delta z}{dx dy} dx dy,$$

we find

$$\delta \iint V dx dy = \int \left\{ \left(V_y - \frac{dV_{y^2}}{dy} - \tfrac{1}{2} \frac{dV_{xy}}{dx} \right) \delta z + \tfrac{1}{2} V_{xy} \frac{d\delta z}{dx} + V_{y^2} \frac{d\delta z}{dy} \right\} dx$$

$$+ \int \left\{ \left(V_x - \frac{dV_{x^2}}{dx} - \tfrac{1}{2} \frac{dV_{xy}}{dy} \right) \delta z + V_{x^2} \frac{d\delta z}{dx} + \tfrac{1}{2} V_{xy} \frac{d\delta z}{dy} \right\} dy \qquad \text{(B)}$$

$$+ \iint \left(Z - \frac{dV_x}{dx} - \frac{dV_y}{dy} + \frac{d^2 V_{x^2}}{dx^2} + \frac{d^2 V_{xy}}{dx dy} + \frac{d^2 V_{y^2}}{dy^2} \right) \delta z dx dy.$$

103. Before proceeding further, it may be well to caution the reader against the supposition that such expressions as

$$\int V_{xy} \frac{d\delta z}{dx} dx$$

admit of being a second time integrated by parts, so as to give

$$\int V_{xy} \frac{d\delta z}{dx} dx = V_{xy} \delta z - \int \frac{dV_{xy}}{dx} \delta z dx.$$

For in all terms which appear with but a single sign of integration, two operations have been already performed, viz.: 1. The operation of simple integration with regard to the variable, y for example, whose differential has disappeared. 2. The substitution of a function of x for y in the result of the integration. Therefore, in the expression

$$\int V_{xy} \frac{d\delta z}{dx} dx,$$

δz is no longer a function of x and y, but a function of x solely, and therefore $\frac{d\delta z}{dx}$, which is a partial differential coefficient taken upon the supposition that δz is a function of *both* variables, is only *part* of the differential coefficient with regard to x, when δz becomes a function of that variable only. In fact, if we denote the complete differential coefficient by $\left(\frac{d\delta z}{dx}\right)$, it is plain that

$$\left(\frac{d\delta z}{dx}\right) = \frac{d\delta z}{dx} + \frac{d\delta z}{dy} \cdot \frac{dy}{dx},$$

the value of $\frac{dy}{dx}$ being derived from the equation

$$\phi(x, y) = 0.$$

It is the more necessary to notice this error, inasmuch as it is one into which Lacroix has fallen, and by which he has been led to give an erroneous expression for the variation of a double integral.*

104. The expression given above for $\delta\iint V dx dy$ consists of two essentially different classes of terms, viz.: 1. A number of terms partially integrated, i. e. with but one integral sign prefixed. 2. A number of terms wholly unintegrated, i. e. with the sign of double integration prefixed. With regard to the latter class, it is evident that they admit of no further reduction, seeing that they involve but one indeterminate variation, δz. But with regard to the former class, two reductions are necessary, before the expression can be applied:—1. The independent variables must be changed, so that in all these terms the integration may be performed with regard to the *same* variable. 2. The number of the variations,

$$\delta z, \ \frac{d\delta z}{dx}, \ \frac{d\delta z}{dy},$$

must be reduced. For, as the terms under a single sign of inte-

* Traite de Calcul. Dif. et Int. tom. ii. p. 785. This misconception of the meaning of the symbol

$$\int V_{xy} \frac{d\delta z}{dx} dx$$

is noticed and guarded against by Poisson, Mem. de l'Institut., tom. xii. p. 296.

gration are extended only to those values of x and y which satisfy the equation

$$\phi(x, y) = 0,$$

these variations are no longer independent. This will appear more readily when we come to consider how this reduction is to be effected.

We now proceed to the first of these reductions, that, namely, which involves the change of the independent variable.

PROP. V.

105. To reduce the single integrals contained in the expression for $\delta\iint V dx dy$ to others in which the integration shall be performed with regard to the same variable.

This is readily effected by means of the formulæ given in the preceding chapter. Adopting the method there given, we shall suppose the independent variable in the reduced integrals to be x. Now since, in general, Art. 97,

$$\int (P_1 - P_0)\, dy = -\int P \frac{dy}{dx} \cdot \frac{\dfrac{d\phi}{dy}}{\sqrt{\left(\dfrac{d\phi^2}{dy^2}\right)}}\, dx,$$

we have

$$\int\left\{\left(V_x - \frac{dV_{x^2}}{dx} - \tfrac{1}{2}\frac{dV_{xy}}{dy}\right)\delta z + V_{x^2}\frac{d\delta z}{dx} + \tfrac{1}{2} V_{xy}\frac{d\delta z}{dy}\right\} dy$$

$$= -\int\left\{\left(V_x - \frac{dV_{x^2}}{dx} - \tfrac{1}{2}\frac{dV_{xy}}{dy}\right)\delta z + V_{x^2}\frac{d\delta z}{dx} + \tfrac{1}{2} V_{xy}\frac{d\delta z}{dy}\right\}\frac{dy}{dx}\, k dx,$$

in which, for the sake of brevity, we have put

$$k = \frac{\dfrac{d\phi}{dy}}{\sqrt{\left(\dfrac{d\phi^2}{dy^2}\right)}}.$$

Hence we find

$$\delta\iint V dx dy = \int\left(\zeta\delta z + \xi\frac{d\delta z}{dx} + \eta\frac{d\delta z}{dy}\right) k dx + \iint \Omega \delta z\, dx dy, \qquad \text{(C)}$$

where

$$\zeta = V_y - \frac{dV_{y^2}}{dy} - \tfrac{1}{2}\frac{dV_{xy}}{dx} - \left(V_x - \frac{dV_{x^2}}{dx} - \tfrac{1}{2}\frac{dV_{xy}}{dy}\right)\frac{dy}{dx},$$

$$\eta = V_{y^2} - \tfrac{1}{2}V_{xy}\frac{dy}{dx}, \quad \xi = \tfrac{1}{2}V_{xy} - V_{x^2}\frac{dy}{dx},$$

$$\Omega = Z - \frac{dV_x}{dx} - \frac{dV_y}{dy} + \frac{d^2V_{x^2}}{dx^2} + \frac{d^2V_{xy}}{dxdy} + \frac{d^2V_{y^2}}{dy^2},$$

the value of $\frac{dy}{dx}$ being derived from the equation

$$\phi(x, y) = 0.$$

PROP. VI.

106. To reduce the number of the variations,

$$\delta z, \quad \frac{d\delta z}{dx}, \quad \frac{d\delta z}{dy},$$

occurring under the sign of simple integration, so that those which remain after the reduction may be independent of one another.

Denoting, as before, the complete differential coefficient of δz with regard to x (taken upon the supposition that x is the only independent variable), by $\left(\frac{d\delta z}{dx}\right)$ we have

$$\left(\frac{d\delta z}{dx}\right) = \frac{d\delta z}{dx} + \frac{d\delta z}{dy}\cdot\frac{dy}{dx};$$

hence, in the terms under a single sign of integration,

$$\frac{d\delta z}{dx} = \left(\frac{d\delta z}{dx}\right) - \frac{dy}{dx}\frac{d\delta z}{dy}.$$

Substituting this value in

$$\int \xi \frac{d\delta z}{dx}\, dx,$$

and integrating by parts the term

$$\int \xi \left(\frac{d\delta z}{dx}\right) dx,$$

we have

$$\int \xi \frac{d\delta z}{dx} dx = \int \left(\frac{d.\xi\delta z}{dx}\right) dx - \int \left(\frac{d\xi}{dx}\right) \delta z dx - \int \xi \frac{dy}{dx} \frac{d\delta z}{dy} dx.$$

Now it is easy to see that any term of the form

$$\int \left(\frac{d\theta}{dx}\right) dx,$$

which occurs in the formula under consideration, must vanish of itself. For it will be remembered that each term in that formula is really double; and that, if the terms were written at full length, we should have, instead of $\frac{d\theta}{dx}$,

$$\left(\frac{d\theta}{dx}\right)_1 - \left(\frac{d\theta}{dx}\right)_0,$$

the former of these two quantities denoting the value of $\frac{d\theta}{dx}$, corresponding to $y = y_1$, and the latter, the value corresponding to $y = y_0$. Now if the limiting values of x, derived from the system of equations,

$$\phi = 0, \quad \frac{d\phi}{dy} = 0,$$

be x_0, x_1, it is plain that a term of the form

$$\int \left(\frac{d\theta}{dx}\right) dx,$$

if written at full length, will be as follows:

$$\int \left(\frac{d\theta}{dx}\right) dx = \int \left(\frac{d\theta}{dx}\right)_1 dx - \int \left(\frac{d\theta}{dx}\right)_0 dx = (\theta_1)_1 - (\theta_1)_0 - (\theta_0)_1 + (\theta_0)_0,$$

where

$(\theta_1)_1$ denotes the value of θ for $y = y_1$, $x = x_1$,
$(\theta_1)_0$ . . . . $y = y_1$, $x = x_0$,
$(\theta_0)_1$ . . . . $y = y_0$, $x = x_1$,
$(\theta_0)_0$ . . . . $y = y_0$, $x = x_0$.

But we have before seen (p. 224) that when $x = x_0$, or $x = x_1$, $y_1 = y_0$. Hence we have

$$(\theta_1)_1 = (\theta_0)_1, \quad (\theta_1)_0 = (\theta_0)_0;$$

and therefore, in general,

$$\int\left(\frac{d\theta}{dx}\right)dx = 0.$$

In the present case, therefore,

$$\int\left(\frac{d.\xi\delta z}{dx}\right)dx = 0.$$

We have, therefore,

$$\int\xi\frac{d\delta z}{dx}dx = -\int\left(\frac{d\xi}{dx}+\frac{d\xi}{dy}\frac{dy}{dx}\right)\delta z dx - \int\xi\frac{dy}{dx}\frac{d\delta z}{dy}dx.$$

Substituting this in the expression (C), replacing ξ, η, ζ, Ω, by their values, and adding the variation resulting from a change in the form of the limiting function, which has been determined in Prop. III., we find, as the complete variation of the given integral,

$$D.\iint V dxdy = \int Vk\delta y dx + \int\left(V_{x^2}\frac{dy^2}{dx^2} - V_{xy}\frac{dy}{dx} + V_{y^2}\right)k\frac{d\delta z}{dy}dx$$

$$+\int\left(V_y - V_x\frac{dy}{dx} + 2\frac{dV_{x^2}}{dx}\frac{dy}{dx} + \frac{dV_{x^2}}{dy}\frac{dy^2}{dx^2} - \frac{dV_{y^2}}{dy} - \frac{dV_{xy}}{dx} + V_{x^2}\frac{d^2y}{dx^2}\right)k\delta z dx$$

$$+\iint\left(Z - \frac{dV_x}{dx} - \frac{dV_y}{dy} + \frac{d^2V_{x^2}}{dx^2} + \frac{d^2V_{xy}}{dxdy} + \frac{d^2V_{y^2}}{dy^2}\right)\delta z dxdy; \qquad \text{(D)}$$

the single integrals in this expression being extended to all systems of values of x and y which satisfy the equation

$$\phi(x, y) = 0.$$

If there be two limiting functions, i. e. if the limits of the given integral be assigned, as stated in p. 216, it is evident that the foregoing formula will be adapted to that case by putting for each of the single integrals which it contains, the difference between two others, one of which is extended to all systems of values of x and y, which satisfy the equation

$$\phi = 0,$$

and the other extended to all systems of values which satisfy the equation

$$\psi = 0.$$

Thus we should have, instead of the single term,

$$\int Vk\delta y dx,$$

the terms

$$\int Vk\delta y dx - \int Vk'\delta y dx,$$

in which

$$k' = \frac{\frac{d\psi}{dy}}{\sqrt{\left(\frac{d\psi^2}{dy^2}\right)}}, \qquad k = \frac{\frac{d\phi}{dy}}{\sqrt{\left(\frac{d\phi^2}{dy^2}\right)}},$$

and similarly for the other terms.

PROP. VII.

107. To find the complete variation of

$$\iiint V dx dy dz,$$

where

$$V = f\left(x, y, z, u, \frac{du}{dx}, \frac{du}{dy}, \frac{du}{dz}, \frac{d^2u}{dx^2}, \frac{d^2u}{dy^2}, \frac{d^2u}{dz^2}, \frac{d^2u}{dxdy}, \frac{d^2u}{dxdz}, \frac{d^2u}{dydz}\right).$$

The complete variation of a triple, like that of a double integral, is found by adding the partial variations which result from a change in each of the elements on which its value depends, namely:—1. The form of the limiting function (or functions, if there be two). 2. The form of the function u.

With regard to the former of these, it is evident from p. 225, that the variation resulting from this cause will be

$$\iint Vk\delta z dx dy,$$

if there be but one limiting function; and

$$\iint Vk\delta z dx dy - \iint Vk'\delta z dx dy,$$

if there be two. Here

$$k = \frac{\frac{d\phi}{dz}}{\sqrt{\left(\frac{d\phi^2}{dz^2}\right)}}, \qquad k' = \frac{\frac{d\psi}{dz}}{\sqrt{\left(\frac{d\psi^2}{dz^2}\right)}}.$$

It remains, then, to consider the variation arising from a change in the form of the function u. If, as before, we denote this partial variation by the symbol δ, we shall evidently have

$$\delta . \iiint V dx dy dz = \iiint \delta V dx dy dz.$$

Substituting for δV its value, as given in Prop. II., we have

$$\delta . \iiint V dx dy dz =$$

$$\iiint \left(U\delta u + V_x \frac{d\delta u}{dx} + V_y \frac{d\delta u}{dy} + V_z \frac{d\delta u}{dz} + V_{x^2} \frac{d^2\delta u}{dx^2} + V_{y^2} \frac{d^2\delta u}{dy^2} + V_{z^2} \frac{d^2\delta u}{dz^2} \right.$$
$$\left. + V_{xy} \frac{d^2\delta u}{dx dy} + V_{xz} \frac{d^2\delta u}{dx dz} + V_{yz} \frac{d^2\delta u}{dy dz} \right) dx dy dz. \qquad \text{(A)}$$

Reducing this expression, as in the case of the double integral, by the method of integration by parts, collecting the different terms, and using, for the sake of symmetry, the semi-sum of the two different expressions which are found for each of the terms

$$\iiint V_{xy} \frac{d^2\delta u}{dx dy} dx dy dz, \quad \iiint V_{xz} \frac{d^2\delta u}{dx dz} dx dy dz, \quad \iiint V_{yz} \frac{d^2\delta u}{dy dz} dx dy dz,$$

we obtain finally,

$$\delta . \iiint V dx dy dz$$

$$= \iint \left\{ \left(V_z - \frac{dV_{z^2}}{dz} - \tfrac{1}{2} \frac{dV_{yz}}{dy} - \tfrac{1}{2} \frac{dV_{xz}}{dx} \right) \delta u \right.$$
$$\left. + \quad \frac{d\delta u}{dz} + \tfrac{1}{2} V_{xz} \frac{d\delta u}{dx} + \tfrac{1}{2} V_{yz} \frac{d\delta u}{dy} \right\} dx dy$$
$$+ \iint \left\{ \left(V_y - \frac{dV_{y^2}}{dy} - \tfrac{1}{2} \frac{dV_{yz}}{dz} - \tfrac{1}{2} \frac{dV_{xy}}{dx} \right) \delta u \right.$$
$$\left. + V_{y^2} \frac{d\delta u}{dy} + \tfrac{1}{2} V_{xy} \frac{d\delta u}{dx} + \tfrac{1}{2} V_{yz} \frac{d\delta u}{dz} \right\} dx dz \qquad \text{(B)}$$
$$+ \iint \left\{ \left(V_x - \frac{dV_{x^2}}{dx} - \tfrac{1}{2} \frac{dV_{xy}}{dy} - \tfrac{1}{2} \frac{dV_{xz}}{dz} \right) \delta u \right.$$
$$\left. + V_{x^2} \frac{d\delta u}{dx} + \tfrac{1}{2} V_{xy} \frac{d\delta u}{dy} + \tfrac{1}{2} V_{xz} \frac{d\delta u}{dz} \right\} dy dz$$
$$+ \iiint \left(U - \frac{dV_x}{dx} - \frac{dV_y}{dy} - \frac{dV_z}{dz} \right.$$
$$\left. + \frac{d^2 V_{x^2}}{dx^2} + \frac{d^2 V_{y^2}}{dy^2} + \frac{d^2 V_{z^2}}{dz^2} + \frac{d^2 V_{xy}}{dx dy} + \frac{d^2 V_{xz}}{dx dz} + \frac{d^2 V_{yz}}{dy dz} \right) \delta u dx dy dz.$$

108. Two reductions must be made in this formula before it can be used, namely:—1. The reduction of the several double integrals which it contains, to others in which the integration shall be performed with regard to the *same* independent variables. 2. The reduction of the several variations,

$$\delta u, \frac{d\delta u}{dx}, \frac{d\delta u}{dy}, \frac{d\delta u}{dz}.$$

occurring under the sign of double integration, to others which shall be independent of one another.

The former of these objects is readily effected, as in the case of a double integral, by means of the formulæ of Art. 95. These formulæ, which are adapted to the present case by putting

$$P = \left(V_x - \frac{dV_{x^2}}{dx} - \frac{1}{2}\frac{dV_{xy}}{dy} - \frac{1}{2}\frac{dV_{xz}}{dz}\right)\delta u$$

$$+ V_{x^2}\frac{d\delta u}{dx} + \frac{1}{2}V_{xy}\frac{d\delta u}{dy} + \frac{1}{2}V_{xz}\frac{d\delta u}{dz};$$

$$Q = \left(V_y - \frac{dV_{y^2}}{dy} - \frac{1}{2}\frac{dV_{yz}}{dz} - \frac{1}{2}\frac{dV_{xy}}{dx}\right)\delta u$$

$$+ V_{y^2}\frac{d\delta u}{dy} + \frac{1}{2}V_{xy}\frac{d\delta u}{dx} + \frac{1}{2}V_{yz}\frac{d\delta u}{dz};$$

$$R = \left(V_z - \frac{dV_{z^2}}{dz} - \frac{1}{2}\frac{dV_{yz}}{dy} - \frac{1}{2}\frac{dV_{xz}}{dx}\right)\delta u$$

$$+ V_{z^2}\frac{d\delta u}{dz} + \frac{1}{2}V_{xz}\frac{d\delta u}{dx} + \frac{1}{2}V_{yz}\frac{d\delta u}{dy};$$

give

$$\delta\iiint V dxdydz = \iint\left(\Upsilon\delta u + \Phi\frac{d\delta u}{dx} + \Theta\frac{d\delta u}{dy} + \Psi\frac{d\delta u}{dz}\right)kdxdy$$

$$+ \iiint \Omega\delta u dxdydz, \qquad \text{(C)}$$

where

$$\Upsilon = V_z - \frac{dV_{z^2}}{dz} - \frac{1}{2}\frac{dV_{yz}}{dy} - \frac{1}{2}\frac{dV_{xz}}{dx}$$

$$- q\left(V_y - \frac{dV_{y^2}}{dy} - \frac{1}{2}\frac{dV_{yz}}{dz} - \frac{1}{2}\frac{dV_{xy}}{dx}\right)$$

$$- p\left(V_x - \frac{dV_{x^2}}{dx} - \frac{1}{2}\frac{dV_{xy}}{dy} - \frac{1}{2}\frac{dV_{xz}}{dz}\right)$$

$$\Phi = -pV_{x^2} - \tfrac{1}{2}qV_{xy} + \tfrac{1}{2}V_{xz},$$
$$\Theta = -qV_{y^2} - \tfrac{1}{2}pV_{xy} + \tfrac{1}{2}V_{yz},$$
$$\Psi = V_{z^2} - \tfrac{1}{2}pV_{xz} - \tfrac{1}{2}qV_{yz},$$
$$\Omega = U - \frac{dV_x}{dx} - \frac{dV_y}{dy} - \frac{dV_z}{dz} +$$
$$\frac{d^2V_{x^2}}{dx^2} + \frac{d^2V_{y^2}}{dy^2} + \frac{d^2V_{z^2}}{dz^2} + \frac{d^2V_{xy}}{dxdy} + \frac{d^2V_{xz}}{dxdz} + \frac{d^2V_{yz}}{dydz}.$$

$$p = \frac{dz}{dx}, \quad q = \frac{dz}{dy}, \quad k = \frac{\frac{d\phi}{dz}}{\sqrt{\left(\frac{d\phi^2}{dz^2}\right)}}.$$

109. It now remains only to reduce, as far as possible, the number of the variations,

$$\delta u, \frac{d\delta u}{dx}, \frac{d\delta u}{dy}, \frac{d\delta u}{dz},$$

which are evidently equivalent to

$$\delta u, \ \delta.\frac{du}{dx}, \quad \delta.\frac{du}{dy}, \quad \delta.\frac{du}{dz}.$$

Now since x, y, z are connected by the equation

$$\phi(x, y, z) = 0,$$

if we denote by $\left(\frac{d\delta u}{dx}\right)$, $\left(\frac{d\delta u}{dy}\right)$, the complete differential coefficients of δu with regard to x and y, after substituting for z its value in terms of those variables, we shall have

$$\left(\frac{d\delta u}{dx}\right) = \frac{d\delta u}{dx} + p\frac{d\delta u}{dz},$$

$$\left(\frac{d\delta u}{dy}\right) = \frac{d\delta u}{dy} + q\frac{d\delta u}{dz};$$

and, therefore,

$$\frac{d\delta u}{dx} = \left(\frac{d\delta u}{dx}\right) - p\frac{d\delta u}{dz},$$

$$\frac{d\delta u}{dy} = \left(\frac{d\delta u}{dy}\right) - q\frac{d\delta u}{dz}.$$

Substituting these values in the expression for

$$\delta . \iiint V dx dy dz,$$

and observing that

$$\iint \Phi \left(\frac{d\delta u}{dx}\right) k dx dy = - \iint \left(\frac{d\Phi}{dx}\right) k dx dy,^*$$

$$\iint \Theta \left(\frac{d\delta u}{dy}\right) k dx dy = - \iint \left(\frac{d\Theta}{dx}\right) k dx dy,$$

we find, ultimately,

$$\delta \iiint V dx dy dz = \iint \left\{ \Upsilon - \left(\frac{d\Phi}{dx}\right) - \left(\frac{d\Theta}{dy}\right) \right\} k \delta u dx dy$$

$$+ \iint (\Psi - p\Phi - q\Theta) \frac{d\delta u}{dz} k dx dy$$

$$+ \iiint \Omega \delta u dx dy dz.$$

This expression, as in the case of a single integral, consists of two distinct parts, viz.:—1. A number of terms depending on the change in the forms of u and its differential coefficients, not for all values of the independent variables, but for those values only which satisfy the equation

$$\phi(x, y, z) = 0.$$

2. A term depending on the change in the general form of the function u.

Adding to this the terms corresponding to the second limiting function, and

$$\iint V k \delta z dx dy - \iint V k' \delta z dx dy,$$

which result from a change in the form of the functions ϕ and ψ, and replacing Υ, Φ, Θ, Ψ, Ω, by their values, we obtain, finally, for the complete variation of a triple integral,

---

* The reader will find no difficulty in proving, by reasoning perfectly analogous to that of p. 231, that any quantity such as

$$\int \Phi \delta u dy,$$

which appears with but a single integral sign, must, in the species of integration with which we are at present concerned, vanish of itself. Hence

$$\iint \Phi \left(\frac{d\delta u}{dx}\right) dx dy = \int \Phi \delta u dy - \iint \left(\frac{d\Phi}{dx}\right) \delta u dx dy = - \iint \left(\frac{d\Phi}{dx}\right) \delta u dx dy.$$

$$\delta.\iiint V dxdydz = \iint V\delta z k dxdy - \iint V\delta z k' dxdy$$

$$+ \iint \Big(V_z - pV_x - qV_y - \frac{dV_{z^2}}{dz} + 2p\frac{dV_{x^2}}{dx} + 2q\frac{dV_{y^2}}{dy} + p^2\frac{dV_{x^2}}{dz} + q^2\frac{dV_{y^2}}{dz}$$

$$+ q\frac{dV_{xy}}{dx} + p\frac{dV_{xy}}{dy} + pq\frac{dV_{xy}}{dz} - \frac{dV_{xz}}{dx} - \frac{dV_{yz}}{dy}$$

$$+ V_{x^2}r + V_{xy}s + V_{y^2}t \Big) \delta u k dxdy \qquad \text{(D)}$$

$$+ \iint (V_{z^2} + p^2 V_{x^2} + q^2 V_{y^2} - pV_{xz} - qV_{yz} + pqV_{xy}) \frac{d\delta u}{dz} k dxdy$$

$$- \iint (V_z - \text{\&c.}) \delta u k' dxdy$$

$$- \iint (V_{z^2} + \text{\&c.}) \frac{d\delta u}{dz} k' dxdy$$

$$+ \iiint \Big(U - \frac{dV_x}{dx} - \frac{dV_y}{dy} - \frac{dV_z}{dz}$$

$$+ \frac{d^2V_{x^2}}{dx^2} + \frac{d^2V_{y^2}}{dy^2} + \frac{d^2V_{xy}}{dxdy} + \frac{d^2V_{xz}}{dxdz} + \frac{d^2V_{yz}}{dydz} \Big) \delta u dxdydz.$$

CHAPTER VII.

ON MAXIMA AND MINIMA OF FUNCTIONS OF TWO OR MORE INDEPENDENT VARIABLES.

110. The general method of investigation to be pursued in this problem is perfectly analogous, in its nature, to that already explained in the case of functions of one independent variable. Whatever be the number of independent variables in the function under consideration, the condition of the existence of a maximum or minimum value is the same, namely, that in any species of variation which the conditions of the question admit of, the linear part of the increment should vanish, and the quadratic part preserve the same sign, negative for a maximum, and positive for a minimum.

If then we denote, as before, by the symbol D, the most general increment of which a function U is susceptible, the existence of a maximum or minimum value of that function requires that the equation

$$DU = 0,$$

should be satisfied without any restriction (except those imposed by the conditions of the question) upon the increments which are assigned to its several varying elements. We shall now proceed to apply this theory successively to the cases of double and triple integrals.

In the following discussion we shall first suppose that there are *two* limiting functions, or, in other words, that the conditions of the question require the given integral to be extended to all systems of values of x and y which satisfy the two *inequalities*,

$$\psi(x, y) > 0, \quad \phi(x, y) < 0;$$

or, in the case of a triple integral,

$$\psi(x, y, z) > 0, \quad \phi(x, y, z) < 0.$$

We shall then examine the case of a *single* limiting function, i. e., the case in which the original integral is extended to all systems of values of x and y which satisfy the single inequality,

$$\phi(x, y) < 0, \text{ or } \phi(x, y, z) < 0.$$

Prop. I.

111. To determine the function z such as to render

$$\iint V dx dy$$

a maximum or minimum, where

$$V = f\left(x, y, z, \frac{dz}{dx}, \frac{dz}{dy}, \frac{d^2z}{dx^2}, \frac{d^2z}{dxdy}, \frac{d^2z}{dy^2}\right).$$

Referring to the expression given in the preceding chapter for the complete variation of a double integral, we find that the equation

$$DU = 0$$

is equivalent to

$$\int \left\{ V\delta y + \left(V_y - \frac{dy}{dx} V_x + 2 \frac{dy}{dx} \frac{dV_{x^2}}{dx} + \frac{dy^2}{dx^2} \frac{dV_{x^2}}{dy} - \frac{dV_{y^2}}{dy} + V_{x^2} \frac{d^2y}{dx^2} \right) \delta z \right.$$
$$\left. + \left(V_{x^2} \frac{dy^2}{dx^2} - V_{xy} \frac{dy}{dx} + V_{y^2} \right) \frac{d\delta z}{dy} \right\} k dx \qquad \text{(A)}$$
$$- \int \left\{ V\delta y + (V_y - \&c.)\, \delta z + \left(V_{x^2} \frac{dy^2}{dx^2} - \&c. \right) \frac{d\delta z}{dy} \right\} k' dx$$
$$+ \iint \left(Z - \frac{dV_x}{dx} - \frac{dV_y}{dy} + \frac{d^2V_{x^2}}{dx^2} + \frac{d^2V_{xy}}{dxdy} + \frac{d^2V_{y^2}}{dy^2} \right) \delta z\, dx dy = 0.$$

In seeking to determine the method of satisfying this equation, it must be remarked, in the first place, that, as in the corresponding case of single integrals, it consists of two classes of terms which are essentially distinct from each other, namely:—1. A number of partially integrated terms, whose value depends upon the forms assigned to the functions δy, δz, not in general, but only for those values of the variables x, y which satisfy the equation

$$\phi(x, y) = 0, \text{ or } \psi(x, y) = 0.$$

2. A number of terms wholly unintegrated, whose values cannot be rendered determinate without fixing the general form of the function δz.

It appears then, by the application of the principle laid down in the case of simple integrals, that the equation (A) can be satisfied only by equating separately to zero these two classes of terms. Assuming, for the sake of brevity,

$$\alpha = V_y - V_x \frac{dy}{dx} + 2\frac{dy}{dx}\frac{dV_{x^2}}{dx} + \&c.$$

$$\beta = V_{x^2}\frac{dy^2}{dx^2} - V_{xy}\frac{dy}{dx} + V_{y^2}$$

$$\Omega = Z - \frac{dV_x}{dx} - \frac{dV_y}{dy} + \&c.$$

we have, in the first place, the three equations,

$$\begin{aligned} &\int\left(V\delta y + \alpha\delta z + \beta\frac{d\delta z}{dy}\right)k'dx = 0, \\ &\int\left(V\delta y + \alpha\delta z + \beta\frac{d\delta z}{dy}\right)kdx = 0, \qquad \text{(B)} \\ &\iint \Omega\delta z dx dy = 0. \end{aligned}$$

Now, it will readily appear on referring to the case of simple integrals, that these equations cannot be satisfied without either limiting the generality of the variations δy, δz, or equating to zero the terms under the sign of integration.

The former of these methods being inadmissible, we have from the other the equations

$$\begin{aligned} &\left(V\delta y + \alpha\delta z + \beta\frac{d\delta z}{dy}\right)_0 = 0, \\ &\left(V\delta y + \alpha\delta z + \beta\frac{d\delta z}{dy}\right)_1 = 0, \qquad \text{(C)} \\ &\Omega = 0, \end{aligned}$$

the first belonging to the limiting function ψ, and the second to the limiting function ϕ. We shall suppose, for the sake of simplicity, that the equation

$$\Omega = 0,$$

which is evidently a partial differential equation of the fourth order, admits of a solution in which the quantity under the functional sign is the same for all the arbitrary functions which it contains.* The number of these functions will in this case be four. In order then to render this solution definite, it is necessary to have some mode of determining these functions, and we shall now proceed to show that the first two equations (C) furnish the means requisite for that purpose. This is to be proved, as in the case of single integrals, by an examination of the different cases which may result from the nature of the data in the particular problem to be solved. Previously, however, to entering upon this examination, we shall enunciate the following principle, which will be found of great use in the application of equation (A), and whose truth is an obvious consequence of the first principles of our science.

If in the general expression for the variation of a multiple integral there be found a term of the form

$$\iint \ldots . U\delta u dx_1\, dx_2 \ldots . dx_n,$$

and if the conditions of the problem, into which this integral enters, be such as to render u a determinate function of $x_1, x_2 \ldots . x_n$, *this term will vanish of itself; and if, in the same expression, there be found two integrals of the form*

$$\iint \ldots . U\delta u dx_1 dx_2 \ldots . dx_n,$$

and

$$\iint \ldots . U'\delta u' dx_1 dx_2 \ldots . dx_n,$$

and that the conditions of the problem into which it enters are such as to render u a determinate function of $x_1, x_2 \ldots . x_n$, *and* u', *these two terms may be reduced to one.*

The first part of this principle is self-evident. For, if u be a determinate function of $x_1, x_2 \ldots .$ it is plain from the meaning of the symbol δ that $\delta u = 0$. The truth of the second part is also apparent. For, if

$$u = f(x_1, x_2 \ldots . u'),$$

* The absence of a perfect analogy between the arbitrary functions which enter into the solution of a partial differential equation, and the arbitrary constants which are found in that of an ordinary differential equation, render this discussion much less satisfactory than that of Chapter III.—*Vide* note on p. 242.

it is plain that we must have (p. 8)

$$\delta u = \frac{df}{du'}\,\delta u',$$

and, therefore, that the two terms mentioned above are reducible to the single term

$$\iint \ldots \left(U\frac{df}{du'} + U'\right)\delta u' dx_1 dx_2 \ldots dx_n.$$

112. The truth of this principle being thus rendered apparent, we shall next proceed to apply it to the several cases which may occur in practice.

(1.) Let the limiting function ϕ be a given determinate function of x and y. In this case, y, as found from the equation

$$\phi(x, y) = 0$$

is a determinate function of x, hence $\delta y = 0$. The other variations,

$$\delta z,\ \frac{d\delta z}{dy},\ \text{ or }\ \delta z,\ \delta q,$$

remaining arbitrary, the first of equations (C) can only be satisfied by equating the coefficients of these variations to zero. Hence we must have at each limit of integration,

$$a = 0,\ \beta = 0.$$

In other words, the first of equations (C) is equivalent to the two equations,

$$a_0 = 0,\ \beta_0 = 0.$$

Similarly, from the other limiting function we have the equations

$$a_1 = 0,\ \beta_1 = 0.$$

These, with the equations

$$\phi(x, y) = 0,\quad \psi(x, y) = 0,$$

are sufficient to determine the four arbitrary functions which enter into the solution of the equation,

$$\Omega = 0.$$

The quantity under the functional sign being supposed the same in all these arbitrary functions, it is evident that the solution of the equation

$$\Omega = 0$$

will be

$$F\{x, y, z, \phi_1(v), \phi_2(v), \phi_3(v), \phi_4(v)\} = 0,$$

v being a determinate function of x, y, z, and which we have supposed to be the same in all the functions. Suppose, to fix our ideas, that

$$v = ax + by + cz.$$

We have then the equations

$$v = ax + by + cz,$$
$$\phi(x, y) = 0,$$
$$a_0 = 0,$$
$$F\{x, y, z, \phi_1(v), \phi_2(v), \phi_3(v), \phi_4(v)\} = 0,$$

Eliminating x, y, z between these equations, we have evidently a result of the form

$$F_1\{v, \phi_1(v), \phi_2(v), \phi_3(v), \phi_4(v)\} = 0. \tag{1}$$

Proceeding in the same way with the remaining equations,

$$a_1 = 0, \quad \beta_0 = 0, \quad \beta_1 = 0,$$

it is plain that the result will be three other equations similar to (1), which we shall denote by

$$F_2\{v, \phi_1(v), \phi_2(v), \phi_3(v), \phi_4(v)\} = 0, \tag{2}$$
$$F_3\{v, \phi_1(v), \phi_2(v), \phi_3(v), \phi_4(v)\} = 0, \tag{3}$$
$$F_4\{v, \phi_1(v), \phi_2(v), \phi_3(v), \phi_4(v)\} = 0. \tag{4}$$

These four equations, (1), (2), (3), (4), determine the four arbitrary functions, ϕ_1, ϕ_2, ϕ_3, ϕ_4, in terms of v. Substituting the values so found in the equation

$$F\{(x, y, z, \phi_1(v), \phi_2(v), \phi_3(v), \phi_4(v)\} = 0,$$

and replacing v by its value,

$$ax + by + cz,$$

we have an equation free from arbitrary functions, and of the form

$$f(x, y, z) = 0.$$

This equation furnishes a complete solution of the problem. If the quantity under the functional sign be not the same in all, the determination of the functions will, in general, present insurmountable difficulties.*

(2.) Suppose that the limiting values of z are also given in terms of x, or in other words, suppose that the limiting values of x, y, z are connected by the determinate equations

$$f(x, y, z) = 0, \quad f'(x, y, z) = 0. \tag{D}$$

The limiting values of z and y being both in this case determinate functions of x, it is plain that $\delta z = 0$, $\delta y = 0$, and that therefore the equations furnished by

$$\left(V\delta y + \alpha\delta z + \beta\frac{d\delta z}{dy}\right)_0 = 0$$

are reduced to one, namely,

$$\beta_0 = 0.$$

Taking, as before, the case in which the quantity under the functional sign is the same in all, and eliminating x, y, z between this equation and

$$v = ax + by + cz$$

$$f(x, y, z) = 0, \quad f'(x, y, z) = 0,$$

$$F\{x, y, z, \phi_1(v), \phi_2(v), \phi_3(v), \phi_4(v)\} = 0,$$

we shall have two equations of the form

$$\left.\begin{aligned} F_1\{v, \phi_1(v), \phi_2(v), \phi_3(v), \phi_4(v)\} &= 0, \\ F_2\{v, \phi_1(v), \phi_2(v), \phi_3(v), \phi_4(v)\} &= 0. \end{aligned}\right\} \tag{E}$$

Two others of the same kind being given by the equation

$$\left(V\delta y + \alpha\delta z + \beta\frac{d\delta z}{dy}\right)_1 = 0,$$

it is plain that the arbitrary functions contained in the general solution are completely determined as before.

* Vid. Lacroix, Traité de Calcul. Dif. et Int., tom. iii. pp. 228–238.

(3.) Similarly, if the limiting value of either* of the differential coefficients,

$$\frac{dz}{dx}, \frac{dz}{dy},$$

be given, it is evident that the equation

$$\left(V\delta y + \alpha\delta z + \beta\frac{d\delta z}{dy}\right)_0 = 0$$

will disappear of itself, inasmuch as

$$\delta y_0 = 0, \quad \delta z_0 = 0, \quad \left(\frac{d\delta z}{dy}\right)_0 = 0,$$

and that we shall have for the determination of the arbitrary functions, in addition to the equations (D), another equation, formed by differentiating the general solution with regard to either x or y, and substituting the given limiting value of the differential coefficient,

$$\frac{dz}{dx}, \text{ or } \frac{dz}{dy}.$$

The number of the ancillary equations remains, therefore, the same as before.

(4.) Let the limiting function $\phi(x, y)$ be an unknown function whose form is to be determined by the elimination of z between the solution of the equation

$$\Omega = 0,$$

and the determinate equation

$$z = f(x, y).$$

---

* If either of these quantities be given it is plain that the other is also given. For, since the equations

$$f(x, y, z) = 0, \quad f'(x, y, x) = 0.$$

are supposed to hold for all the limiting values of x and y, we have

$$\left\{\frac{df}{dx} + \frac{df}{dz}\frac{dz}{dx} + \frac{dy}{dx}\left(\frac{df}{dy} + \frac{df}{dz}\frac{dz}{dy}\right)\right\}_0 = 0,$$

$$\left\{\frac{df'}{dx} + \frac{df'}{dz}\frac{dz}{dx} + \frac{dy}{dx}\left(\frac{df'}{dy} + \frac{df'}{dz}\frac{dz}{dy}\right)\right\}_0 = 0.$$

Eliminating $\left(\frac{dy}{dx}\right)_0$, and substituting the given value of either $\left(\frac{dz}{dx}\right)_0$ or $\left(\frac{dz}{dy}\right)_0$, we have the value of the other.

No term of the equations (C) disappears in this case, but as the limiting values of x, y, and z are connected by a determinate equation, it appears, from the second part of the above-stated principle, that the terms containing their variations are consolidated into one. In fact, if we denote the solution of the equation

$$\Omega = 0 \text{ by } z = \chi(x, y),$$

we shall have

$$f(x, y) = \chi(x, y),$$

and therefore, taking the complete variation of each side of this equation,

$$\frac{df}{dy}\delta y = \frac{d\chi}{dy}\delta y + \delta\chi = \frac{d\chi}{dy}\delta y + \delta z;$$

or, denoting $\frac{df}{dy}, \frac{d\chi}{dy}$, by q', q,

$$q'\delta y = q\delta y + \delta z,$$

and, therefore,

$$\delta z = (q' - q)\delta y.$$

In this case, then, the equation

$$\left(V\delta y + \alpha\delta z + \beta\frac{d\delta z}{dy}\right)_0 = 0,$$

is equivalent to

$$V_0 + \alpha_0(q' - q) = 0, \quad \beta_0 = 0.$$

These, with the equation

$$z = f(x, y),$$

furnish two relations between the arbitrary functions which enter into the general solution. Two others are furnished in a similar manner by the conditions which are supposed to hold at the other limit.

And, in general, it is plain, as in the case of a single integral, that when any condition diminishes the number of equations furnished by (C), by annulling or consolidating any of its terms, it will, at the same time, supply a number of additional equations sufficient to make good the deficiency.

113. It appears, therefore, from the foregoing discussion, that in the case which we have been considering, that, namely, in which there are two limiting functions, the problem is in general

possible and determinate, the number of the ancillary equations being the same as that of the arbitrary functions which enter into the general solution. There remains, however, one important case to be considered, namely, the case of a *single* limiting function, or that in which the given integral is supposed to be extended to all values of x and y, which satisfy the condition

$$\phi(x, y) < 0.$$

The number of the equations by which the arbitrary functions are determined being, in this case, reduced by one-half, it may naturally be expected that the problem will, in general, be indeterminate. In certain cases, however, the following consideration will render it determinate as before.

It is an essential condition of the process of definite integration that the element of the integral, i. e. any one of the quantities of whose sum the given integral is the limit, should not become infinite within the limits of integration. Hence it is plain that after determining, as far as possible, the arbitrary functions which enter into the integral of the equation

$$\Omega = 0,$$

it will be necessary to reject from the result any solution which renders the element of the integral infinite, for any system of values of x and y which is found within the limits of integration. If this condition leaves any of the arbitrary functions still undetermined, we infer that the problem is in its nature indeterminate.

114. The above-stated theory is, as might naturally be expected, subject to exceptions, similar in their nature to those which are to be found in the application of the Calculus of Variations to integrals of the first order. Of these exceptions, the most remarkable is of a nature analogous to that stated in Art. 28, (1). The existence of that exception results, as will be seen upon referring to the Article quoted, from the fact that in certain cases the differential equation is of an order less than $2n$, and that therefore its solution does not contain a sufficient number of arbitrary constants to satisfy the conditions furnished by the terms which are found without the sign of integration.

Similarly, in the case of a double integral, if the partial differential equation,

$$\Omega = 0,$$

be of an order less than double that of the highest coefficient contained in V, its solution will not in general contain a sufficient number of arbitrary functions to satisfy the conditions furnished by the terms which appear with but a single sign of integration. The conditions necessary to the existence of such an exception are, therefore, to be investigated by considering in what case the terms of the highest order disappear from the equation

$$\Omega = 0.$$

Confining our attention for the present to the case in which V contains no differential coefficients of an order higher than the second, we shall have

$$\Omega = Z - \frac{dV_x}{dx} - \frac{dV_y}{dy} + \frac{d^2V_{x^2}}{dx^2} + \frac{d^2V_{xy}}{dxdy} + \frac{d^2V_{y^2}}{dy^2}.$$

In this expression it is plain that the differential coefficients of the highest, i. e. of the fourth order, are contained exclusively in the last three terms. Putting

$$p = \frac{dz}{dx}, \quad q = \frac{dz}{dy}, \quad r = \frac{d^2z}{dx^2}, \quad s = \frac{d^2z}{dxdy}, \quad t = \frac{d^2z}{dy^2},$$

and neglecting all terms except those of the highest order, we have

$$\left.\begin{aligned} \frac{d^2V_{x^2}}{dx^2} &= \frac{d^2V}{dr^2}\frac{d^4z}{dx^4} + \frac{d^2V}{drds}\frac{d^4z}{dx^3dy} + \frac{d^2V}{drdt}\frac{d^4z}{dx^2dy^2} \\ \frac{d^2V_{xy}}{dxdy} &= \frac{d^2V}{drds}\frac{d^4z}{dx^3dy} + \frac{d^2V}{ds^2}\frac{d^4z}{dx^2dy^2} + \frac{d^2V}{dsdt}\frac{d^4z}{dxdy^3} \\ \frac{d^2V_{y^2}}{dy^2} &= \frac{d^2V}{drdt}\frac{d^4z}{dx^2dy^2} + \frac{d^2V}{dsdt}\frac{d^4z}{dxdy^3} + \frac{d^2V}{dt^2}\frac{d^4z}{dy^4}. \end{aligned}\right\} \quad (F)$$

Now if the terms of the highest order disappear from Ω, it is evident that the coefficients of the several quantities,

$$\frac{d^4z}{dx^4}, \frac{d^4z}{dx^3dy}, \frac{d^4z}{dx^2dy^2}, \frac{d^4z}{dxdy^3}, \frac{d^4z}{dy^4},$$

must vanish of themselves. Adding, therefore, the equations (F), and equating the coefficients of these quantities separately to zero, we have the equations

$$\frac{d^2V}{dr^2}=0,\quad \frac{d^2V}{drds}=0,\quad \frac{d^2V}{ds^2}+2\frac{d^2V}{drdt}=0,\quad \frac{d^2V}{dsdt}=0,\quad \frac{d^2V}{dt^2}=0.$$

Integrating the first two of these equations, we find

$$V=rf(x, y, z, p, q, t)+F(x, y, z, p, q, s, t).$$

Similarly from the last two,

$$V=tf'(x, y, z, p, q, r)+F'(x, y, z, p, q, s, r).$$

Identifying these two expressions, it is plain that we must have

$$V=Art+Br+Dt+F+\phi(s, p, q, x, y, z), \tag{G}$$

where A, B, D, F are functions of x, y, z, p, q. Hence we have

$$\frac{d^2V}{drdt}=A,\quad \frac{d^2V}{ds^2}=\frac{d^2\phi}{ds^2}.$$

Substituting these values in the equation

$$\frac{d^2V}{ds^2}+2\frac{d^2V}{drdt}=0,$$

we have

$$\frac{d^2\phi}{ds^2}=-2A,$$

and therefore

$$\phi=-As^2+2Cs+C'.$$

Substituting this value in equation (G), we find, ultimately,

$$V=A(rt-s^2)+Br+2Cs+Dt+E. \tag{H}$$

If, therefore, V be of this form, the equation

$$\Omega=0$$

cannot be of an order higher than the third.

We shall now proceed to show that in this case the above-mentioned equation cannot rise above the second order, or, in other words, that if the terms of the fourth order disappear, those of the third order will disappear also.

We have, from equation (H),

$$V_{x^2} = At + B, \quad V_{xy} = -2As + 2C, \quad V_{y^2} = Ar + D,$$

$$V_x = (rt - s^2)\frac{dA}{dp} + r\frac{dB}{dp} + 2s\frac{dC}{dp} + t\frac{dD}{dp},$$

$$V_y = (rt - s^2)\frac{dA}{dq} + r\frac{dB}{dq} + 2s\frac{dC}{dq} + t\frac{dD}{dq},$$

neglecting the term E, which will evidently give no differential coefficients of an order higher than the second. Hence we find by differentiation,

$$\frac{dV_{x^2}}{dx} = A\frac{dt}{dx} + t\left(r\frac{dA}{dp} + s\frac{dA}{dq} + p\frac{dA}{dz} + \frac{dA}{dx}\right)$$
$$+ r\frac{dB}{dp} + s\frac{dB}{dq} + p\frac{dB}{dz} + \frac{dB}{dx}.$$

Similarly,

$$\frac{1}{2}\frac{dV_{xy}}{dy} = -A\frac{ds}{dy} - s\left(s\frac{dA}{dp} + t\frac{dA}{dq} + q\frac{dA}{dz} + \frac{dA}{dy}\right)$$
$$+ s\frac{dC}{dp} + t\frac{dC}{dq} + q\frac{dC}{dz} + \frac{dC}{dy}.$$

Hence, if we neglect terms of an order lower than the second,

$$\frac{dV_{x^2}}{dx} + \frac{1}{2}\frac{dV_{xy}}{dy} - V_x$$
$$= t\left(\frac{dC}{dq} - \frac{dD}{dp} + \frac{dA}{dx} + p\frac{dA}{dz}\right) + s\left(\frac{dB}{dq} - \frac{dC}{dp} - \frac{dA}{dy} - q\frac{dA}{dz}\right)$$
$$= mt - ns \text{ (suppose).}$$

In the same way we find

$$\frac{dV_{y^2}}{dy} + \frac{1}{2}\frac{dV_{xy}}{dx} - V_y$$
$$= r\left(\frac{dC}{dp} - \frac{dB}{dq} + \frac{dA}{dy} + q\frac{dA}{dz}\right) + s\left(\frac{dD}{dp} - \frac{dC}{dq} - \frac{dA}{dx} - p\frac{dA}{dz}\right)$$
$$= nr - ms.$$

But since

$$\Omega = Z + \frac{d}{dx}\left(\frac{dV_{x^2}}{dx} + \tfrac{1}{2}\frac{dV_{xy}}{dy} - V_x\right) + \frac{d}{dy}\left(\frac{dV_{y^2}}{dy} + \tfrac{1}{2}\frac{dV_{xy}}{dx} - V_y\right)$$

$$= Z + \frac{d}{dx}(mt - ns) + \frac{d}{dy}(nr - ms),$$

it is plain that if the differentiations be actually performed, all terms of the third order will disappear from **Ω**.

115. It is necessary to notice here a remarkable difference which exists between the present case and the corresponding exception, which was noticed in the case of single integrals. The existence of such an exception in the latter case was shewn to result from the fact that the integral to which the calculus of variations was applied had not been previously reduced to its lowest terms. In fact, if V be a linear function of the highest differential coefficient which it contains, we have seen (p. 47) that

$$\int V dx = \phi + \int \phi' dx,$$

in which ϕ' contains no differential coefficients of an order higher than $n-1$.

No such reduction, however, is practicable in the present case. For, if we assume

$$\iint V dx dy = \phi + \iint \phi' dx dy. \qquad \text{(I)}$$

ϕ being a quantity consisting of single integrals and referring solely to the limits of integration, and take the variation of both sides according to the foregoing rules, it is evident that the coefficients of δz under the sign of double integration must be the same in each. Now, if V be of the second order, ϕ' will be of the first, and, therefore, in its variation the coefficient of δz under the sign of double integration will be a *linear* function of r, s, t. Hence, it is plain that the assumption (I) will be impossible if Ω contain any powers of these quantities higher than the first.

But it is easily seen that if V be of the form (H), Ω will in general contain a term of the form

$$M(rt - s^2).$$

The reduction of the given integral to another of the first order is, therefore, in general impossible.

116. As an example of the foregoing theory, let it be required to determine the form of the function Z such as to render the double integral

$$\iint (px + qy - z)^m \, dxdy,$$

a maximum or minimum. Here we have

$$Z = - m(px + qy - z)^{m-1},$$
$$V_x = m(px + qy - z)^{m-1} x,$$
$$V_y = m(px + qy - z)^{m-1} y,$$
$$V_{x^2} = 0, \; V_{xy} = 0, \; V_{y^2} = 0,$$

$$\frac{dV_x}{dx} = m \, (px + qy - z)^{m-1} + m.\overline{m-1} \, (px + qy - z)^{m-2} \, (rx^2 + sxy),$$

$$\frac{dV_y}{dy} = m \, (px + qy - z)^{m-1} + m.\overline{m-1} \, (px + qy - z)^{m-2} \, (sxy + ty^2).$$

Substituting these values in the equation

$$\Omega = 0,$$

it becomes

$$rx^2 + 2sxy + ty^2 + \frac{3}{m-1} (px + qy - z) = 0. \qquad \text{(a)}$$

To integrate this equation, assume

$$u = px + qy - z.$$

Differentiating successively with regard to x and y, we find

$$\frac{du}{dx} = rx + sy, \quad \frac{du}{dy} = sx + ty \; ;$$

hence

$$rx^2 + 2sxy + ty^2 = x \frac{du}{dx} + y \frac{du}{dy}.$$

Substituting this value, as also that of $px + qy - z$ in equation (a), and putting n for $\frac{3}{1 - m}$, we find

$$x \frac{du}{dx} + y \frac{du}{dy} - nu = 0.$$

Integrating in the ordinary way this partial differential equation of the first order, we have

$$u = x^n F\left(\frac{y}{x}\right),$$

or, replacing u by its value, $px + qy - z$,

$$px + qy - z = x^n F\left(\frac{y}{x}\right).$$

This, which is also a partial differential equation of the first order, admits of being integrated by the ordinary method, which gives, therefore, as the complete integral of equation (a),

$$z = xf\left(\frac{y}{x}\right) + x^n F\left(\frac{y}{x}\right). \qquad \text{(b)}$$

We shall now proceed to determine the arbitrary functions in this solution, according to the various data of the problems which may occur.

(1.) Let the forms of the limiting function, ϕ, ψ, be given, e. g. let it be supposed that the limits of integration are so taken that the definite integral may represent the limit of the sum of the elements,

$$(px + qy - z)^m dxdy,$$

for all systems of values of x and y which render the signs of the functions

$$(x - a)^2 + (y - b)^2 - c^2,$$
$$(x - a)^2 + (y - b)^2 - c'^2,$$

different.

Here $\delta y = 0$, and the first of equations (C) becomes

$$V_y - V_x \frac{dy}{dx} = 0; \qquad \text{(c)}$$

or putting for V_y, V_x their values, and for $\frac{dy}{dx}$ its value derived from either of the equations

$$(x - a)^2 + (y - b)^2 - c^2 = 0,$$
$$(x - a)^2 + (y - b)^2 - c'^2 = 0,$$

we find the following equation, which must hold for each limit of integration,

$$m(px+qy-z)^{m-1}\left(y+\frac{x.\overline{x-a}}{y-b}\right)=0.$$

This equation may be satisfied either by making

$$px+qy-z=0,$$

or by making

$$y+\frac{x.\overline{x-a}}{y-b}=0.$$

With regard to the latter of these suppositions, it must be remembered that equation (c) is supposed to be true independently of x, and to require merely that y should have one of the two values,

$$b\pm\sqrt{\{c^2-(x-a)^2\}}.$$

This supposition, therefore, which would establish a *second* equation between y and x, is inadmissible. Neglecting this, then, let us consider the remaining equation,

$$px+qy-z=0.$$

Substituting for p and q their values derived from equation (b), we find that the equation

$$px+qy-z=0$$

is equivalent to

$$F\left(\frac{y}{x}\right)=0.$$

This equation of condition would reduce equation (b) to

$$z=xf\left(\frac{y}{x}\right),$$

which would give, generally,

$$px+qy-z=0,$$

and therefore

$$V=0.$$

And it will readily be seen that this conclusion does not depend upon the particular form of the limiting function which we have selected, but merely upon the fact that it is a function of x and y

only. It appears, therefore, that if the form of the limiting function be given, the limiting values of z remaining indeterminate, the given definite integral does not admit of a finite maximum or minimum.

(2.) Let the limiting values of z also be given, e. g. let the limits of integration be supposed to be so taken, that the definite integral may represent the sum of the elements for all values of x and y which render the signs of the functions $x^2 + y^2 - a^2$ and $x^2 + y^2 - a'^2$ different, and let it be supposed that, for all values of x and y which cause the first of these functions to vanish, z has the constant value c, and that, for all values which cause the second to vanish, it has the constant value c'. Assuming $t = \frac{y}{x}$, we have for the first system the equations

$$z = xf(t) + x^n F(t)$$
$$z = c, \quad y = tx,$$
$$x^2 + y^2 = a^2.$$

Eliminating x, y, z between these equations, we find

$$c = \frac{a}{\sqrt{(1+t^2)}} f(t) + \left(\frac{a}{\sqrt{(1+t^2)}}\right)^n F(t).$$

Similarly from the equations

$$z = xf(t) + x^n F(t), \quad y = tx,$$
$$z = c', \quad x^2 + y^2 = a'^2,$$

we find

$$c' = \frac{a'}{\sqrt{(1+t^2)}} f(t) + \left(\frac{a'}{\sqrt{(1+t^2)}}\right)^n F(t).$$

Solving these equations for $f(t)$ and $F(t)$, we have

$$f(t) = \frac{ca'^n - c'a^n}{aa'^n - a'a^n} \sqrt{(1+t^2)},$$

$$F(t) = \frac{c'a - ca'}{aa'^n - a'a^n} (1+t^2)^{\frac{n}{2}}.$$

Substituting these values in the equation

$$z = xf(t) + x^n F(t),$$

and replacing t by its value, we have, finally,

$$(aa'^n - a'a^n)\, z = (ca'^n - c'a^n)\,\sqrt{(y^2 + x^2)} + (c'a - ca')\,(y^2 + x^2)^{\frac{n}{2}}. \quad \text{(a)}$$

We shall next proceed to consider the case in which there is but one limiting function. Suppose, for example, that it were required by the conditions of the problem that the given integral should be extended to all systems of values of x and y, which render the function

$$x^2 + y^2 - a^2$$

negative.

Writing for n its value $\dfrac{3}{1-m}$, the general solution (b) will be

$$z = xf\left(\frac{y}{x}\right) + x^{\frac{3}{1-m}} F\left(\frac{y}{x}\right).$$

Forming from this equation the value of

$$V = (px + qy - z)^m,$$

we find

$$V = \left(\frac{2+m}{1-m}\right)^m x^{\frac{3m}{1-m}} \left\{ F\left(\frac{y}{x}\right)\right\}^m$$

Now, according to the principle of p. 248, every solution must be rejected which would render this value infinite for any values of y and x within the limits of integration. But if $\dfrac{3m}{1-m}$ be negative, i. e., if m be either between $-\infty$ and 0, or between 1 and $+\infty$, it is plain that unless $F\left(\dfrac{y}{x}\right)$ be zero for all values of $\dfrac{y}{x}$, we shall have $V = \infty$ when $y = 0$, $x = 0$. In this case, therefore, the part of the solution involving $F\left(\dfrac{y}{x}\right)$ being neglected, equation (b) becomes

$$z = xf\left(\frac{y}{x}\right);$$

and it is easy to see that the condition, that when $x^2 + y^2 - a^2 = 0$, z shall $= c$, reduces this to

$$az = c\sqrt{(y^2 + x^2)}.$$

If, on the other hand, $\frac{3m}{1-m}$ be positive, i. e. if m lie between 0 and 1, the quantity $x^{\frac{3m}{1-m}}$ will not become infinite for any value of x within the limits of integration, and we may, therefore, assume for $F\left(\frac{y}{x}\right)$ any function which does not become infinite within those limits. Such a value being assumed for $F\left(\frac{y}{x}\right)$, the other arbitrary function is determined, as before, by the condition that when $z = c$, $x^2 + y^2 = a^2$. This would give

$$z = \frac{\sqrt{(y^2 + x^2)}}{a}\left\{c - \frac{a^n x^n}{(y^2 + x^2)^{\frac{n}{2}}} F\left(\frac{y}{x}\right)\right\} + x^n F\left(\frac{y}{x}\right), \qquad \text{(e)}$$

putting n for $\frac{3}{1-m}$. This result is the only one which furnishes a real finite solution.

(3.) As V does not contain any differential coefficients of an order higher than the first, the term involving δq will vanish of itself from the first of equations (C). Case 3 is therefore here inapplicable.

(4.) Let it be supposed that the forms of the limiting function or functions are not directly given, but to be determined by the elimination of z between the solution of equation (a) and the equations

$$\begin{aligned} z &= ax + by + c, \\ z &= a'x + b'y + c'. \end{aligned} \qquad \text{(f)}$$

The equation

$$\{V + a(q' - q)\}_0 = 0,$$

or

$$\{V + V_x(p' - p) + V_y(q' - q)\}_0 = 0,$$

of p. 247, becomes, in this case, by the substitution of the values of V, V_x, V_y,

$$(px + qy - z)^m + m(px + qy - z)^{m-1}\{(p' - p)x + (q' - q)y\} = 0;$$

or, neglecting the factor $(px + qy - z)^{m-1}$,

$$(1 - m)(px + qy - z) + m(p'x + q'y - z) = 0.$$

Substituting for p, q their values derived from the general solution (b), and for p', q' their values derived from the first of equations (f), this becomes

$$(2+m)x^n F\left(\frac{y}{x}\right) = mc.$$

Making, as before, $\frac{y}{x} = t$, we have

$$z = xf(t) + x^n F(t),$$
$$z = (a+bt)\,x + c,$$
$$(m+2)\,x^n F(t) = mc.$$

Putting, for the sake of brevity,

$$u = f(t), \quad v = F(t),$$

and eliminating z and vx^n, we find

$$x = \frac{2c}{m+2}\,\frac{1}{u-a-bt}.$$

Substituting this for x in the third of the above equations, it becomes

$$(u-a-bt)^n = \frac{2^n}{m}\,(m+2)^{1-n}\,c^{n-1}\,v.$$

Similarly, from the second of equations (f),

$$(u-a'-b't)^n = \frac{2^n}{m}\,(m+2)^{1-n}\,c'^{n-1}\,v.$$

From these equations we find readily the values of u and v, namely,

$$u = \frac{c'^{\frac{n-1}{n}}(a+bt) - c^{\frac{n-1}{n}}(a'+b't)}{c'^{\frac{n-1}{n}} - c^{\frac{n-1}{n}}},$$

$$v = m\,(m+2)^{n-1}\left\{\frac{(a+bt)-(a'+b't)}{2\left(c'^{\frac{n-1}{n}} - c^{\frac{n-1}{n}}\right)}\right\}^n;$$

and hence

$$z = \frac{c'^{\frac{n-1}{n}}(ax+by) - c^{\frac{n-1}{n}}(a'x+b'y)}{c'^{\frac{n-1}{n}} - c^{\frac{n-1}{n}}},$$

$$+\,m\,(m+2)^{n-1}\left\{\frac{(ax+by)-(a'x+b'y)}{2\left(c'^{\frac{n-1}{n}} - c^{\frac{n-1}{n}}\right)}\right\}^n.$$

It is not difficult to see that this solution may, by transformation of co-ordinates, be put under the form

$$(x+y)^n = A^{n-1} z.$$

The general solution (b) fails when $n = -1$, as it will then contain but one arbitrary function. But by integrating, with special regard to this case, it is easy to see that the real solution is

$$z = xf\left(\frac{y}{x}\right) + xlxF\left(\frac{y}{x}\right).$$

This case corresponds to

$$V = (px + qy - z)^{-2},$$

and the reader will find no difficulty in applying to it, according to the conditions of the problem, the mode of determining the arbitrary function given above. If the limiting functions be

$$x^2 + y^2 - a^2, \quad x^2 + y^2 - a'^2,$$

and the limiting values of z, c, and c', as in case 2, the result will be of the form

$$Az = B\sqrt{(y^2 + x^2)}\, l\left(\frac{y^2 + x^2}{C^2}\right).$$

117. As a more general problem of the same kind, we may suppose that it were required to determine the function z of such a form as to render

$$\iint f(px + qy - z)\, dxdy$$

a maximum or minimum, f denoting any function whatever.

In this case, putting $u = px + qy - z$, we have

$$Z = -f'(u), \quad V_x = xf'u, \quad V_y = yf'(u).$$

The equation

$$\Omega = 0, \text{ or } Z - \frac{dV_x}{dx} - \frac{dV_y}{dy} = 0,$$

becomes, therefore,

$$x\frac{du}{dx} + y\frac{du}{dy} + 3\frac{f'(u)}{f''(u)} = 0.$$

Integrating this equation in the ordinary way, we find

$$x^3 f'(u) = \phi\left(\frac{y}{x}\right), \text{ or } u = f'^{-1}\left\{\frac{1}{x^3}\cdot\phi\left(\frac{y}{x}\right)\right\};$$

or finally, replacing u by its value,

$$px + qy - z = f'^{-1}\left\{\frac{1}{x^3}\phi\left(\frac{y}{x}\right)\right\}.$$

This equation, which admits of being integrated as a partial differential equation of the first order, gives

$$z = x\psi\frac{y}{x} + x\int f'^{-1}\left(\frac{a}{x^3}\right)\frac{dx}{x^2}, \tag{a}$$

in which, after the integration with regard to x is performed, a must be eliminated by the equation

$$a = \phi\left(\frac{y}{x}\right).$$

Thus, for example, if the given integral were

$$\iint l\,(px + qy - z)dxdy,$$

it is easy to see that the solution would be

$$z = x^3\phi\left(\frac{y}{x}\right) + x\psi\left(\frac{y}{x}\right). \tag{b}$$

This result is not contained in the general solution of the problem given in p. 254. For, in order to make equation (b) identical with the general solution,

$$z = x\psi\left(\frac{y}{x}\right) + x^n\,\phi\left(\frac{y}{x}\right),$$

it would be necessary to have $n = 3$, which would give

$$m = \frac{n+3}{n} = 0,$$

a value obviously inadmissible.

118. As another example, let μ be any function of x, y, z, and let it be required to determine z such as to render

$$\iint \mu\,(px + qy - z)\,dxdy$$

a maximum or minimum. Here

$$Z = (px + qy - z)\frac{d\mu}{dz} - \mu,$$

$$V_x = \mu x, \quad V_y = \mu y.$$

The equation

$$\Omega = 0$$

becomes, therefore,

$$x\frac{d\mu}{dx} + y\frac{d\mu}{dy} + z\frac{d\mu}{dz} = -3\mu.$$

If the given function μ were a homogeneous function of the nth order, this equation would be equivalent to

$$(n + 3)\mu = 0,$$

and could therefore be satisfied only by making $\mu = 0$, unless when $n = -3$, in which case the equation would become identical.*

119. Again, let it be required to assign such a form to the function z as will render

$$\iint \sqrt{(p^2 + q^2)}\, dxdy$$

a maximum or minimum. Here

$$V_x = \frac{p}{\sqrt{(p^2 + q^2)}}, \quad V_y = \frac{q}{\sqrt{(p^2 + q^2)}},$$

and the equation

$$\Omega = 0$$

becomes

$$q^2 r - 2pqs + p^2 t = 0. \qquad \text{(a)}$$

The integral of this equation is, as is well known,

$$y = xfz + Fz. \qquad \text{(b)}$$

(1.) Suppose that the forms of the limiting functions, ϕ, ψ, are given, e. g. let it be supposed that the integral is extended to all systems of values of y and x which give different signs to the functions

$$x^2 + y^2 - a^2 \text{ and } x^2 + y^2 - a'^2.$$

---

* For an explanation of this case, *vid.* Chapter X.

Since $\delta y = 0$, the first equation (C) is reduced to

$$V_y - V_x \frac{dy}{dx} = 0;$$

or substituting for V_x, V_y, and $\frac{dy}{dx}$, their values,

$$\frac{p}{\sqrt{(p^2+q^2)}}, \quad \frac{q}{\sqrt{(p^2+q^2)}}, \text{ and } -\frac{x}{y},$$

$$px + qy = 0.$$

Substituting for p and q their values derived from the equation

$$y = xf(z) + F(z),$$

this condition becomes equivalent to

$$Fz = 0.$$

The general solution is therefore in this case reduced to

$$y = xfz, \qquad \text{(c)}$$

the function f remaining indeterminate, inasmuch as it is plain that the condition,

$$V_y - V_x \frac{dy}{dx} = 0,$$

is satisfied also at the second limit of integration. Here, therefore, the conditions furnished by the nature of the problem are not independent. But if the second function were of a form differing from that of the first, the equation would again become determinate. Thus, for example, if the second limiting function were

$$\left(\frac{x}{a}\right)^2 + \left(\frac{y}{b}\right)^2 - 1,$$

it would be easy to show, by putting the general solution (c) under the form

$$z = F_1\left(\frac{y}{x}\right),$$

to which it is reducible by the condition $F = 0$, that the conditions of the question are only satisfied by the equation

$$z = c.$$

As, however, this equation renders $V = 0$, it does not give a finite maximum or minimum.

A similar conclusion would be arrived at if we supposed the integral to be extended to *all* values of x and y which render

$$\phi_1 (= x^2 + y^2 - a^2)$$

negative. For if we derive from the equation

$$z = f_1 \left(\frac{y}{x}\right),$$

to which, as before, the general solution is reducible, the values of p and q, we shall have

$$\sqrt{(p^2 + q^2)} = \frac{f'_1}{x} \sqrt{\left(1 + \frac{y^2}{x^2}\right)}.$$

It appears, then, by the reasoning of pp. 248, 258, that $f'_1 = 0$, and therefore that

$$f_1 = z = c.$$

(2.) Next suppose that the limiting values of z are also given, e. g. let it be granted that when

$$x^2 + y^2 - a^2 = 0, \quad z = m\frac{y}{x},$$

and that when

$$x^2 + y^2 - a'^2 = 0, \quad z = m'\frac{y}{x}.$$

Eliminating y and x from the general solution,

$$y = xf(z) + Fz,$$

by means of the first two of the above equations we have the relations,

$$\frac{az}{\sqrt{(m^2 + z^2)}} = \frac{am}{\sqrt{(m^2 + z^2)}} f(z) + F(z).$$

Similarly from the second two,

$$\frac{a'z}{\sqrt{(m'^2 + z^2)}} = \frac{a'm'}{\sqrt{(m'^2 + z^2)}} f(z) + F(z).$$

Solving these equations for $\phi(z)$ and $\psi(z)$, we find

$$f(z) = \frac{a\sqrt{(m'^2+z^2)} - a'\sqrt{(m^2+z^2)}}{am\sqrt{(m'^2+z^2)} - a'm'\sqrt{(m^2+z^2)}}\,z,$$

$$F(z) = \frac{aa'(m-m')\,z}{am\sqrt{(m'^2+z^2)} - a'm'\sqrt{(m^2+z^2)}}.$$

Hence the complete solution of this case is given by the equation

$$\{am\sqrt{(m'^2+z^2)} - a'm'\sqrt{(m^2+z^2)}\}\,y$$
$$= \{a\sqrt{(m'^2+z^2)} - a'\sqrt{(m^2+z^2)}\}\,zx + aa'(m-m')\,z.$$

120. Problems of relative maxima and minima are to be resolved on precisely the same principles as those which have been already applied to expressions involving single integrals. Thus if it were required to assign such a form to the function z as to render one integral,

$$\iint Vdxdy,$$

a maximum or minimum, and another,

$$\iint V'dxdy,$$

equal to a given constant, we should obtain the solution by assigning such a form to the function as to render

$$\iint Vdxdy + m\iint V'dxdy$$

a maximum or minimum, determining, as before, the arbitrary constant, m, by the given value of the second of the above integrals. The remainder of the investigation is precisely similar to the case of absolute maxima or minima.

The principle upon which it is shown that the indeterminate coefficient, by which the second integral is multiplied, must be a *constant*, is also perfectly analogous to that of Art. 61. It results in every case from the fact that, as the equation which is to be satisfied is of the form

$$\iint \ldots V'dxdy \ldots = c,$$

the indeterminate coefficient by which this equation is multiplied is found *without* the sign of integration. For, since the final result of the process of definite integration is a constant quantity, i. e. one which does not involve any of the variables, x, y, . . .

which are found in the function to be integrated, it is plain that an equation of the form

$$\iint \dots W dx dy \dots \quad + \lambda \iint \dots W' dx dy \dots = 0.$$

cannot be satisfied by any other than a constant value of λ.

The same thing will also appear by considering the definite integral as the *sum* of a number of elements. The equation of condition will then be

$$V'_1 dx_1 dy_1 \dots + V'_2 dx_2 dy_2 \dots + V'_3 dx_3 dy_3 \dots + \text{\&c.} = \text{const.}$$

V'_1, V'_2, &c., being the values of V' corresponding to the several systems of values,

$$x_1, y_1, \dots x_2, y_2, \dots \text{\&c.}$$

Multiplying this equation by an indeterminate coefficient, λ, we shall have

$$\lambda (V'_1 dx_1 dy_1 \dots + V'_2 dx_2 dy_2 \dots + V'_3 dx_3 dy_3 \dots + \text{\&c.} - c) = 0.$$

Hence it is plain that the factor λ, which is evidently identical with m in

$$\iint \dots V dx dy \dots + m \iint \dots V' dx dy,$$

is the same for all the quantities,

$$V'_1, V'_2, V'_3, \text{\&c.},$$

i. e. for all systems of values of $x, y, \dots$ It is therefore constant.

121. Another species of relative maxima and minima is that in which the integral or integrals which are to have a given value relate only to those values of the independent variable which satisfy the limiting equation,

$$\phi (x, y) = 0,$$

and are therefore of an order which is necessarily inferior by one to that of the integral which is to be made a maximum or minimum.

There is no essential difference between the mode of treating this and any other case of relative maxima and minima. The variation of the integral whose value is given is to be multiplied by an indeterminate constant, and added to the terms outside the highest sign of integration in the variation of the original integral. The problem is then to be treated as in the ordinary case, and the arbitrary constant determined by means of the integral whose value is given.

122. As an example of relative maxima and minima, let it be required to find, among all the functions which render the definite integral,

$$\int \sqrt{(p^2 + q^2)}\, dx,$$

equal to a given quantity, that one which shall render

$$\int (z - px - qy)\, dx$$

a maximum or minimum. Here

$$V = z - px - qy + n\sqrt{(p^2 + q^2)},$$

and, therefore,

$$Z = 1, \quad V_x = -x + \frac{np}{\sqrt{(p^2 + q^2)}}, \quad V_y = -y + \frac{nq}{\sqrt{(p^2 + q^2)}}.$$

Deriving from these the values of $\frac{dP}{dx}$ and $\frac{dQ}{dy}$, substituting them in the equation

$$Z - \frac{dV_x}{dx} - \frac{dV_y}{dy} = 0,$$

and putting, for the sake of brevity, $m = -\frac{3}{n}$, we have

$$q^2 r - 2pqs + p^2 t = m(p^2 + q^2)^{\frac{3}{2}}. \qquad \text{(a)}$$

This equation, which is easily integrable by the ordinary method, gives

$$\{x + \psi_1(z)^2\} + \{y + \psi_2(z)^2\} = \frac{1}{m^2}. \qquad \text{(b)}$$

Among the different cases which may present themselves for the determination of the arbitrary function, we shall select that in which there are supposed to be two limiting functions determined by the elimination of z between equation (b) and the two equations

$$\left.\begin{array}{c} x^2 + y^2 + z^2 = A^2, \\ (x + z)^2 + (y + z)^2 = B^2, \end{array}\right\} \qquad \text{(c)}$$

respectively. Substituting the values of V, V_x, V_y in the equation

$$V + V_x(p' - p) + V_y(q' - q) = 0,$$

it becomes

$$z - p'x - q'y + \frac{n(pp' + qq')}{\sqrt{(p^2 + q^2)}} = 0. \qquad \text{(d)}$$

But from the first of equations (c),

$$p' = -\frac{x}{z}, \quad q' = -\frac{y}{z};$$

hence

$$z - p'x - q'y = \frac{A^2}{z}.$$

Again, equation (b) gives

$$p = -\frac{x + \psi_1(z)}{\{x + \psi_1(z)\}\psi'_1(z) + \{y + \psi_2(z)\}\psi'_2(z)},$$

$$q = -\frac{y + \psi_2(z)}{\{x + \psi_1(z)\}\psi'_1(z) + \{y + \psi_2(z)\}\psi'_2(z)}.$$

Making these substitutions in equation (d), it becomes

$$A^2 = 3\{x + \psi_1(z)\}x + 3\{y + \psi_2(z)\}y,$$

or

$$x\psi_1(z) + y\psi_2(z) = z^2 - \tfrac{2}{3}A^2.$$

From this and the equation

$$\{x + \psi_1(z)\}^2 + \{y + \psi_2(z)\}^2 = \frac{1}{m^2},$$

it is easy to see that we have between ψ_1 and ψ_2 the relation

$$\{\psi_1(z)\}^2 + \{\psi_2(z)\}^2 = \frac{1}{m^2} + \tfrac{1}{3}A^2 - z^2.$$

Adopting the same method with the second of equations (c), and supposing, for the sake of simplicity, that $A = B$, we find

$$\psi_1(z) + \psi_2(z) = \tfrac{1}{2}z.$$

Determining from these equations the values of $\psi_1(z)$ and $\psi_2(z)$, and substituting them in equation (b), we find, ultimately,

$$\{4x + z + \sqrt{(m'^2 - 9z^2)}\}^2 + \{4y + z - \sqrt{(m'^2 - 9z^2)}\}^2 = \frac{16}{m^2},$$

where

$$m'^2 = \frac{8}{m^2} + \frac{8A^2}{3}.$$

The constant m is of course determined by means of the given value of the integral

$$\iint \sqrt{(p^2 + q^2)}\, dxdy.$$

PROP. II.

123. If z be a function of x and y, satisfying the equation

$$D.\iint V dxdy = 0,$$

to determine what other conditions are necessary, in order that the corresponding value of the integral may be a maximum or minimum.

The general theory of maxima and minima requires, as in the case of a single integral, that the second variation should preserve the same sign, whatever form (consistent with the conditions of the question) may be assigned to the variation δz. Distinguishing, as in Prop. V., Chap. III., between the problem which concerns the variation of the limiting values of z and its differential coefficients, and that which concerns the variation in the general form of the function, and confining our attention to the latter question, we shall suppose the limiting values of

$$y,\ z,\ \text{and}\ \frac{dz}{dx},\ \text{or}\ \frac{dz}{dy},$$

to be given. This condition will annul any terms which are found outside the sign of double integration. Now the second variation of the given integral will evidently be

$$\begin{aligned}
\delta^2 \iint V dxdy = \iint \Big(& \frac{d^2V}{dr^2}\delta r^2 + \frac{d^2V}{ds^2}\delta s^2 + \frac{d^2V}{dt^2}\delta t^2 + 2\frac{d^2V}{drds}\delta r\delta s \\
& + 2\frac{d^2V}{dsdt}\delta s\delta t + 2\frac{d^2V}{drdt}\delta r\delta t + 2\frac{d^2V}{drdp}\delta r\delta p + 2\frac{d^2V}{drdq}\delta r\delta q \\
& + 2\frac{d^2V}{drdz}\delta r\delta z + 2\frac{d^2V}{dsdp}\delta s\delta p + 2\frac{d^2V}{dsdq}\delta s\delta q + 2\frac{d^2V}{dsdz}\delta s\delta z \qquad \text{(A)} \\
& + 2\frac{d^2V}{dtdp}\delta t\delta p + 2\frac{d^2V}{dtdq}\delta t\delta q + 2\frac{d^2V}{dtdz}\delta t\delta z + \frac{d^2V}{dp^2}\delta p^2 \\
& + \frac{d^2V}{dq^2}\delta q^2 + 2\frac{d^2V}{dpdq}\delta p\delta q + 2\frac{d^2V}{dpdz}\delta p\delta z + 2\frac{d^2V}{dqdz}\delta q\delta z \\
& + \frac{d^2V}{dz^2}\delta z^2 \Big) dxdy.
\end{aligned}$$

The general principle upon which this method is founded is, that in consequence of the indeterminate nature of the variations which are found in (A), such values may be assigned to these quantities (by giving a proper form to δz) as to render (A) either positive or negative, unless one of the two following conditions be fulfilled, namely:—1. That the sign of the element or quantity under the integral sign in (A) be independent of the variations δr, δs, &c. 2. That the integral (A) be capable of being reduced to another, in which the sign of the element shall be independent of these variations. In the first case the element of the integral in (A), considered as a quadratic function of six independent variables, must preserve the same sign, independently of the particular values of the variables which it contains. It must, therefore, be capable of being resolved into a function of six squares of the form

$$\begin{aligned} A(\delta z + a'\delta p + a''\delta q + \&c.)^2 + B(\delta p + \beta'\delta q + \&c.)^2 \\ + \&c. \\ + I\delta t^2, \end{aligned} \tag{B}$$

in which the coefficients $A, B, \ldots I$ have all the same sign.*

In the second case, in which the integral (A) is reduced to another which satisfies the first condition, it is plain that the element of the second integral will differ from that of the first by a quantity which can be made to vanish in the process of definite integration, i. e. by a quantity which is in itself *integrable.*

We infer, therefore, that it is necessary to the fulfilment of this condition, that the quantity under the integral sign in (A) should be capable of being resolved into three others, one of which is integrable with regard to x, another integrable with regard to y, and the third such as to satisfy the first condition.

The application of this principle to the case before us leading to formulæ of great length, we shall content ourselves with giving the principal steps and results of this method.

---

* Théorie des Fonctions Analytiques, pp. 267–269. The nature of the equations derived from this condition, and the mode of deducing them, will be found in the Article quoted. The present discussion being rather curious than useful, I have not thought it necessary to enumerate them.

(1.) Assume, in accordance with the principle stated above,

$$\frac{d^2 V}{dr^2}\delta r^2 + \frac{d^2 V}{ds^2}\delta s^2 + \frac{d^2 V}{dt^2}\delta t^2 + \&c. = \frac{dP}{dx} + \frac{dQ}{dy} + K. \qquad (C)$$

Then it readily appears that P, Q must be of the forms

$$P = \alpha\delta q^2 + \beta\delta q\delta p + \gamma\delta q\delta z + \varepsilon\delta p^2 + \zeta\delta p\delta z + \eta\delta z^2,$$
$$Q = \alpha'\delta q^2 + \beta'\delta q\delta p + \gamma'\delta q\delta z + \varepsilon'\delta p^2 + \zeta'\delta p\delta z + \eta'\delta z^2.$$

(2.) Differentiating the first of these with regard to x, the second with regard to y, and substituting in (C), we find

$$K = \frac{d^2 V}{dr^2}\delta r^2 + \frac{d^2 V}{ds^2}\delta s^2 + \frac{d^2 V}{dt^2}\delta t^2 + 2\frac{d^2 V}{drds}\delta r\delta s$$
$$+ 2\frac{d^2 V}{dsdt}\delta s\delta t + 2\frac{d^2 V}{drdt}\delta r\delta t + \lambda\delta r\delta p + \mu\delta r\delta q$$
$$+ \&c.$$

in which it will be remarked that the coefficients of the squares and products of the variations of the highest order are the same as in (A).

(3.) As the sign of K is independent of the variations δr, δs, &c., which it contains, it must be capable of being resolved into a linear function of six squares similar to (B). This will give five conditions, two of which are evidently independent of the indeterminate quantities, α, β, &c., α', β', &c., and express the fact that the sign of the polynomial,

$$\frac{d^2 V}{dr^2}\delta r^2 + \frac{d^2 V}{ds^2}\delta s^2 + \frac{d^2 V}{dt^2}\delta t^2$$
$$+ 2\frac{d^2 V}{drds}\delta r\delta s + 2\frac{d^2 V}{dsdt}\delta s\delta t + 2\frac{d^2 V}{drdt}\delta r\delta t, \qquad (D)$$

is independent of the variations δr, δs, δt.

(4.) If this condition be fulfilled, it appears from (3) that there will remain three others to be satisfied by means of the indeterminate quantities, α, β, &c., α', β', &c. If it be possible to assign values to α, β, &c., α', β', &c., such as to satisfy these three conditions, without rendering K infinite within the limits of integration, the value of the given integral will be a maximum or minimum, according as the sign of the polynomial (D) is negative or positive. The same method may be extended to cases in which

V contains differential coefficients of any order. But it is so rarely possible to integrate the equation

$$\Omega = 0,$$

of Prop. I., that it is, for any practical purpose, useless to pursue the subject further.*

PROP. III.

124. Let V be a function of the independent variables x, y, z, and of the function u and its differential coefficients, as far as the second order inclusively, and let it be required to determine the form of the function, such as to render the definite integral,

$$\iiint V dx dy dz \ (= U),$$

a maximum or minimum; the limits of integration being supposed to be so assigned that the definite integral may extend to all values of the variables x, y, z, which satisfy the conditions

$$\phi(x, y, z) < 0, \quad \psi(x, y, z) > 0.$$

It appears by reasoning precisely similar to that employed in the cases of single and double integrals, that to the fulfilment of this condition it is necessary that the complete variation of the first order should vanish, or, in other words, that the equation

$$DU = 0$$

should be satisfied. Substituting for DU its value found in the preceding Chapter, and putting, for the sake of brevity,

$$\Theta = V_z - pV_x - qV_y - \frac{dV_{z^2}}{dz} + 2p\frac{dV_{x^2}}{dx} + 2q\frac{dV_{y^2}}{dy} + p^2\frac{dV_{x^2}}{dz} + q^2\frac{dV_{y^2}}{dz}$$

$$+ q\frac{dV_{xy}}{dx} + p\frac{dV_{xy}}{dy} + pq\frac{dV_{xy}}{dz} - \frac{dV_{xz}}{dx} - \frac{dV_{yz}}{dy}$$

$$+ V_{x^2}r + V_{xy}s + V_{y^2}t,$$

and

$$\Theta' = V_{z^2} + p^2V_{x^2} + q^2V_{y^2} - pV_{xz} - qV_{yz} + pqV_{xy},$$

* The preceding discussion is taken from the Memoir of M. Delaunay, Journal de l'Ecole Polyt., tom. xvii. pp. 92–97.

$$\Omega = U - \frac{dV_x}{dx} - \frac{dV_y}{dy} - \frac{dV_z}{dz}$$

$$+ \frac{d^2 V_{x^2}}{dx^2} + \frac{d^2 V_{y^2}}{dy^2} + \frac{d^2 V_{z^2}}{dz^2} + \frac{d^2 V_{xy}}{dxdy} + \frac{d^2 V_{xz}}{dxdz} + \frac{d^2 V_{yz}}{dydz},$$

the equation

$$DU = 0$$

becomes

$$\iint \left(V\delta z + \Theta \delta u + \Theta' \frac{d\delta u}{dz} \right) k dx dy$$

$$- \iint \left(V\delta z + \Theta \delta u + \Theta' \frac{d\delta u}{dz} \right) k' dx dy + \iiint \Omega \delta u dx dy dz = 0.$$

This equation is evidently equivalent to

$$\left(V\delta z + \Theta \delta u + \Theta' \frac{d\delta u}{dz} \right)_1 = 0,$$

$$\left(V\delta z + \Theta \delta u + \Theta' \frac{d\delta u}{dz} \right)_0 = 0, \qquad \text{(B)}$$

$$\Omega = 0;$$

the first of these equations being supposed to hold for all values of x, y, z, which satisfy the equation

$$\phi(x, y, z) = 0;$$

the second for all values which satisfy the equation

$$\psi(x, y, z) = 0;$$

and the third for all values satisfying the conditions

$$\phi(x, y, z) < 0, \quad \psi(x, y, z) > 0.$$

The method of treating these equations is precisely analogous to that of the foregoing Proposition. The equation

$$\Omega = 0,$$

which is evidently a partial differential equation of the fourth order, gives the general relation between the function u and the independent variables x, y, z, and the remaining equations serve to determine the arbitrary functions which enter into its solution. In considering the mode of determining these functions, we shall, as in Prop. I., confine our attention to the case in which the equation

$$\Omega = 0$$

admits of a solution containing arbitrary functions of the *same* quantities. This supposition will, as before, render the *number* of the arbitrary functions *complete*, i. e. equal to the order of the partial differential equation.

(1.) Let the form of the limiting function ϕ be given. This condition will give, as in the case of a double integral,

$$\delta z = 0.$$

The first of equations (B) gives, therefore,

$$\Theta_1 = 0, \quad \Theta'_1 = 0. \qquad \text{(C)}$$

These, with the equation

$$\phi(x, y, z) = 0, \qquad \text{(D)}$$

will furnish one equation between the arbitrary functions which enter into the solution of the equation

$$\Omega = 0.$$

This is proved exactly as in the case of double integrals. Let the solution of the differential equation be

$$u = F\{x, y, z, \phi_1(v, v'), \phi_2(v, v'), \phi_3(v, v'), \phi_4(v, v')\}, \qquad \text{(E)}$$

v, v' being given functions of x, y, z. Suppose

$$v = ax + by + cz, \quad v' = a'x + b'y + c'z. \qquad \text{(F)}$$

Eliminating u, x, y, z between the six equations, (C), (D), (E), (F), we have two equations of the form

$$f_1\{v, v', \phi_1(v, v'), \phi_2(v, v'), \phi_3(v, v'), \phi_4(v, v')\} = 0,$$
$$f_2\{v, v', \phi_1(v, v'), \phi_2(v, v'), \phi_3(v, v'), \phi_4(v, v')\} = 0.$$

Two others being found in a similar manner from the other limiting function ψ, we have altogether four equations for the determination of the four arbitrary functions, ϕ_1, ϕ_2, ϕ_3, ϕ_4. Substituting the values of these functions in (E), we have finally a determinate result of the form

$$f(u, x, y, z) = 0.$$

(2.) If the value of u, corresponding to the limiting function ϕ, be also given, the preceding discussion will be modified in a manner perfectly analogous to that of Art. 112. The equation

$$\Theta_1 = 0$$

will disappear, and be replaced by an equation resulting from the given value of u. Similarly, if the limiting value of any one of the coefficients,

$$\frac{du}{dx}, \frac{du}{dy}, \frac{du}{dz},$$

be also given, the equation

$$\Theta'_1 = 0$$

will also disappear, and be replaced by an equation resulting from the given value of the differential coefficient.

(3.) Finally, if the form of the limiting function ϕ be determined by the elimination of u between the equation (E) and a given determinate equation,

$$u = f(x, y, z),$$

the variations δu, δz will be connected by the equations

$$\frac{df}{dz}\delta z = \frac{dF}{dz}\delta z + \delta F = \frac{dF}{dz}\delta z + \delta u.$$

Hence, as in the case of double integrals, the first of equations (B) is equivalent to

$$V_1 + \Theta_1\left(\frac{df}{dz} - \frac{dF}{dz}\right) = 0,$$

$$\Theta'_1 = 0$$

being the same in number as before. It is unnecessary to pursue this discussion further, its principles being perfectly analogous to the corresponding case of double integrals. Any condition which reduces the number of equations furnished by (B) will, as in the former case, introduce a sufficient number of new equations to make up the deficiency.

CHAPTER VIII.

APPLICATION OF THE CALCULUS OF VARIATIONS TO GEOMETRY.

II.—*Theory of Surfaces.*

125. THE applications of the Calculus of Variations to the theory of surfaces are necessarily of a very limited character. Excluding, according to the principle stated in p. 138, problems in which the quantity represented by the integral under consideration is of physical nature, the number of those which remain will be very small; and even in the case of these problems, the imperfection of the means which we possess for the integration of partial differential equations will generally prevent our arriving at satisfactory solutions. But it must be remembered that this is an imperfection in the Integral Calculus, not in the Calculus of Variations. The rules of this latter science will in every case indicate, by an equation or equations, the distinguishing property which marks the surface, or class of surfaces, which we seek for. The deduction of the equation of the surface in finite terms, as a function of the co-ordinates, properly belongs to the Integral Calculus.

PROP. I.

126. To find the surface which will render $\iint \mu dS$ a maximum or minimum, dS being the element of the superficial area, and μ a given function of the co-ordinates x, y, z.

Putting for dS its value,

$$\sqrt{(1+p^2+q^2)}\, dxdy,$$

the given integral becomes

$$\iint \mu \sqrt{(1+p^2+q^2)}\, dxdy.$$

Here, therefore,

$$V = \mu \sqrt{(1+p^2+q^2)},$$

whence

$$Z = \sqrt{(1+p^2+q^2)}\,\frac{d\mu}{dz},\quad V_x = \frac{\mu p}{\sqrt{(1+p^2+q^2)}},\quad V_y = \frac{\mu q}{\sqrt{(1+p^2+q^2)}},$$

$$V_{x^2} = 0,\quad V_{xy} = 0,\quad V_{y^2} = 0;$$

we find, therefore, by differentiation,

$$\frac{dV_x}{dx} = \frac{p}{\sqrt{(1+p^2+q^2)}}\left(\frac{d\mu}{dx} + p\,\frac{d\mu}{dz}\right) + \mu\,\frac{(1+q^2)r - pqs}{(1+p^2+q^2)^{\frac{3}{2}}},$$

$$\frac{dV_y}{dy} = \frac{p}{\sqrt{(1+p^2+q^2)}}\left(\frac{d\mu}{dy} + q\,\frac{d\mu}{dz}\right) + \mu\,\frac{(1+p^2)t - pqs}{(1+p^2+q^2)^{\frac{3}{2}}}.$$

The equation

$$Z - \frac{dV_x}{dx} - \frac{dV_y}{dy} = 0,$$

p. 241, becomes, therefore,

$$\frac{(1+q^2)r - 2pqs + (1+p^2)t}{(1+p^2+q^2)^{\frac{3}{2}}} + \frac{1}{\mu\sqrt{(1+p^2+q^2)}}\left(p\,\frac{d\mu}{dx} + q\,\frac{d\mu}{dy} - \frac{d\mu}{dz}\right) = 0. \quad \text{(A)}$$

This equation is susceptible of a geometrical interpretation, analogous to that given to the corresponding equation, p. 140. For, if R, R' be the principal radii of curvature of the surface, it is known that

$$\frac{1}{R} + \frac{1}{R'} = -\frac{(1+q^2)r - 2pqs + (1+p^2)t}{(1+p^2+q^2)^{\frac{3}{2}}}.$$

If then we denote by α, β, γ the *acute* angles which the normal makes with the axes of co-ordinates, the equation just found may be put under the form

$$\frac{1}{R} + \frac{1}{R'} = -\frac{1}{\mu}\left(\cos\alpha\,\frac{d\mu}{dx} + \cos\beta\,\frac{d\mu}{dy} + \cos\gamma\,\frac{d\mu}{dz}\right). \quad \text{(B)}$$

From this equation we may readily deduce theorems analogous to those which have been already established in the case of curves. Thus if μ be a homogeneous function of the degree m, we shall arrive at a theorem similar to that of p. 152.

Let, as before, a surface be described whose equation is $\mu = \text{const.}$, and let a plane be drawn through the origin conjugate (with regard to this surface) to the direction of the line drawn from the origin to any point on the surface possessing the property ef rendering $\iint \mu dS$ a maximum or minimum. Produce the normal to this surface till it meet the conjugate plane, and let the line so produced be called n. Then, by proceeding exactly as in p. 152, we shall find

$$\frac{1}{R} + \frac{1}{R'} = \frac{m}{n}.$$

Hence,

If μ *be a homogeneous function of* x, y, z, *the surface which renders* $\iint \mu dS$ *a maximum or minimum is such, that the sum of the reciprocals of the radii of curvature is equal to m times the reciprocal of the normal* (drawn as above described).

To find the conditions which are to be satisfied at the limits of integration, let us suppose that the required surface is bounded by two given surfaces whose equations are

$$dz = p'dx + q'dy,$$
$$dz = p''dx + q''dy.$$

Then since (p. 241)

$$\alpha_0 = V_y - V_x \frac{dy}{dx} = \frac{\mu_0\left(q - p\dfrac{dy}{dx}\right)}{\sqrt{(1 + p^2 + q^2)}},$$

and

$$\frac{dy}{dx} = \frac{p' - p}{q - q'},$$

the equation

$$V_0 + \alpha_0 (q' - q) = 0$$

becomes

$$\mu_0 \sqrt{(1 + p^2 + q^2)} + \mu_0 \frac{q(q' - q) + p(p' - p)}{\sqrt{(1 + p^2 + q^2)}} = 0,$$

or, clearing of fractions, and neglecting the supposition $\mu_0 = 0$,

$$1 + pp' + qq' = 0;$$

and in the same way we should find

$$1 + pp'' + qq'' = 0.$$

The required surface, therefore, cuts its bounding surfaces at right angles.

Example.

127. To find a surface such that the portion of its superficial area, which is included between two given surfaces, may be a minimum.

Here $$\mu = 1, \quad \frac{d\mu}{dx} = 0, \quad \frac{d\mu}{dy} = 0, \quad \frac{d\mu}{dz} = 0.$$

The general equation becomes, then, in this case,

$$\frac{1}{R} + \frac{1}{R'} = 0,$$

giving either

$$R = \infty, \ R' = \infty, \text{ or } R + R' = 0.$$

The first of these suppositions gives, as the required surface, a plane. This, however, is not the general solution, inasmuch as it would evidently be impossible to make it fulfil, in all cases, the conditions at the limits. We must, therefore, have recourse to the other supposition,

$$R + R' = 0.$$

From which we infer that "*The surface of a minimum area is in general such, that its principal radii of curvature at every point of the surface are equal and of contrary signs.*"*

PROP. II.

128. To find the surface which will render

$$\iint (\mu dS + \mu' dx dy)$$

a maximum or minimum, dS being the element of the surface, and μ, μ' being given functions of x, y, z.

In this case

$$V = \mu \sqrt{(1 + p^2 + q^2)} + \mu',$$

whence

* For a discussion of this surface, the reader is referred to Monge, Application de l'Analyse a la Geometrie, p. 184.

$$Z = \sqrt{(1+p^2+q^2)}\,\frac{d\mu}{dz} + \frac{d\mu'}{dz},$$

$$V_x = \frac{\mu p}{\sqrt{(1+p^2+q^2)}}$$

$$\frac{dV_x}{dx} = \frac{p}{\sqrt{(1+p^2+q^2)}}\left(\frac{d\mu}{dx} + p\frac{d\mu}{dz}\right) + \mu\,\frac{(1+q^2)r - pqs}{(1+p^2+q^2)^{\frac{3}{2}}},$$

$$V_y = \frac{\mu q}{\sqrt{(1+p^2+q^2)}}$$

$$\frac{dV_y}{dy} = \frac{q}{\sqrt{(1+p^2+q^2)}}\left(\frac{d\mu}{dy} + q\frac{d\mu}{dz}\right) + \mu\,\frac{(1+p^2)t - pqs}{(1+p^2+q^2)^{\frac{3}{2}}}.$$

The equation

$$Z - \frac{dV_x}{dx} - \frac{dV_y}{dy} = 0$$

becomes, therefore,

$$\frac{(1+q^2)r - 2pqs + (1+p^2)t}{(1+p^2+q^2)^{\frac{3}{2}}}$$

$$+\frac{1}{\mu\sqrt{(1+p^2+q^2)}}\left(p\frac{d\mu}{dx} + q\frac{d\mu}{dy} - \frac{d\mu}{dz}\right) - \frac{1}{\mu}\frac{d\mu'}{dz} = 0. \qquad \text{(A)}$$

Introducing, as in the foregoing proposition, R, R', α, β, γ, we have

$$\frac{1}{R} + \frac{1}{R'} = -\frac{1}{\mu}\left(\cos\alpha\frac{d\mu}{dx} + \cos\beta\frac{d\mu}{dy} + \cos\gamma\frac{d\mu}{dz} + \frac{d\mu'}{dz}\right). \qquad \text{(B)}$$

The terms outside the sign of integration will evidently be

$$\int\left\{(\mu\sqrt{(1+p^2+q^2)} + \mu')\,\delta y + \frac{\mu q - \mu p\dfrac{dy}{dx}}{\sqrt{(1+p^2+q^2)}}\,\delta z\right\}dx.$$

Treating this, as in the former Proposition, under the supposition that the surface is limited by two given surfaces, we have at each limit of the equation,

$$\mu(1+pp'+qq') + \mu'\sqrt{(1+p^2+q^2)} = 0.$$

Hence, if ω_0, ω_1 be the angles at which the surface cuts its bounding surfaces, we have

$$\cos \omega_0 = -\frac{\mu'_0}{\mu_0} \cos \theta_0,$$

$$\cos \omega_1 = -\frac{\mu'_1}{\mu_1} \cos \theta_1; \qquad \text{(C)}$$

θ_0, θ_1 being the angles made with the plane of xy by the tangent planes to the two bounding surfaces.

Example.

129. To find a surface such that, under a given superficial area, it may contain the greatest possible volume.

It is evident, from the general principle of isoperimetrical problems, that the integral which is here to be made a maximum is

$$\iint \{z - a\sqrt{(1 + p^2 + q^2)}\} \, dxdy.$$

The solution, therefore, may be deduced from the general proposition by making

$$\mu = -a, \quad \mu' = z.$$

This will give

$$\frac{d\mu}{dx} = 0, \quad \frac{d\mu}{dy} = 0, \quad \frac{d\mu}{dz} = 0, \quad \frac{d\mu'}{dz} = 1,$$

and thus reduce the equations (A) and (B) to

$$(1 + q^2)r - 2pqs + (1 + p^2)t + \frac{1}{a}(1 + p^2 + q^2)^{\frac{3}{2}} = 0,$$

$$\frac{1}{R} + \frac{1}{R'} = \frac{1}{a}. \qquad \text{(D)}$$

Hence, *The surface which, under a given superficial area, contains a maximum volume, is such that the sum of its curvatures at every point is constant.*

130. The equations furnished by the terms outside the sign of double integration will, of course, depend upon the particular form in which the question is given. If it be required to determine a surface such that the portion of it which is bounded by the curves in which it intersects a given surface and the plane of xy may be given, and that the volume included between the surface so found, the projecting cylinder of the first curve, and the

plane of xy, may be a maximum, the first equation (C), p. 281, will be reduced to

$$z_1 \cos\theta_1 = a \cos\omega_1. \qquad \text{(a)}$$

Thus, for example, if the first bounding curve were situated in a plane parallel to xy at the distance b, we should have

$$z_1 = b, \quad \cos\theta_1 = 1,$$

and, therefore,

$$\cos\omega_1 = \frac{b}{a}. \qquad \text{(b)}$$

Hence we infer that the required surface cuts the bounding plane at a constant angle. The curve of intersection is in this case a line of curvature. The second equation (C) becomes, under the same circumstances,

$$\cos\omega_0 = 0.$$

The surface, therefore, cuts the plane of xy at right angles.

If $b = a$, the equation (b) becomes

$$\cos\omega_1 = 1, \text{ or } \omega_1 = 0.$$

It is easy to see, therefore, that the surface will touch the bounding plane in a point, the nature of the problem evidently excluding the idea of a *curve* of contact. If $b > a$, the equation (b) becomes impossible.

131. As another instance, let us suppose that it were required to join, by a surface of given area, two curves of given length, situated, the first in the plane of xy, the second in a plane parallel to xy, so that the included volume may be a maximum.*

Denoting by

$$(\iint z\,dx\,dy),$$

the value of the integral,

$$\iint z\,dx\,dy,$$

taken through the entire space included within the first of the foregoing curves, and by

$$[\iint z\,dx\,dy],$$

the value of the same integral taken through the entire space included within the projection of the second, it is plain that the volume which is to be made a maximum will be represented by

* This problem is taken from M. Delaunay's Memoir, p. 111.

$$(\iint z dx dy) - [\iint z dx dy] + z_0 \int y_0 dx,$$

the single integral being extended through the same space as

$$[\iint z dx dy].$$

Again, the surface which bounds this volume laterally will be represented on the same system of notation by

$$(\iint \sqrt{(1+p^2+q^2)} dx dy) - [\iint \sqrt{(1+p^2+q^2)} dx dy].$$

Finally, the lengths of the limiting curves will be expressed by

$$\int \sqrt{\left(1 + \frac{dy_1^2}{dx_1^2}\right)} dx, \quad \int \sqrt{\left(1 + \frac{dy_0^2}{dx_0^2}\right)} dx, \qquad \text{(a)}$$

respectively. Hence, according to the general principle of relative maxima and minima, the function which is to be made a maximum will be

$$\iint \{z - a\sqrt{(1+p^2+q^2)}\} dx dy$$

$$+ \int \left\{ y_0 z_0 + m_0 \sqrt{\left(1 + \frac{dy_0^2}{dx_0^2}\right)} \right\} dx + m_1 \int \sqrt{\left(1 + \frac{dy_1^2}{dx_1^2}\right)} dx, \quad \text{(b)}$$

the double integral being extended through the whole space lying between the first curve and the projection of the second. The complete variation of this expression will be

$$\iint \left\{ 1 + a\left(\frac{(1+q^2)r - 2pqs + (1+p^2)t}{(1+p^2+q^2)^{\frac{3}{2}}}\right) \right\} \delta z dx dy$$

$$+ \int \left\{ a\left(\frac{q - p\dfrac{dy}{dx}}{\sqrt{(1+p^2+q^2)}}\right)_1 \delta z_1 + \{z - a\sqrt{(1+p^2+q^2)}\}_1 \delta y_1 \right\} dx$$

$$- \int \left\{ a\left(\frac{q - p\dfrac{dy}{dx}}{\sqrt{(1+p^2+q^2)}}\right)_0 \delta z_0 + \{z - a\sqrt{(1+p^2+q^2)}\}_0 \delta y_0 \right\} dx \qquad \text{(c)}$$

$$- m_1 \int \frac{\dfrac{d^2 y_1}{dx_1^2}}{\left(1 + \dfrac{dy_1^2}{dx_1^2}\right)^{\frac{3}{2}}} \delta y_1 dx$$

$$+ Dz_0 \int y_0 dx + \int \left(z_0 - m_0 \frac{\dfrac{d^2 y_0}{dx_0^2}}{\left(1 + \dfrac{dy_0^2}{dx_0^2}\right)^{\frac{3}{2}}} \right) \delta y_0 dx.$$

Now since one of the limiting curves is situated in the plane of xy, it is plain that

$$Dz_1 = \delta z_1 + q_1 \delta y_1 = 0.$$

The second curve being situated in a plane parallel to xy, the increment Dz_0 is evidently constant through the entire of the upper limiting curve. Eliminating, therefore, δz_1, δz_0 by the equations

$$\delta z_1 + q_1 \delta y_1 = 0, \quad \delta z_0 + q_0 \delta y_0 = Dz_0,$$

and equating to zero the coefficients of the remaining variations, δz, δy_1, δy_0, Dz_0, we have the equations

$$(1+q^2)r - 2pqs + (1+p^2)t + \frac{1}{a}(1+p^2+q^2)^{\frac{3}{2}} = 0,$$

$$a\frac{1+p_1^2+p_1q_1\dfrac{dy_1}{dx_1}}{\sqrt{(1+p_1^2+q_1^2)}} - m_1\frac{\dfrac{d^2y_1}{dx_1^2}}{\left(1+\dfrac{dy_1^2}{dx_1^2}\right)^{\frac{3}{2}}} = 0,$$

$$a\frac{1+p_0^2+p_0q_0\dfrac{dy_0}{dx_0}}{\sqrt{(1+p_0^2+q_0^2)}} + m_0\frac{\dfrac{d^2y_0}{dx_0^2}}{\left(1+\dfrac{dy_0^2}{dx_0^2}\right)^{\frac{3}{2}}} = 0, \qquad \text{(d)}$$

$$\int y_0 dx - a\int\frac{q_0 - p_0\dfrac{dy_0}{dx_0}}{\sqrt{(1+p_0^2+q_0^2)}}dx = 0;$$

recollecting that any term which appears without any sign of integration must vanish of itself (p. 231). Now since in passing from one point to another of either of the limiting curves, z remains unaltered, we have

$$p_1 + q_1\frac{dy_1}{dx_1} = 0, \quad p_0 + q_0\frac{dy_0}{dx_0} = 0.$$

Again, if we represent by ρ_1, ρ_0 the radii of curvature of these curves respectively, we have

$$\frac{1}{\rho_1} = -\frac{\dfrac{d^2y_1}{dx_1^2}}{\left(1+\dfrac{dy_1^2}{dx_1^2}\right)^{\frac{3}{2}}}, \quad \frac{1}{\rho_0} = -\frac{\dfrac{d^2y_0}{dx_0^2}}{\left(1+\dfrac{dy_0^2}{dx_0^2}\right)^{\frac{3}{2}}}.$$

Making these substitutions in the second and third of the preceding equations, we have

$$\frac{a}{\sqrt{(1+p_1^2+q_1^2)}}+\frac{m_1}{\rho_1}=0,\quad \frac{a}{\sqrt{(1+p_0^2+q_0^2)}}-\frac{m_0}{\rho_0}=0; \quad \text{(e)}$$

or, if we denote by θ_1, θ_0, the angles which the tangent planes at any two points of the bounding curves respectively make with the plane of xy,

$$\rho_1\cos\theta_1=-\frac{m_1}{a},\quad \rho_0\cos\theta_0=\frac{m_0}{a}.$$

Hence we infer that at each point of either of the bounding curves the projection of the radius of curvature of that curve upon the tangent plane to the surface is constant.

We shall next consider the signification of the fourth equation (d), namely,

$$\int y_0 dx - a\int\frac{q_0-p_0\dfrac{dy_0}{dx_0}}{\sqrt{(1+p_0^2+q_0^2)}}dx=0, \quad \text{(f)}$$

the integrals being supposed to be extended through the whole of the upper curve.

Now we have, in general,

$$\iint\left(\frac{1}{R}+\frac{1}{R'}\right)dxdy=-\int\frac{q-p\dfrac{dy}{dx}}{\sqrt{(1+p^2+q^2)}}dx,* \quad \text{(g)}$$

the double integral being extended through the entire space enclosed by a given curve, and the single integral being taken through the circumference of the bounding curve. Hence if we describe a closed curve upon a surface whose mean curvature is constant, and denote the single integral in (g), when extended through the entire circumference, by

$$\int\frac{q_0-p_0\dfrac{dy_0}{dx_0}}{\sqrt{(1+p_0^2+q_0^2)}}dx,$$

we shall have

---

* For a proof of this theorem, which is, as far as I am aware, due to Laplace, see Chap. X.

$$\int \frac{q_0 - p_0 \frac{dy_0}{dx_0}}{\sqrt{(1 + p_0^2 + q_0^2)}} dx = - \iint \left(\frac{1}{R} + \frac{1}{R'} \right) dx dy = \frac{1}{a} \iint dx dy = \frac{1}{a} \int y_0 dx.$$

Making this substitution in (f), it becomes identically zero, and therefore furnishes no additional condition.

Now it is evidently necessary to the validity of the foregoing reasoning that the portion of the surface situated at one side of the plane of the limiting curve should be *closed*, as the theorem expressed in equation (g) would not otherwise be true. It is necessary also, that if a line be drawn perpendicular to the plane of xy, it should only intersect this part of the surface once. The failure of this condition would also introduce a modification in equation (g). If these conditions be fulfilled, the fourth equation (f) becomes, as we have seen, identical.

If these conditions be not fulfilled, this equation will furnish a condition for the determination of one of the arbitrary constants which enters into the solution of equation (D). Thus, for example, if we take as a particular solution of (D) a surface of revolution round an axis parallel to the axis of z, it may be proved from this equation that the required surface is spherical.*

---

* Delaunay, p. 116. As the case of a surface of revolution has been considered before (p. 161), I do not think it necessary to dwell upon it here.

CHAPTER IX.

APPLICATION OF THE CALCULUS OF VARIATIONS TO MECHANICS.

132. The applications of the Calculus of Variations to Mechanics are of two kinds. In that science, as in Geometry, we meet with problems of maxima and minima, which involve functions of variable form, and which, therefore, require for their solution the aid of the Calculus of Variations.

But a more important application of our science was made by Lagrange, namely, to the deduction of the equations of equilibrium or motion of a system whose constitution is known. We shall now proceed to give examples of both these classes of problems, commencing with the following very general case of the celebrated problem of the brachystochrone.

Prop. I.

133. Let a material point be constrained to move on a given surface, and let it be supposed that the forces which act upon it are such as to render the expression

$$Xdx + Ydy + Zdz$$

a perfect differential; find the curve along which it should be constrained to move, in order that the time of passing from one point to another may be a minimum.

Let ds be the element of the path, v the tangential velocity, and T the time; then it is evident that

$$T = \int \frac{ds}{v}.$$

We have, moreover,

$$v^2 = 2\int (Xdx + Ydy + Zdz) = \phi(x, y, z).$$

This question is therefore a case of the more general problem discussed at the commencement of the preceding Chapter, to which it may be reduced by putting

$$\mu = \frac{1}{v}.$$

Hence we have

$$\frac{d\mu}{dx} = -\frac{1}{v^2}\cdot\frac{dv}{dx} = -\frac{X}{v^3},$$

$$\frac{d\mu}{dy} = -\frac{1}{v^2}\cdot\frac{dv}{dy} = -\frac{Y}{v^3},$$

$$\frac{d\mu}{dz} = -\frac{1}{v^2}\cdot\frac{dv}{dz} = -\frac{Z}{v^3}.$$

The first of equations (F), p. 180, becomes, therefore,

$$\frac{1}{\rho^2} - \frac{1}{\rho'^2} = \frac{1}{v^4}(X\cos\alpha + Y\cos\beta + Z\cos\gamma)^2. \qquad \text{(A)}$$

From this equation may be deduced the following general properties of the curve in question:

(1.) Let R be the resultant of the forces which act upon the material point, and ω the angle which that resultant makes with the perpendicular to the plane of the normal section which passes through the tangent to the path; also let θ be the angle between the plane of this normal section and the osculating plane to the path; then we have

$$\frac{1}{\rho^2} - \frac{1}{\rho'^2} = \frac{\sin^2\theta}{\rho^2}.$$

We have also

$$X\cos\alpha + Y\cos\beta + Z\cos\gamma = R\cos\omega.$$

Making these substitutions in equation (A), it becomes

$$\frac{\sin^2\theta}{\rho^2} = \frac{R^2\cos^2\omega}{v^4}, \text{ or } \frac{v^4}{\rho^2}\sin^2\theta = R^2\cos^2\omega;$$

or, since the signs of the two sides of the equation are the same,*

$$\frac{v^2}{\rho}\sin\theta = R\cos\omega.$$

* If the signs were different the pressure on the curve would vanish, and the motion would not be constrained. This supposition applies to another problem, which is solved by the same equation.—Vid. *infra*, p. 293.

This equation expresses the fact that the resolved part of the centrifugal force, perpendicular to the plane of the normal section, i. e. along the tangent to the perpendicular section of the surface, is equal to the resolved part of R in the same direction. Now it is evident that the pressure on the curve in any direction is equal to the sum of the resolved parts (in that direction) of the resultant and the centrifugal force; if, therefore, the total pressure be resolved into two, one along the normal to the surface, the other in the tangent plane, the value of this latter component is

$$\frac{v^2}{\rho}\sin\theta + R\cos\omega = 2R\cos\omega.$$

The total force on the point may be resolved into three, namely:—1. A force along the tangent to the path, which acts solely in increasing the velocity. 2. A force along the normal to the surface, which acts altogether in producing a pressure on the surface. 3. A force perpendicular to these, which may be termed the deflecting force, inasmuch as it is that part of the total force which tends to deflect the material point from the path in which it is constrained to move.

Adopting this definition, the result arrived at above may be stated thus:

If a material point, whose motion is restricted to a given surface, be constrained to move in a groove of such a form, that the time of passing from one point to another may be a minimum, the deflecting force, or pressure on the side of the groove, is double what it would be if the particle were at rest, i. e. if the tangential component were destroyed.

(2.) If the resultant be in the plane of the normal section, the required curve is a geodesic line. For in this case

$$X\cos\alpha + Y\cos\beta + Z\cos\gamma = 0;$$

hence

$$\rho = \rho', \quad \theta = 0.$$

The osculating plane is therefore perpendicular to the surface, &c.

(3.) If gravity be the only force acting upon the point, and if the axis of z be taken perpendicular to the horizon, we find from equation (A), since $X = 0$, $Y = 0$, $Z = -g$,

$$\frac{1}{\rho^2} - \frac{1}{\rho'^2} = \frac{1}{v^4} \cdot g^2 \cos^2\gamma.$$

Again,

$$v^2 = 2\int(Xdx + Ydy + Zdz) = 2g(h - z);$$

h being the initial height. Hence

$$\frac{1}{\rho^2} - \frac{1}{\rho'^2} = \frac{\cos^2\gamma}{4(h-z)^2}. \qquad \text{(a)}$$

Let P (Fig. 14) be the place of the material point at any instant, PN a normal to its path drawn in the tangent plane to the surface, and produced to meet the horizontal plane through the point of departure of the particle, PZ a perpendicular to this plane. We have then

$$PZ = h - z;$$

and, since PN is evidently perpendicular to the plane of the normal section,

$$NPZ = \gamma.$$

Hence

$$h - z = PZ = PN . \cos\gamma;$$

and if this value be substituted in the equation

$$\frac{1}{\rho^2} - \frac{1}{\rho'^2} = \frac{\cos^2\gamma}{4(h-z)^2},$$

we find

$$\frac{1}{\rho^2} - \frac{1}{\rho'^2} = \frac{1}{4PN^2}. \qquad \text{(b)}$$

If the tangent plane to the surface at the point P be horizontal,

$$PN = \infty, \text{ and } \frac{1}{\rho^2} - \frac{1}{\rho'^2} = 0.$$

Hence it appears that whenever the tangent plane is horizontal, the osculating plane of the curve of quickest descent at that point coincides with that of a geodetic line.

Again, if θ be the angle which PN makes with the osculating plane, we have, by Meunier's theorem,

$$\rho = \rho' \sin\theta;$$

hence

$$\frac{1}{\rho^2} - \frac{1}{\rho'^2} = \frac{\cos^2\theta}{\rho^2} = \frac{1}{4PN^2};$$

or

$$\rho = 2PN . \cos\theta. \qquad \text{(c)}$$

Hence at every point of a line of quickest descent on any surface, the radius of curvature is double the projection of the normal (drawn as stated above) on the osculating plane.

Let the given surface be a sphere whose radius is unity, and let r be the spherical radius of curvature,* then

$$\rho = \sin r, \quad \rho' = 1, \quad \frac{1}{\rho^2} - \frac{1}{\rho'^2} = \cot^2 r.$$

Let HH' (Fig. 15) be the great circle whose plane is horizontal, and let it be supposed that the body starts from some point in this circle without initial velocity. Let the origin of spherical co-ordinates be taken at o, the pole of this circle, and let p be the place of the body at any instant, op the radius vector, oq the perpendicular on the tangent, and pn the normal. Assume

$$\mu = op, \quad \nu = pn, \quad \pi = oq.$$

Then it is plain that

$$\cos\gamma = \sin\pi, \quad PN = \tan\nu, \quad h - z = \cos\mu.$$

Making these substitutions in equation (a), we find

$$\tan r = 2\,\frac{\cos\mu}{\sin\pi}.$$

But

$$\tan r = -\frac{\sin\mu d\mu}{\cos\pi d\pi},\dagger$$

(the negative sign being taken, inasmuch as the curve is convex towards the origin); hence

$$\tan\mu d\mu = -2\cot\pi d\pi,$$

or, by integration,

$$\sin^2\pi = m\cos\mu. \tag{d}$$

From equation (b) we find

$$\tan r = 2\tan\nu \tag{e}$$

expressing a property obviously analogous to one of the cycloid.

* See Graves' edition of Chasles, pp. 93, 95.

† Ibid., p. 95.

(4.) Let the force acting upon the material point be a single central force, varying as the nth power of the distance. Then if we take as origin the centre of force, we have

$$X = r^{n-1}\,x, \quad Y = r^{n-1}\,y, \quad Z = r^{n-1}\,z,$$

$$v^2 = 2\int(Xdx + Ydy + Zdz) = \frac{2}{n+1}\,r^{n+1}.*$$

Making these substitutions in (A), we have

$$\frac{1}{\rho^2} - \frac{1}{\rho'^2} = \left(\frac{n+1}{2}\right)^2 \left(\frac{x\cos\alpha + y\cos\beta + z\cos\gamma}{r^2}\right)^2;$$

or, if we denote by p the perpendicular from the origin upon the plane of the normal section,

$$\frac{1}{\rho^2} - \frac{1}{\rho'^2} = \left(\frac{n+1}{2}\right)^2 \frac{p^2}{r^4}. \qquad \text{(f)}$$

Hence also if θ be the acute angle between the osculating plane and the plane of the normal section,

$$\frac{\sin\theta}{\rho} = \frac{\tan\theta}{\rho'} = \pm\,\frac{n+1}{2}\,\frac{p}{r^2}, \qquad \text{(g)}$$

the upper or lower sign being taken according as the force is repulsive or attractive.

(5.) In general if the central force be denoted by R, we shall have from (A),

$$\frac{v^2}{\rho}\sin\theta = \pm\,(X\cos\alpha + Y\cos\beta + Z\cos\gamma) = \pm\,R\cos\omega,$$

where ω is the angle between R and the normal to the curve, drawn in the tangent plane. Hence

$$\pm\,R = \frac{v^2}{\rho\cos\omega\,\operatorname{cosec}\theta}. \qquad \text{(h)}$$

This formula may be expressed geometrically as follows:

Draw through the centre of absolute curvature a line perpendicular to the osculating plane. From the point where this perpendicular cuts the tangent plane as centre, describe a sphere

* It is here supposed, for the sake of simplicity, that the constant vanishes. The reader will find no difficulty in interpreting this restriction.

through the point in the curve. Then if c denote the chord of this sphere which passes through the centre of force, we shall have

$$R = \pm \frac{2v^2}{c}. \qquad \text{(i)}$$

The principle stated in (6) shows that the same expression applies to the motion of a particle, unconstrained (except by the given surface), and subject to the action of a central force.

(6.) We have seen (p. 180) that the equation

$$\frac{1}{\rho^2} - \frac{1}{\rho'^2} = \frac{1}{\mu^2}\left(\cos\alpha \frac{d\mu}{dx} + \cos\beta \frac{d\mu}{dy} + \cos\gamma \frac{d\mu}{dz}\right)^2$$

contains the solution of two problems, namely:—1. To find the curve which renders $\int \mu ds$ a minimum. 2. To find the curve which renders $\int \frac{ds}{\mu}$ a minimum. Hence, in the present case, the equation

$$\frac{1}{\rho^2} - \frac{1}{\rho'^2} = \frac{1}{v^4}(X\cos\alpha + Y\cos\beta + Z\cos\gamma)^2$$

contains not only the solution of the problem of the brachystochrone, corresponding to the integral T, p. 287, but also that of determining the actual path which the body would describe under the action of the given forces, and merely constrained to move on the given surface. For this latter problem is (in accordance with the principle of least action) solved by determining the path of the body such as to render

$$\int v ds$$

a minimum. The two questions are therefore connected, as in p. 180. Hence,

The actual path of a material particle, restricted to a given surface, and urged by a system of forces, which make

$$Xdx + Ydy + Zdz$$

a perfect differential, and the curve of quickest passage from one point to another, are represented by the same differential equation,

$$\rho^2 = f\left(x, y, \frac{dy}{dx}\right).$$

(7.) If the material particle be wholly unrestricted, the equation $u = 0$, of p. 177, will disappear, and the equations (C) will become

$$\frac{d\mu}{dx} - \mu\frac{d^2x}{ds^2} - \frac{dx}{ds}\frac{d\mu}{ds} = 0,$$

$$\frac{d\mu}{dy} - \mu\frac{d^2y}{ds^2} - \frac{dy}{ds}\frac{d\mu}{ds} = 0, \qquad \text{(k)}$$

$$\frac{d\mu}{dz} - \mu\frac{d^2z}{ds^2} - \frac{dz}{ds}\frac{d\mu}{ds} = 0.$$

Let α, β, γ be the angles made with the axes of co-ordinates by any line, l, in the normal plane to the curve. Multiply the foregoing equations by $\cos\alpha$, $\cos\beta$, $\cos\gamma$, respectively. Now

$$\cos\alpha\frac{d^2x}{ds^2} + \cos\beta\frac{d^2y}{ds^2} + \cos\gamma\frac{d^2z}{ds^2} = -\frac{1}{\rho}\cos\omega,$$

ω being the angle between the radius of curvature and the line l. Also,

$$\cos\alpha\frac{dx}{ds} + \cos\beta\frac{dy}{ds} + \cos\gamma\frac{dz}{ds} = 0.$$

Hence we have, in general,

$$\frac{\mu}{\rho}\cos\omega = -\left(\cos\alpha\frac{d\mu}{dx} + \cos\beta\frac{d\mu}{dy} + \cos\gamma\frac{d\mu}{dz}\right);$$

and in the particular case before us, where $\mu = \frac{1}{v}$,

$$\frac{v^2}{\rho}\cos\omega = X\cos\alpha + Y\cos\beta + Z\cos\gamma = R\cos\omega', \qquad (1)$$

ω' being the angle between R and l.

Let ϕ be the angle between R and the normal plane, and η the angle between l and the projection of R on the normal plane. Then

$$\cos\omega' = \cos\phi\cos\eta.$$

Hence

$$\frac{v^2}{\rho}\cos\omega = R\cos\omega' = R\cos\phi\cos\eta;$$

and as the line l is drawn arbitrarily in the normal plane, this equation can only be satisfied by making

$$\omega = \eta, \quad \frac{v^2}{\rho} = R \cos \phi.$$

Hence we infer,

If a material point, acted on by any system of forces which satisfy the condition

$$Xdx + Ydy + Zdz = d\Pi,$$

be constrained to move in a path such that the time of passing from one given point to another may be a minimum, the radius of absolute curvature will coincide with the projection of the resultant force upon the normal plane, and the pressure on the curve will be double what it would be if the point were reduced to rest by the destruction of the tangential force.

For example, let this point be acted on by a single central force. Let the origin be taken at the centre of force. Then if we multiply the first of equations (k) by y, and the second by x, and subtract them, we shall have

$$y\frac{d\mu}{dx} - x\frac{d\mu}{dy} + \mu\left(x\frac{d^2y}{ds^2} - y\frac{d^2x}{ds^2}\right) + \frac{d\mu}{ds}\left(x\frac{dy}{ds} - y\frac{dx}{ds}\right) = 0. \quad \text{(m)}$$

But since

$$\frac{d\mu}{dx} = -\frac{X}{v^3}, \quad \frac{d\mu}{dy} = -\frac{Y}{v^3},$$

it is plain that

$$y\frac{d\mu}{dx} - x\frac{d\mu}{dy} = 0.$$

Equation (m) becomes therefore integrable, giving

$$\mu\left(x\frac{dy}{ds} - y\frac{dx}{ds}\right) = c.$$

Similarly,

$$\mu\left(z\frac{dx}{ds} - x\frac{dz}{ds}\right) = b.$$

Eliminating μ, and integrating, we find

$$ax + by + cz = 0.$$

The curve is therefore plane. Hence we have

$$\frac{v^2}{\rho} = R\cos\phi,$$

ϕ being the angle between the normal to the curve and the radius vector drawn from the centre of force. This equation gives

$$R = \frac{v^2}{\rho \cos \phi} = \frac{v^2}{\text{semichord of curvature}}.$$

This is the well-known expression for the central force in the case of unconstrained motion. The preceding analysis shows that this expression applies equally to the case of constrained motion, provided that the curve to which the particle is constrained be such as to render the time of passing from one point to another a minimum.

If in any of the preceding cases the path of the material particle be limited not by two given points, but by two given curves, it appears from p. 181 that the required curve will cut its bounding curves at right angles.*

134. As an example of the preceding case, suppose the material particle to be acted upon solely by gravity, and to have, at the commencement of its motion, a given velocity, and let it be required to determine the curve of quickest descent from one given curve to another.

Let h be the height due to the initial velocity, and y_0 the ordinate of the point from which the motion begins. Then if the axis of y be taken vertical, and in the direction of gravity, we shall have

$$v^2 = 2g(y - y_0 + h). \tag{a}$$

Equation (1) becomes, therefore,

$$\frac{2(y - y_0 + h)}{\rho} = \cos \beta.$$

If then the origin be transferred to a point in the axis of $-y$ (i. e. vertically *above* the origin at first assumed), at a distance equal to $y_0 - h$, equation (a) will give

$$\text{rad. of curv.} = 2N,$$

N being the normal terminated in the new axis of x. This is a well-known property of the cycloid whose axis coincides with

* The suppositions $\mu_1 = 0$, $\mu_0 = 0$, would give $v_1 = \infty$, $v_0 = \infty$, and are therefore inadmissible.

the axis of x. The curve of quickest descent is therefore a cycloid described upon the horizontal line, from which the particle may be supposed to start *without* initial velocity.

With regard to the conditions to be observed at the limits, it is evident, from the general principle of Art. 133, that the cycloid will cut its bounding curves at right angles.

135. Again, suppose the particle to be attracted to a fixed centre by a force varying inversely as the nth power of the distance, and to have an initial velocity equal to that which would be acquired in falling from infinity. In this case we have

$$v^2 = -2\int R dr = \frac{2f}{n-1} r^{-(n-1)},$$

$$X\cos\alpha + Y\cos\beta = R\cos\phi = fr^{-(n+1)}(x\cos\alpha + y\cos\beta) = fr^{-(n+1)}p,$$

p being the perpendicular from the centre of force upon the tangent. Making these substitutions in the equation of p. 295, and putting for ρ its value in terms of r and p, we have

$$2rdp + (n-1)pdr = 0.$$

Integrating, and adding an arbitrary constant, we find

$$r^{n-1}p^2 = a^{n+1}.$$

Substituting for p its value in terms of r and ω, and putting

$$m = \frac{n+1}{2},$$

we have

$$d\omega = a^m \frac{dr}{r\sqrt{(r^{2m} - a^{2m})}}.$$

Hence we find easily

$$r^m \cos m\omega = a^m,$$

the polar equation of the required curve.

Prop. II.

136. To find the line of quickest descent for a material particle moving in a medium resisting as any function of the velocity, and acted on solely by gravity.

The present question is evidently not included in that which forms the subject of the foregoing Proposition, inasmuch as any system of forces which includes the resistance of a medium will not satisfy the criterion,

$$Xdx + Ydy + Zdz = d\Pi.$$

It will therefore be necessary to consider it separately.

Assume $\theta = v^2$, and let Θ $\{= f(\theta)\}$ be the function which expresses the resistance of the medium. Then if, as in Chapter IV., we take s as the independent variable, we shall have the time of descent expressed by

$$\int \frac{ds}{\sqrt{\theta}},$$

the functions x, y, θ being connected by the equations

$$\left.\begin{aligned} L &= \frac{dx^2}{ds^2} + \frac{dy^2}{ds^2} - 1 = 0, \\ L_1 &= \frac{d\theta}{ds} - 2g\frac{dy}{ds} + \Theta = 0, \end{aligned}\right\} \quad \text{(A)}$$

the axis of y being vertical, and the positive ordinates being measured *downwards*. Hence, according to the principle of the method of Lagrange, we have the equation

$$\int \left(\delta \frac{1}{\sqrt{\theta}} + \lambda \delta L + \lambda_1 \delta L_1\right) ds = 0. \quad \text{(B)}$$

But

$$\delta \frac{1}{\sqrt{\theta}} = -\tfrac{1}{2}\theta^{-}\ \delta\theta, \quad \delta L = 2\frac{dx}{ds}\frac{d\delta x}{ds} + 2\frac{dy}{ds}\frac{d\delta y}{ds},$$

$$\delta L_1 = \frac{d\delta\theta}{ds} - 2g\frac{d\delta y}{ds} + \frac{d\Theta}{d\theta}\delta\theta.$$

Substituting these values in (B), integrating by parts the terms involving the several quantities,

$$\frac{d\delta x}{ds}, \ \frac{d\delta y}{ds}, \ \frac{d\delta\theta}{ds},$$

and equating to zero the coefficients of the variations δx, δy, $\delta\theta$, we have

$$\frac{d}{ds} \cdot \lambda \frac{dx}{ds} = 0,$$

$$\frac{d}{ds} \cdot \lambda \frac{dy}{ds} - g \frac{d\lambda_1}{ds} = 0, \qquad \text{(C)}$$

$$\tfrac{1}{2} \theta^{-\frac{3}{2}} + \frac{d\lambda_1}{ds} - \lambda_1 \frac{d\Theta}{d\theta} = 0.$$

Integrating the first two, we have

$$\lambda \frac{dx}{ds} = a, \quad \lambda \frac{dy}{ds} - g\lambda_1 = b, \qquad \text{(D)}$$

Assume

$$t = \theta^{-\frac{1}{2}} + \lambda_1 \Theta.$$

Differentiating this expression, we find

$$\left(\tfrac{1}{2}\theta^{-\frac{3}{2}} - \lambda_1 \frac{d\Theta}{d\theta}\right) \frac{d\theta}{ds} = \Theta \frac{d\lambda_1}{ds} - \frac{dt}{ds}.$$

Hence, and from the third equation (C), we find

$$\frac{d\lambda_1}{ds} \left(\frac{d\theta}{ds} + \Theta\right) = \frac{dt}{ds};$$

or, eliminating θ by means of the equation $L_1 = 0$,

$$2g \frac{d\lambda_1}{ds} \frac{dy}{ds} = \frac{dt}{ds}. \qquad \text{(E)}$$

But if the first two equations (C) be multiplied by $\frac{dx}{ds}, \frac{dy}{ds}$, respectively, and added, we shall have

$$g \frac{d\lambda_1}{ds} \frac{dy}{ds} = \frac{d\lambda}{ds}.$$

Substituting this in (E), and integrating,

$$t + c = 2\lambda.$$

It is easy to show, by reasoning similar to that of Prop. I., Chap. IV., that if the length of the curve be not given, the constant c must be neglected. For the addition of an arbitrary constant to t would be equivalent to supposing that the original integral was

$$\int \left(\frac{1}{\sqrt{\theta}} + c\right) ds$$

Neglecting c, therefore, and eliminating λ by means of the equation

$$\lambda \frac{dx}{ds} = a,$$

we have

$$(\theta^{-\frac{1}{2}} + \lambda_1 \Theta) \frac{dx}{ds} = 2a;$$

or eliminating λ_1 by means of the equations (D),

$$\left(\theta^{-\frac{1}{2}} - \frac{b}{g}\Theta\right)\frac{dx}{ds} + \frac{a}{g}\Theta\frac{dy}{ds} = 2a. \qquad \text{(F)}$$

Eliminating $\frac{dx}{ds}, \frac{dy}{ds}$ between (A) and (F), we find

$$\left(\theta^{-\frac{1}{2}} - \frac{b}{g}\Theta\right)^2 \left\{1 - \frac{1}{4g^2}\left(\frac{d\theta}{ds} + \Theta\right)^2\right\}$$
$$- a^2\left\{2 - \frac{1}{2g^2}\left(\frac{d\theta}{ds} + \Theta\right)\Theta\right\}^2 = 0. \qquad \text{(G)}$$

This equation determines θ in terms of s, and if the value so found be substituted in the second equation (A), an equation will be obtained in y and s, which will be that of the curve required.

If the resistance of the medium vanish, we have $\Theta = 0$, and the equations $L_1 = 0$ and (G) become

$$\frac{d\theta}{ds} - 2g\frac{dy}{ds} = 0,$$

$$\frac{1}{\theta}\left(1 - \frac{1}{4g^2}\frac{d\theta^2}{ds^2}\right) - 4a^2 = 0.$$

Integrating the first of these, we find

$$\theta - 2gy = c.$$

Eliminating θ between this and the second, and solving for $\frac{dy}{ds}$, we have

$$\frac{dy}{ds} = 2a\sqrt{\left(\frac{1}{4a^2} - 2gy - c\right)},$$

the equation of a cycloid, as before.

PROP. III.

137. Let a flexible cord of variable thickness be attached to two fixed points; find the curve which it ought to form in order that the centre of gravity may be lowest.

The cord being supposed to be inextensible, it is plain that its thickness at any point will be a given function of the arc, s. Let this function be denoted by S. Then the distance of the centre of gravity from the axis of x will be represented by

$$\frac{\int ySds}{\int Sds}.$$

But as the integrals in this expression are taken through the whole extent of the curve, it is evident that

$$\int Sds = \text{volume of cord} = \text{const.}$$

Hence the integral which is to be rendered a minimum is

$$\int ySds.$$

Treating this expression according to the method stated in Chapter IV., we shall have the two equations,

$$\left.\begin{aligned} 2\frac{d}{ds}\cdot\lambda\frac{dx}{ds} &= 0, \\ 2\frac{d}{ds}\cdot\lambda\frac{dy}{ds} - S &= 0. \end{aligned}\right\} \quad \text{(A)}$$

Integrating these equations, and assuming

$$S_1 = \int Sds,$$

we shall have

$$\left.\begin{aligned} 2\lambda\frac{dx}{ds} &= a, \\ 2\lambda\frac{dy}{ds} - S_1 &= b. \end{aligned}\right\} \quad \text{(B)}$$

Hence we find

$$2\lambda = \sqrt{\{a^2 + (b + S_1)^2\}}.$$

Substituting this value in the equations (B), and integrating, we have

$$\left.\begin{aligned} x + c &= \int \frac{a ds}{\sqrt{\{a^2 + (b + S_1)^2\}}}, \\ y + d &= \int \frac{(b + S_1)\, ds}{\sqrt{\{a^2 + (b + S_1)^2\}}}. \end{aligned}\right\} \quad (C)$$

If s be eliminated between these two equations the result will be the equation of the required curve. The solution contains, as will be seen, four arbitrary constants: of these, two are determined by the given position of the extremities of the cord, one by its length, and the fourth by supposing that the arc s, which has been taken for the independent variable, is reckoned from one extremity of the curve. This constant will evidently disappear in the elimination of s. If the thickness of the curve be constant we shall have

$$S = 1, \quad S_1 = s.$$

Substituting these values in (C), and performing the integrations, we find

$$x + c = al\left[\frac{s + b}{a} + \sqrt{\left\{1 + \left(\frac{s + b}{a}\right)^2\right\}}\right],$$

$$y + d = a\sqrt{\left\{1 + \left(\frac{s + b}{a}\right)^2\right\}};$$

or, eliminating s,

$$x + c = al\left[\frac{y + d}{a} + \sqrt{\left\{\left(\frac{y + d}{a}\right)^2 - 1\right\}}\right].$$

the equation of a catenary.

If the extremities of the cord be not fixed, but merely restricted to two given curves,

$$u_0 = 0, \quad u_1 = 0,$$

the terms free from the sign of integration will give the equations

$$\left(\frac{dx}{ds}\right)_0 \delta x_0 + \left(\frac{dy}{ds}\right)_0 \delta y_0 = 0, \quad \left(\frac{dx}{ds}\right)_1 \delta x_1 + \left(\frac{dy}{ds}\right)_1 \delta y_1 = 0,$$

the variations δx_0, δy_0, δx_1, δy_1, being connected by the equations

$$\frac{du_0}{dx_0}\delta x_0 + \frac{du_0}{dy_0}\delta y_0 = 0, \quad \frac{du_1}{dx_1}\delta x_1 + \frac{du_1}{dy_1}\delta y_1 = 0.$$

Hence we have

$$\frac{du_0}{dx_0}\left(\frac{dy}{ds}\right)_0 - \frac{du_0}{dy_0}\left(\frac{dx}{ds}\right)_0 = 0, \quad \frac{du_1}{dx_1}\left(\frac{dy}{ds}\right)_1 - \frac{du_1}{dy_1}\left(\frac{dx}{ds}\right)_1 = 0;$$

showing that the required curve cuts its bounding curves at right angles.

It is evident, from the principles of Mechanics, that the curve determined in this Proposition is that which the cord will assume when in equilibrium under the influence of the force of gravity only. This principle is assumed in the following Proposition.

Prop. IV.

138. If in the preceding Proposition the cord be placed upon a given surface, determine its position of equilibrium.

Let the axis of z be vertical, and let the equation of the given surface be

$$u = 0.$$

Then it is evident that the method of Lagrange will give, as in Prop. V., Chap. IV., the equations;

$$\begin{aligned} &\lambda' \frac{du}{dx} - \frac{d}{ds}.\lambda\frac{dx}{ds} = 0, \\ &\lambda' \frac{du}{dy} - \frac{d}{ds}.\lambda\frac{dy}{ds} = 0, \\ &\lambda' \frac{du}{dz} - \frac{d}{ds}.\lambda\frac{dz}{ds} + S = 0. \end{aligned} \qquad (A)$$

Multiplying these equations by $\frac{dx}{ds}, \frac{dy}{ds}, \frac{dz}{ds}$, respectively, and adding them, we find

$$\frac{d\lambda}{ds} = S, \text{ whence } \lambda = S_1 + a.$$

Again, if we denote by α, β, γ the angles which the plane of the tangent normal section makes with the co-ordinate planes, and proceed as in pp. 179, 180, we shall readily obtain the equation

$$\frac{1}{\rho^2} - \frac{1}{\rho'^2} = \frac{S^2 \cos^2\gamma}{(S_1 + a)^2}. \tag{B}$$

This equation defines the required curve. If the extremities of the cord be not fixed, it is easy to see that it will cut at right angles the curves to which these extremities are restricted.

Prop. V.

139. To determine the form of the *elastic curve*, i. e. the curve of equilibrium of an elastic spring, whose extreme points are given, or restricted to two given curves, and which is not acted on by any external forces.*

Adopting the principle of Daniel Bernouilli, we shall define the elastic curve to be that in which the integral

$$\int \frac{ds}{\rho^2},$$

taken through the whole extent of the curve, is a minimum. This problem is therefore included in the more general one discussed in Prop. IV. Chap. IV., which is applied to the present case by putting

$$\mu = \frac{1}{\rho^2}, \quad \mu' = -\frac{2}{\rho^3}.$$

(1.) Let the extreme points be given, and let the line joining these points be taken for the axis of x. Equation (E), p. 169, becomes (putting $-\frac{4}{b^2}$ for a)

$$\frac{1}{\rho} = \frac{2y}{b^2}.$$

Hence in p. 170 we have

$$Y = \frac{2y}{b^2}, \quad Y_1 = \frac{y^2}{b^2}.$$

---

* Euler, Meth. Inven., p. 245.

Making these substitutions in equation (F), we have

$$x + f = \int \frac{\left(\frac{y^2}{b^2} + e\right) dy}{\sqrt{\left\{1 - \left(\frac{y^2}{b^2} + e\right)^2\right\}}}. \qquad \text{(A)}$$

This is the equation of the elastic curve if the position of the extreme tangents be not given.

If the position of the extreme tangents be given (in the same plane), we must recur to the general equation (C), p. 168,

$$\mu' \rho^2 = ay - bx + c.$$

But as this equation may always be reduced by transformation of co-ordinates to the form (E), it is evident that the elastic curve may in all cases be represented by the equation (A), the position of the axes of co-ordinates depending upon the given position of the extreme tangents. If a general solution, not requiring any particular position of the axes, be required, it is immediately deduced from (A) by a transformation of co-ordinates. It may also be easily deduced from the general equation (C) as follows:

If we substitute for μ' in equation (C) its value $-\frac{2}{\rho^3}$, we shall have

$$-\frac{2}{\rho} = ay - bx + c.$$

Substituting for $\frac{1}{\rho}$ its value,

$$-(1 + p^2)^{-\frac{3}{2}} \frac{pdp}{dy},$$

and integrating with regard to y, we find

$$\frac{2}{\sqrt{(1 + p^2)}} = \tfrac{1}{2} ay^2 - b \int xdy + cy + c'.$$

Again, putting for $\frac{1}{\rho}$ its value $-(1 + p^2)^{-\frac{3}{2}} \frac{dp}{dx}$, and integrating with regard to x, we have

$$\begin{aligned} \frac{2p}{\sqrt{(1 + p^2)}} &= \tfrac{1}{2} bx^2 - a \int ydx - cx + c'', \\ &= \tfrac{1}{2} bx^2 - axy + a \int xdy - cx + c''. \end{aligned}$$

Eliminating $\int x\,dy$, and arranging, we have

$$\frac{4(a+bp)}{\sqrt{(1+p^2)}} = (ay - bx + c)^2 + e, \qquad \text{(B)}$$

where

$$e = 2ac' + 2bc'' - c^2.$$

Hence it appears that the general solution contains five arbitrary constants; of these four are determined by the conditions which must be satisfied at the limiting points, and the fifth by the given length of the elastic spring.

If the extreme points be merely limited to two given curves, it appears, from the principle of p. 171, that the line joining the extreme points is perpendicular to the tangents to the two limiting curves.

Finally, let it be supposed that the elastic spring is so adjusted as to *touch* the limiting curve at each extremity.

The general equations* for the solution of the present case are,

$$\left.\begin{aligned} a + b\left(\frac{df_1}{dx}\right)_1 - (ay_1 - bx_1 + c)\left(\frac{d^2f_1}{dx^2}\right)_1\left(\frac{dx^2}{ds^2}\right)_1 = 0, \\ a + b\left(\frac{df_0}{dx}\right)_0 - (ay_0 - bx_0 + c)\left(\frac{d^2f_0}{dx^2}\right)_0\left(\frac{dx^2}{ds^2}\right)_0 = 0, \end{aligned}\right\} \qquad \text{(C)}$$

which, with the equations

$$y_1 = f_1(x_1),\quad y_0 = f_0(x_0),\quad \left(\frac{dy}{dx}\right)_1 = \left(\frac{df_1}{dx}\right)_1,\quad \left(\frac{dy}{dx}\right)_0 = \left(\frac{df_0}{dx}\right)_0,$$

are sufficient to determine four of the arbitrary constants, a, b, c, e, f. The fifth is determined by the given length of the spring.

Suppose, for example, that the bounding curves were the right lines,

$$y = mx + p,$$
$$y = nx + q.$$

These equations will give

$$\left(\frac{df_1}{dx}\right)_1 = m,\quad \left(\frac{df_0}{dx}\right)_0 = n,\quad \left(\frac{d^2f_1}{dx^2}\right)_1 = 0,\quad \left(\frac{d^2f_0}{dx^2}\right)_0 = 0.$$

---

* The discussion of this case has been inadvertently omitted in the text. It will be found in the note on p. 171.

Making these substitutions in equations (C), we have

$$a + mb = 0, \quad a + nb = 0,$$

or (unless the given lines be parallel)

$$a = 0, \quad b = 0.$$

The general equation (C), p. 168, becomes

$$\rho = \text{const.}$$

The elastic curve is, therefore, in this case, a circle.

PROP. VI.

140. Given the volume of a surface of revolution, to determine its form so that the attraction upon a point in its axis may be a maximum.*

If the solid be divided by planes perpendicular to the axis, the attraction of an indefinitely thin lamina included between two of these planes is

$$2\pi dx \left(1 - \frac{x}{\sqrt{(x^2 + y^2)}}\right),$$

the axis of revolution being taken as axis of x. The attraction of the entire solid will, therefore, be

$$2\pi \int \left(1 - \frac{x}{\sqrt{(x^2 + y^2)}}\right) dx.$$

The volume of the solid will be

$$\pi \int y^2 dx.$$

Hence the integral which is to be made an absolute maximum is

$$\int V dx,$$

where

$$V = 1 - \frac{x}{\sqrt{(x^2 + y^2)}} + ay^2,$$

a being an indeterminate constant. The equation

$$N - \frac{dP_1}{dx} + \&c. = 0,$$

* Airy's Mathematical Tracts, p. 247.

of p. 38, is therefore reduced to $N=0$, or

$$\frac{xy}{(x^2+y^2)^{\frac{3}{2}}} + 2ay = 0. \qquad \text{(A)}$$

The factor $y=0$ being inadmissible, we have, as the equation of the generating curve,

$$x + 2a(x^2+y^2)^{\frac{3}{2}} = 0,$$

the constant a being determined from the given volume.

Prop. VII.

141. Given the superficial area of an attracting surface, to determine its form so that its *potential*, with regard to a given point, may be a maximum.

If the given point be taken as origin, and the law of the attraction be represented by $\psi(r)$, the potential of the attracting surface will be

$$\iint \phi(r)dS,$$

where dS is the element of the surface, and

$$\phi(r) = -\int\psi(r)dr.$$

Hence, as the superficial area is given, the integral which is to be made an absolute maximum is

$$\iint \{\phi(r)+a\}\, dS.$$

The present question is therefore a case of Prop. I., Chap. VIII., in which

$$\mu = \phi(r)+a, \quad \frac{d\mu}{dx} = \frac{d\phi}{dr}\frac{x}{r}, \quad \frac{d\mu}{dy} = \frac{d\phi}{dr}\frac{y}{r}, \quad \frac{d\mu}{dz} = \frac{d\phi}{dr}\frac{z}{r}.$$

Substituting these values in equation (B), p. 277, and putting for $\frac{d\phi}{dr}$ its value $-\psi(r)$, we have

$$\frac{1}{R} + \frac{1}{R'} = \frac{\psi(r)}{\phi(r)+a}\left(\frac{x\cos\alpha + y\cos\beta + z\cos\gamma}{r}\right), \qquad \text{(A)}$$

or, if P be the perpendicular from the origin upon the tangent plane,

$$\frac{1}{R} + \frac{1}{R'} = \frac{P}{r} \cdot \frac{\psi(r)}{\phi(r) + a}. \qquad \text{(B)}$$

If the portion of surface with which we are concerned be bounded by two given surfaces, it is evident, from the general principle of p. 278, that unless

$$\phi(r_1) + a = 0, \qquad \text{(C)}$$

the surface must cut the first limiting surface at right angles, and that unless

$$\phi(r_0) + a = 0, \qquad \text{(D)}$$

it must so cut the other limiting surface. The equations (C) and (D) evidently denote that the limiting curves of the several surfaces which solve the problem are situated upon spheres whose centre is the attracted point, and whose radii are the several roots of the equation

$$\phi(r) + a = 0.$$

142. Suppose, for example,

$$\psi(r) = r^n,$$

then

$$\phi(r) = -\frac{1}{n+1} r^{n+1}.$$

Making these substitutions in (B), we find

$$\frac{1}{R} + \frac{1}{R'} = \frac{P r^{n-1}}{a - \frac{r^{n+1}}{n+1}}.$$

Equations (C) and (D) would become, on the same supposition,

$$r_1^{n+1} = (n+1)\, a = r_0^{n+1}.$$

The two limiting curves would therefore be situated upon the same sphere. The general equation (B) would give for each of these curves either

$$\frac{1}{R} + \frac{1}{R'} = \infty, \text{ or } P = 0,$$

denoting that the curve is either a *singular* curve, or the curve of contact of the circumscribing cone. But it is easy to see that the hypotheses,

$$\mu_1 = 0, \quad \mu_0 = 0,$$

would not, in general, give a real maximum. For we have seen (p. 271) that it is necessary to the existence of a maximum or minimum value that the quantity

$$\frac{d^2V}{dp^2}\delta p^2 + 2\frac{d^2V}{dpdq}\delta p\delta q + \frac{d^2V}{dq^2}\delta q^2$$

should not change its sign within the limits of integration. Now if

$$V = \mu\sqrt{(1 + p^2 + q^2)},$$

we have

$$\frac{d^2V}{dp^2} = \frac{\mu(1 + q^2)}{(1 + p^2 + q^2)^{\frac{3}{2}}}, \quad \frac{d^2V}{dpdq} = -\frac{\mu pq}{(1 + p^2 + q^2)^{\frac{3}{2}}}, \quad \frac{d^2V}{dq^2} = \frac{\mu(1 + p^2)}{(1 + p^2 + q^2)^{\frac{3}{2}}}.$$

Hence,

$$\frac{d^2V}{dp^2}\delta p^2 + 2\frac{d^2V}{dpdq}\delta p\delta q + \frac{d^2V}{dq^2}\delta q^2$$

$$= \mu\frac{(1 + q^2)\delta p^2 - 2pq\delta p\delta q + (1 + p^2)\delta p^2}{(1 + p^2 + q^2)^{\frac{3}{2}}}$$

$$= \mu\frac{\delta p^2 + \delta q^2 + (q\delta p - p\delta q)^2}{(1 + p^2 + q^2)^{\frac{3}{2}}}.$$

The sign of this quantity is evidently the same as that of μ. Hence, unless μ be a function which passes through zero without changing its sign, the supposition that μ vanishes within the limits of integration ought to be inadmissible. We must, therefore, in general, recur to the other supposition, namely, that the surface intersects its limiting surfaces at right angles.

ON THE APPLICATION OF THE CALCULUS OF VARIATIONS TO THE DEDUCTION OF EQUATIONS OF EQUILIBRIUM AND MOTION.

143. Hitherto the applications of the Calculus of Variations, to which our attention has been directed, have been in strict accordance with the definition of that science as laid down in Chapter I. So long as the increments with which we are concerned are of a nature purely mathematical, it is plain that the

value of a function can be changed in two, and but two, ways, namely, either by a change in one or more of the independent variables, or by a change in the form of the function. And if we wish to give to the symbol of the Calculus of Variations, δ, a distinct signification, we must be careful to apply it only to those increments which result from a change in *form.* If, therefore, x denote an independent variable, δx, considered as a symbol distinct from dx, is, mathematically speaking, unmeaning. As long as we confine ourselves to purely mathematical conceptions, an independent variable admits of but one species of increment. But the case is materially altered by the introduction of the mechanical conception of *motion.* The symbol δ no longer denotes the increment which is produced by a change in form, but the increment which is produced by a change in position, by the motion of a particle from one point of space to another. In fact the increment denoted by δ is not a *variation,* but a *displacement;* and although the science which treats of such increments is, generally speaking, identical in its rules with the Calculus of Variations, it is, nevertheless, in its fundamental conceptions, essentially distinct; and no small obscurity has been thrown over the purely mathematical science by the introduction of principles which properly belong only to the mechanical science. Our first object, then, must be to consider how far this fundamental difference of conception introduces a difference in the rules which have been laid down in the preceding part of this work. One of these differences has been already alluded to. The symbol δ, as applied to an independent variable, ceases to be identical with d, the latter having reference to the distance between one molecule and another, and the former denoting the displacement of the *same* molecule. Again, in the purely mathematical, or, to speak more properly, geometrical science, we have not found it necessary to make use of such symbols as

$$\delta.dx, \quad \delta.dxdy, \quad \delta.dxdydz, \text{ \&c.},$$

the quantities with which we have been hitherto concerned not being *elements*, but *aggregates* of elements, or, in other words, *integrals.* Thus in investigating the variation of a definite integral, such as

$$\iiint V dxdydz,$$

it is plainly unmeaning to assign a variation to the element $dxdydz$, inasmuch as the value of the expression does not in any way depend upon this element, but solely upon the limits of integration and the form of the function to be integrated, V.

But in the investigation of the equations of equilibrium or motion of a continuous system, the *variation* of the *element*, that is to say, the change in geometrical magnitude which it undergoes in consequence of the displacement of the molecules which compose it, cannot be neglected. We shall find that this species of variation is introduced into mechanical problems in two ways, namely:—1. By the nature of molecular force. 2. By the nature of the equations of condition which subsist in a continuous system.

The *moment* of a force, as defined by Lagrange, is measured by the product of the force and the effect which it tends to produce. Thus, if F be the force, ω the quantity which it tends to change, and δ a symbol denoting the change, the moment of the force F will be

$$F\delta\omega.$$

Now certain forces (as, for example, the force of elasticity) are defined by the change which they tend to produce in an *element* of the continuous system under consideration, and cannot be adequately defined with respect to a change in any finite portion. Thus, although the force of elasticity does tend to alter the magnitude of a finite portion of the system, it cannot be adequately defined with respect to such a change, inasmuch as the same effect would be produced by other kinds of force wholly different in their nature, as, for example, by forces whose operation was limited to certain regions only of the finite portion. The essential difference of the force of elasticity is that it acts in *every* element of the system. It is evident, therefore, that the mathematical expression of the moment of a force, such as the force of elasticity, must include the variation (as above defined) of an element.

Again, it may so happen that one of the conditions by which the system is restricted may not be capable of being expressed otherwise than by an equation consisting of elements or infinitesimals. Thus, for example, the incompressibility of a fluid can only be represented by an equation of the form

$$dm = \text{const.}$$

For it is evident that an equation such as

$$M = \text{const.},$$

M being any finite portion of the fluid, would not denote that the fluid was incompressible, but merely that the compression of part of M was equal to the dilatation of the remainder. Now since the application of the method of Lagrange requires the variation of the equations of condition, it is plain that if one of these equations be of the form stated above, its variation will involve the variation of an element.

If, then, we distinguish by the name of *mechanical* variation a variation produced by displacement, the investigation of the mechanical variation of an element will be an essential preliminary to our present purpose.

Again, in the application of the principles of Lagrange to the equilibrium or motion of a continuous system, it is frequently necessary to determine the mechanical variation of functions similar in their nature to those which have been already investigated mathematically. Thus, for example, if ξ, η, ζ be the actual displacements of a molecule in a continuous system, and if

$$V = f\left(x, y, z, \frac{d\xi}{dx}, \frac{d\xi}{dy}, \frac{d\xi}{dz}, \frac{d\eta}{dx} \cdots \frac{d\zeta}{dx} \cdots\right)$$

it is frequently necessary to determine the mechanical variation of V, i. e. the change which V undergoes in consequence of a *virtual* displacement of the molecule. It will, therefore, be necessary to inquire whether and how far this *mechanical* variation is identical with that which has been deduced by purely mathematical considerations. We shall proceed to consider these questions severally.

Prop. VIII.

144. To find the mechanical variation of the element

$$dxdydz,$$

or, in other words, the change in the volume of an elementary parallelopiped of a continuous body, produced by the displacement of the molecules of which it is composed.*

* Mécanique Analytique, pp. 191–4.

Let x, y, z be the co-ordinates of a molecule in the body, and suppose that after the displacement of the molecule these co-ordinates become

$$x+\xi, \quad y+\eta, \quad z+\zeta,$$

the complete displacement being thus made up of three partial displacements parallel to the axes of x, y, z respectively.

It is evidently necessary to the application of mathematics to a problem like the present, that we should suppose the displacements to follow some regular law, connecting by a determinate relation the displacement of any given molecule with its original position. We assume, therefore, that each of the quantities ξ, η, ζ is a function of the co-ordinates x, y, z.

Let $ABCD$ (Fig. 16) be one of the faces of the parallelopiped $dxdydz$, as, for example, the face $dxdy$. Then it is plain that the co-ordinates of the several points are

$$A,\ x,\ y,\ z, \qquad B,\ x+dx,\ y,\ z,$$
$$C,\ x,\ y+dy,\ z, \quad D,\ x+dx,\ y+dy,\ z.$$

Now suppose the several molecules to be displaced, and let the new position of $ABCD$ be $A'B'C'D'$. Then the co-ordinates of the points $A'B'C'D'$ are:

Of A', $x+\xi$, $y+\eta$, $z+\zeta$.

Of B', $x+dx+\xi+\frac{d\xi}{dx}dx$, $y+\eta+\frac{d\eta}{dx}dx$, $z+\zeta+\frac{d\zeta}{dx}dx$.

(A)

Of C', $x+\xi+\frac{d\xi}{dy}dy$, $y+dy+\eta+\frac{d\eta}{dy}dy$, $z+\zeta+\frac{d\zeta}{dy}dy$.

Of D', $x+dx+\xi+\frac{d\xi}{dx}dx+\frac{d\xi}{dy}dy$, $y+dy+\eta+\frac{d\eta}{dx}dx+\frac{d\eta}{dy}dy$,

$$z+\zeta+\frac{d\zeta}{dx}dx+\frac{d\zeta}{dy}dy.$$

Hence we have evidently

$$A'B' = dx\sqrt{\left\{\left(1+\frac{d\xi}{dx}\right)^2+\frac{d\eta^2}{dx^2}+\frac{d\zeta^2}{dz^2}\right\}} = C'D'.$$

(B)

$$A'C' = dy\sqrt{\left\{\frac{d\xi^2}{dy^2}+\left(1+\frac{d\eta}{dy}\right)^2+\frac{d\xi^2}{dy^2}\right\}} = B'D'.$$

The figure $A'B'C'D'$ is therefore a parallelogram. Neglecting quantities of the second order, we have

$$A'B' = \left(1 + \frac{d\xi}{dx}\right) dx, \quad A'C' = \left(1 + \frac{d\eta}{dy}\right) dy.$$

Proceeding in the same way with the remaining faces of the parallelopiped $dxdydz$, we shall easily see that the figure which it assumes is still a parallelopiped whose three edges are

$$\left(1 + \frac{d\xi}{dx}\right) dx, \quad \left(1 + \frac{d\eta}{dy}\right) dy, \quad \left(1 + \frac{d\zeta}{dz}\right) dz.$$

Neglecting terms of an order higher than the first, the product of the three edges will be

$$\left(1 + \frac{d\xi}{dx} + \frac{d\eta}{dy} + \frac{d\zeta}{dz}\right) dxdydz. \qquad \text{(C)}$$

To find the change in the angle at which two edges, $A'B'$, $A'C'$, are inclined to each other, we have, denoting the angle between $A'B'$ and $A'C'$ by γ',

$$2A'B'.A'C' \cos\gamma' = A'B'^2 + A'C'^2 - B'C'^2. \qquad \text{(D)}$$

But from expressions (A) we find

$$B'C'^2 = \left\{\left(1 + \frac{d\xi}{dx}\right) dx - \frac{d\xi}{dy} dy\right\}^2$$

$$+ \left\{\left(1 + \frac{d\eta}{dy}\right) dy - \frac{d\eta}{dx} dx\right\}^2 + \left(\frac{d\zeta}{dx} dx - \frac{d\zeta}{dy} dy\right)^2.$$

Substituting in (D) the several values of $A'B'$, $A'C'$, $B'C'$, and neglecting quantities of an order higher than the first, we have

$$\cos\gamma' = \frac{d\xi}{dy} + \frac{d\eta}{dx},$$

and in the same way

$$\cos\beta' = \frac{d\zeta}{dx} + \frac{d\xi}{dz},$$

$$\cos\alpha' = \frac{d\eta}{dz} + \frac{d\zeta}{dy}.$$

From these expressions it is plain that the angles of the new parallelopiped will differ infinitely little from right angles, and therefore, that the volume of the parallelopiped will differ from the expression (C) by a quantity of an higher order than any of the terms in that expression. Hence it is evident that the variation of $dxdydz$ will be

$$\left(\frac{d\xi}{dx} + \frac{d\eta}{dy} + \frac{d\zeta}{dz}\right) dxdydz.$$

If the displacements be of the species denominated *virtual*, it is usual to denote them by the symbol δ. Employing this notation we shall have

$$\delta . dxdydz = \left(\frac{d\delta x}{dx} + \frac{d\delta y}{dy} + \frac{d\delta z}{dz}\right) dxdydz.$$

We shall, for the sake of uniformity, confine the use of the symbol δ to displacements of this nature, using ξ, η, ζ to denote *real* displacements.

Prop. IX.

145. To find the mechanical variation of V, where

$$V = f\left(x, y, z, \xi, \eta, \zeta, \frac{d\xi}{dx}, \frac{d\xi}{dy}, \text{\&c.}, \frac{d\eta}{dx}, \text{\&c.}, \frac{d\xi}{dx}, \text{\&c.}\right),$$

containing differential coefficients of ξ, η, ζ of any order.

If we assume, as in p. 314, that the displacements follow some regular law, and, therefore, that ξ, η, ζ are functions of x, y, z, it will immediately appear that V can be varied only in one of two ways, namely:—1. By a change in some one of the quantities x, y, z. 2. By a change in the form of some one of the functions ξ, η, ζ. The variables x, y, z are, as is evident from p. 311, capable of two species of increment, namely, the increment which relates to the *distance* between two points or molecules in the body, and the increment which refers to the *displacement* of an individual molecule. The former species being excluded from the present problem by the signification of the term *variation*, which denotes a change referring throughout to the *same* molecule, we shall have, by the principles of Chap. I.,

$$\delta V = \left(\frac{df}{dx} + \frac{df}{d\xi}\frac{d\xi}{dx} + \frac{df}{d \cdot \frac{d\xi}{dx}}\frac{d^2\xi}{dx^2} + \&c. + \frac{df}{d\eta}\frac{d\eta}{dx} + \&c.\right)\delta x$$

$$+ \left(\frac{df}{dy} + \frac{df}{d\xi}\frac{d\xi}{dy} + \&c.\right)\delta y + \left(\frac{df}{dz} + \frac{df}{d\xi}\frac{d\xi}{dz} + \&c.\right)\delta z$$

$$+ \frac{df}{d\xi}\delta\xi + \frac{df}{d \cdot \frac{d\xi}{dx}}\frac{d\delta\xi}{dx} + \&c.$$

$$+ \frac{df}{d\eta}\delta\eta + \&c. + \frac{df}{d\zeta}\delta\zeta + \&c.$$

Or, as it may be more briefly written,

$$\left.\begin{aligned}\delta V = \left(\frac{dV}{dx}\right)\delta x + \left(\frac{dV}{dy}\right)\delta y + \left(\frac{dV}{dz}\right)\delta z \\ + \frac{dV}{d\xi}\delta\xi + \frac{dV}{d \cdot \frac{d\xi}{dx}}\frac{d\delta\xi}{dx} + \&c.\end{aligned}\right\} \quad (A)$$

denoting by

$$\left(\frac{dV}{dx}\right), \left(\frac{dV}{dy}\right), \left(\frac{dV}{dz}\right),$$

the complete differential coefficients of V considered as a function implicit as well as explicit, of x, y, z. It is evident from this expression that the *mathematical* variation, which is

$$\frac{dV}{d\xi}\delta\xi + \frac{dV}{d \cdot \frac{d\xi}{dx}}\frac{d\delta\xi}{dx} + \&c., \quad (B)$$

will not be identical with the mechanical unless

$$\left(\frac{dV}{dx}\right)\delta x + \left(\frac{dV}{dy}\right)\delta y + \left(\frac{dV}{dz}\right)\delta z = 0.$$

But in the ordinary use of the symbols x, y, z, which denote the co-ordinates not of any really existing molecules, but of former positions of the molecules, no such symbol as δx can occur. In this case, therefore, δV is given by (B).

We shall now proceed to give some examples of the application of the method of Lagrange to problems of equilibrium and motion, confining our attention to the case of a *continuous* system, partly because it is to this case that the method of Lagrange is most advantageously applied, and partly because it is only in this case that the peculiar rules of the Calculus of Variations, as distinguished from any other method of arbitrary increments, become necessary. The method to be pursued being, in its nature, very uniform, a small number of examples will suffice to illustrate it.

Prop. X.

146. To find the conditions of equilibrium of a flexible surface, extensible or inextensible, acted on by any system of forces.*

Let dS be the element of the surface. Then, if this surface be inextensible, the principle of virtual velocities gives the equation

$$\iint (X\delta x + Y\delta y + Z\delta z)\, dS = 0;$$

and the inextensibility of the surface gives the equation of condition,

$$\delta dS = 0.$$

Hence, according to the principle of Lagrange, the equation, which is to be treated by the rules of the Calculus of Variations, is

$$\iint (X\delta x + Y\delta y + Z\delta z)\, dS + \iint \lambda \delta dS = 0. \qquad \text{(A)}$$

Assuming

$$U = \surd(1 + p^2 + q^2),$$

we have

$$dS = U dx dy,$$

and therefore

$$\delta . dS = U\delta . dx dy + \delta U dx dy.$$

Now it is evident from Prop. VIII. that

$$\delta . dx dy = \left(\frac{d\delta x}{dx} + \frac{d\delta y}{dy}\right) dx dy;$$

also,

$$\delta U = \frac{dU}{dx}\delta x + \frac{dU}{dy}\delta y + \frac{1}{U}\left(p\frac{d\delta' z}{dx} + q\frac{d\delta' z}{dy}\right);$$

* Mécanique Analytique, p. 148.

denoting here by $\delta' z$ the *mathematical* variation of z, that is to say, the increment which it receives in consequence of a change of form only. Substituting for $\delta' z$ its value,

$$\delta z - p\delta x - q\delta y,$$

integrating by parts, and putting, for the sake of brevity,

$$V = \frac{d \cdot \frac{\lambda p}{U}}{dx} + \frac{d \cdot \frac{\lambda q}{U}}{dy},$$

we have

$$\begin{aligned} \int\{U\delta y + \frac{q}{U}(\delta z - p\delta x - q\delta y)\}\lambda dx + \int\{U\delta x + \frac{p}{U}(\delta z - p\delta x - q\delta y)\}\lambda dy \\ + \iint(UX + \lambda\frac{dU}{dx} - \frac{d\lambda U}{dx} + Vp)\,\delta x dx dy \\ + \iint(UY + \lambda\frac{dU}{dy} - \frac{d.\lambda U}{dy} + Vq)\,\delta y dx dy \\ + \iint(UZ - V)\delta z dx dy = 0. \end{aligned} \qquad (B)$$

Equating to zero the coefficients of δx, δy, δz, under the sign of double integration, and reducing, we find

$$\begin{aligned} U\left\{X - \left(\frac{d\lambda}{dx}\right)\right\} + Vp = 0, \\ U\left\{Y - \left(\frac{d\lambda}{dy}\right)\right\} + Vq = 0, \\ UZ - V = 0, \end{aligned} \qquad (C)$$

where

$$\left(\frac{d\lambda}{dx}\right) = \frac{d\lambda}{dx} + p\frac{d\lambda}{dz},$$

$$\left(\frac{d\lambda}{dy}\right) = \frac{d\lambda}{dy} + q\frac{d\lambda}{dz}.$$

Multiplying these equations by dx, dy, dz, respectively, and adding, we have

$$\frac{d\lambda}{dx}dx + \frac{d\lambda}{dy}dy + \frac{d\lambda}{dz}dz = Xdx + Ydy + Zdz. \qquad (D)$$

If the forces which act upon the surface be such as to satisfy the condition

$$Xdx + Ydy + Zdz = d\Pi,$$

we have, from equation (D),

$$d\lambda = d\Pi, \quad \lambda = \Pi + a.$$

Substituting this value in the third of equations (C) we have, as the equation of the surface,

$$\frac{d \cdot \dfrac{(\Pi + a)\,p}{U}}{dx} + \frac{d \cdot \dfrac{(\Pi + a)\,q}{U}}{dy} - UZ = 0; \qquad \text{(E)}$$

or if the equation be transformed, as in Prop. I., Chap. VIII.,

$$\frac{1}{R} + \frac{1}{R'} = -\frac{1}{\Pi + a}(X\cos\alpha + Y\cos\beta + Z\cos\gamma). \qquad \text{(F)}$$

Now it appears from Chap. VIII. that this equation also furnishes the solution of an ordinary mathematical problem of maxima and minima, namely:

Given the superficial area of a surface, to determine it so that

$$\iint \Pi\, dx dy$$

may be a maximum or minimum.

It is evident, therefore, that the mechanical problem of equilibrium is in this case identical with a purely mathematical question of maxima and minima. We shall see afterwards that this is a case of a more general theorem.

We shall now proceed to consider the conditions to be fulfilled at the limits of integration.

Suppose the given flexible surface to be bounded by two flexible threads, each capable of motion along a given surface, with which it is everywhere in contact, but incapable of motion in the substance of the membrane itself. The fulfilment of this last condition being evidently implied in the signification of the symbol δ, which excludes the idea of change from one point of the substance to another, it is unnecessary to consider it further.

Let the equation of one of the limiting surfaces be

$$dz = p'dx + q'dy.$$

We have then, from the signification of δ,

$$\delta z = p'\delta x + q'\delta y. \tag{G}$$

The terms under the sign of single integration in (B) give, as in p. 247, for each limit of integration,

$$\{U\delta y + \frac{q}{U}(\delta z - p\delta x - q\delta y)\} - \{U\delta x + \frac{p}{U}(\delta z - p\delta x - q\delta y\}\left(\frac{dy}{dx}\right) = 0.$$

Substituting for δz its value derived from (G), eliminating $\frac{dy}{dx}$ by the equation

$$\frac{dy}{dx} = -\frac{p' - p}{q' - q},$$

and equating severally to zero the coefficients of δy, δx, we find the single equation,

$$1 + pp' + qq' = 0.$$

The surface of the membrane, therefore, cuts its bounding surface at right angles. The same is manifestly true for the second bounding surface.

147. The reduction of the equations derived from the coefficients of δx, δy to one, is evidently essential to the possibility of fulfilling the conditions of the problem. It is not peculiar to the present case, and may be proved generally as follows:

It will readily appear, from the composition of the terms under the single sign of integration, that the increments δx, δy are introduced into these terms in two ways, namely:—1. By the mechanical variation of the element $dxdy$. 2. By the substitution of

$$\delta z - p\delta x - q\delta y, \text{ or } (p' - p)\,\delta x + (q' - q)\,\delta y$$

for the mathematical variation $\delta' z$. The terms introduced by the variation of $dxdy$ are evidently of the form

$$\int\Theta\delta x dy + \int\Theta\delta y dx, \tag{H}$$

the coefficients of δx, δy being the same in both. With regard to the terms introduced by $\delta' z$, if these terms be represented by

$$\int\Omega\delta' z dx + \int\Omega_1\delta' z dy,$$

it is evident that the part which contains δx, δy will be

$$\int \Omega\{(p'-p)\delta x + (q'-q)\delta y\}dx + \int \Omega_1\{(p'-p)\delta x + (q'-q)\delta y\}dy. \quad \text{(I)}$$

Adding the expressions (H) and (I), changing the independent variable as before, and equating to zero the coefficients of δx, δy, we find

$$-\Theta\frac{dy}{dx} + \Omega(p'-p) - \Omega_1(p'-p)\frac{dy}{dx} = 0,$$

$$\Theta + \Omega(q'-q) - \Omega_1(q'-q)\frac{dy}{dx} = 0.$$

If now we substitute for $\frac{dy}{dx}$ its value $-\frac{p'-p}{q'-q}$ these equations become identical, and equivalent to the single equation

$$\Theta + \Omega(q'-q) + \Omega_1(p'-p) = 0.$$

If the membrane be flexible and extensible, the condition

$$\delta dS = 0$$

will disappear, and a new force be introduced, whose moment is

$$F\delta dS.$$

Hence it is evident that the preceding discussion is adapted to the present case by simply changing λ into F. The conclusions arrived at are therefore the same.

148. As an example of this proposition, let us suppose the membrane to be urged by a force which is at all points perpendicular to its surface. This condition will evidently give

$$Xdx + Ydy + Zdz = 0,$$

and therefore $\lambda = a$. Substituting this value in the third equation (C), we find

$$\frac{d\,.\,\frac{p}{U}}{dx} + \frac{d\,.\,\frac{q}{U}}{dy} = \frac{F}{a},$$

where F is the acting force. Equation (F) becomes, in the same case,

$$\frac{1}{R} + \frac{1}{R'} = -\frac{F}{a}.$$

If the force be constant, the surface of equilibrium will have its mean curvature constant, a property which, as we have seen in p. 281, belongs to the surface which contains a maximum volume under a given superficial area. This is evidently the case of a flexible surface subjected to the pressure of a homogeneous elastic fluid.

If the acting force were that of gravity solely, we should have in (F)

$$\Pi = gz,$$

and, therefore,

$$\frac{1}{R} + \frac{1}{R'} = -\frac{g\cos\gamma}{gz + a};$$

or, if N be the portion of the normal intercepted between the surface and the plane whose equation is

$$z + \frac{a}{g} = 0,$$

$$\frac{1}{R} + \frac{1}{R'} = -\frac{1}{N}$$

expressing a property analogous to that of the catenary.

Prop. XI.

149. If any continuous body be acted on by a system of forces which satisfy the condition

$$Xdx + Ydy + Zdz = d\Pi,$$

as also by a force tending to increase or diminish the element of the body, the investigation of its equilibrium may be reduced to a question of ordinary maxima and minima.

Let dm be the element of the body. The moment of the force which tends to augment dm being

$$F\delta dm,$$

the equation furnished by the principle of virtual velocities will be

$$\int(X\delta x + Y\delta y + Z\delta z)\,dm + \int F\delta dm = 0. \qquad \text{(A)}$$

Substituting for dm its value, which will be of the form

$$Vdx,\ Vdxdy,\ \text{or}\ Vdxdydz,$$

according as the given body is a line, a surface, or a solid, integrating by parts, and equating to zero the coefficients of δx, δy, δz, severally, we find the three equations of equilibrium. We shall consider specially the case of a solid body, as being the most comprehensive. We have then

$$dm = Vdxdydz,$$

where

$$V = f\left(x, y, z, u, \frac{du}{dx}, \&c.\right),$$

u being a quantity depending upon the aggregation of the molecules, heat, or any other cause which affects the density of the body, which is, in this case, represented by V. We have, therefore,

$$\delta dm = \left\{ V\left(\frac{d\delta x}{dx} + \frac{d\delta y}{dy} + \frac{d\delta z}{dz}\right) + \delta V\right\} dxdydz;$$

or, since

$$\delta V = \left(\frac{dV}{dx}\right)\delta x + \left(\frac{dV}{dy}\right)\delta y + \left(\frac{dV}{dz}\right)\delta z + \delta' V$$

$$= \left(\frac{dV}{dx}\right)\delta x + \left(\frac{dV}{dy}\right)\delta y + \left(\frac{dV}{dz}\right)\delta z + U\delta' u + V_x\frac{d\delta' u}{dx} + \&c.$$

$$\delta dm = \left\{ V\left(\frac{d\delta x}{dx} + \frac{d\delta y}{dy} + \frac{d\delta z}{dz}\right) + \left(\frac{dV}{dx}\right)\delta x + \left(\frac{dV}{dy}\right)\delta y + \left(\frac{dV}{dz}\right)\delta z\right.$$

$$\left. + U\delta' u + V_x\frac{d\delta' u}{dx} + \&c.\right\} dxdydz,$$

where U, V_x, &c., have the same signification as in Prop. III. Chap. VII., and

$$\left(\frac{dV}{dx}\right), \left(\frac{dV}{dy}\right), \left(\frac{dV}{dz}\right)$$

denote the complete differential coefficients of V considered as a function, implicit as well as explicit, of x, y, z. Substituting this value in (A), integrating by parts, and equating to zero the coefficients of δx, δy, δz, $\delta' u$, we have

$$VX + F\left(\frac{dV}{dx}\right) - \left(\frac{d.FV}{dx}\right) = 0,$$

$$VY + F\left(\frac{dV}{dy}\right) - \left(\frac{d.FV}{dy}\right) = 0, \qquad \text{(B)}$$

$$VZ + F\left(\frac{dV}{dz}\right) - \left(\frac{d.FV}{dz}\right) = 0.$$

$$FU - \left(\frac{d.FV_x}{dx}\right) - \&c. = 0.$$

If the first three equations be multiplied by dx, dy, dz respectively, and added, we shall have

$$\frac{dF}{dx}dx + \frac{dF}{dy}dy + \frac{dF}{dz}dz = Xdx + Ydy + Zdz = d\Pi;$$

or, by integration,

$$F = \Pi + a.$$

If this value of F be substituted in (A) that equation will become

$$\int\{\delta\Pi dm + (\Pi + a)\delta dm\} = 0,$$

or

$$\delta\int\Pi dm + a\delta\int dm = 0.$$

Hence it is evident that the condition of equilibrium is found by determining, among all forms of the function u which leave the mass of the body unaltered, that form which renders

$$\int\Pi dm$$

a maximum or minimum. The simplification which this introduces into the general problem, in permitting us to disregard the variations of the elements, is too obvious to require further notice.

Prop. XII.

150. To find the equations of motion of an elastic body, whose several molecules have undergone an indefinitely small displacement from their original position of equilibrium.

In applying the method of Lagrange to the present case, we commence with two important assumptions, namely: 1. That the sum of the internal moments of the body may be represented by the variation (mechanical) of a single function, V. 2. That V is a function of the first differential coefficients of the displacements of the molecules.

The first of these assumptions is evidently true for a body whose molecules attract or repel according to any law of the distance between them. But recent experiments show that there are bodies whose constitution is not of this nature, and the assumption of the general truth of (1) can only be regarded as a hypothesis whose degree of probability will depend upon the number of experimentally true results which can be deduced from it.

The reason for the second assumption is grounded upon the nature of molecular force, which appears to depend upon *relative*, not *absolute* displacements of the molecules.

We shall now proceed to investigate the equations of motion for an elastic body not acted on by any external forces, and with a given form of the function V. The method being in its nature very uniform, it will be sufficient to exemplify it by a simple case. For more ample information upon this important problem the reader is referred to Professor Mac Cullagh's memoir on the Undulatory Theory of Light;* Mr. Green's memoir on the same subject;† and especially to Mr. Haughton's two Memoirs, on the Equilibrium and Motion of Solid and Fluid Bodies,‡ and on the Classification of Elastic Media.§

Let the co-ordinates of any molecule, in its original position, be x, y, z; and suppose that after the displacement they become

$$x+\xi, \quad y+\eta, \quad z+\zeta.$$

Then we shall assume that the nature of the body is such that the sum of the moments of the forces developed by the displacement may be represented by

$$\iiint \delta V dx dy dz,$$

where

$$V = \tfrac{1}{2} U \left(\frac{d\xi}{dx} + \frac{d\eta}{dy} + \frac{d\zeta}{dz} \right)^2, \qquad \text{(A)}$$

---

* Transactions of the Royal Irish Academy, vol. xxi.

† Transactions of the Cambridge Philosophical Society, vol. vii. p. 11.

‡ Transactions of the Royal Irish Academy, vol. xxi.

§ Ibid. vol. xxii. Part. i.

U being any function of x, y, z. The general equation of motion being

$$\iiint\left(\frac{d^2\xi}{dt^2}\,\delta\xi + \frac{d^2\eta}{dt^2}\,\delta\eta + \frac{d^2\zeta}{dt^2}\,d\zeta\right)\varepsilon dx dy dz = \iiint\delta V dx dy dz, \qquad \text{(B)}$$

our first object in adapting it to the present case must be to determine the mechanical variation denoted by the symbol δV. Now in performing the operation denoted by δ, it must be recollected that this symbol refers to a virtual displacement of the molecules from the actual position in which they are placed by the conditions of the problem. We have, therefore, in this operation, no concern with any position which these molecules may have previously had. Hence it is evident that the symbol δ does not apply to x, y, z, which are the co-ordinates not of any really existing molecules, but merely of the *positions* which these molecules *had* in their original state of equilibrium. The mechanical variation of V therefore coincides with its mathematical variation, and results solely from a change in the form of the functions ξ, η, ζ. We have, therefore,

$$\delta V = \tfrac{1}{2}\,\delta . U\left(\frac{d\xi}{dx} + \frac{d\eta}{dy} + \frac{d\zeta}{dz}\right)^2$$

$$= U\left(\frac{d\xi}{dx} + \frac{d\eta}{dy} + \frac{d\zeta}{dz}\right)\left(\frac{d\delta\xi}{dx} + \frac{d\delta\eta}{dy} + \frac{d\delta\zeta}{dz}\right).$$

Substituting this value in the second member of equation (B), integrating by parts, and equating to zero the coefficients of $\delta\xi, \delta\eta, \delta\zeta$, under the sign of triple integration, we have the three equations of motion,

$$-\,\varepsilon\frac{d^2\xi}{dt^2} = \frac{d}{dx}\cdot U\left(\frac{d\xi}{dx} + \frac{d\eta}{dy} + \frac{d\zeta}{dz}\right)$$

$$-\,\varepsilon\frac{d^2\eta}{dt^2} = \frac{d}{dy}\cdot U\left(\frac{d\xi}{dx} + \frac{d\eta}{dy} + \frac{d\zeta}{dz}\right) \qquad \text{(C)}$$

$$-\,\varepsilon\frac{d^2\zeta}{dt^2} = \frac{d}{dz}\cdot U\left(\frac{d\xi}{dx} + \frac{d\eta}{dy} + \frac{d\zeta}{dz}\right);$$

or, if we assume

$$\omega = \frac{d\xi}{dx} + \frac{d\eta}{dy} + \frac{d\zeta}{dz},$$

$$\frac{d^2\xi}{dt^2} = -\frac{U}{\varepsilon}\frac{d\omega}{dx} - \frac{\omega}{\varepsilon}\frac{dU}{dx}$$

$$\frac{d^2\eta}{dt^2} = -\frac{U}{\varepsilon}\frac{d\omega}{dy} - \frac{\omega}{\varepsilon}\frac{dU}{dy} \qquad \text{(D)}$$

$$\frac{d^2\zeta}{dt^2} = -\frac{U}{\varepsilon}\frac{d\omega}{dz} - \frac{\omega}{\varepsilon}\frac{dU}{dz}.$$

An equation, with a single dependent variable, may be deduced from these by adding them after differentiating respectively with regard to x, y, z. Performing this operation, we easily find

$$\frac{d^2\omega}{dt^2} = -\frac{U}{\varepsilon}\left(\frac{d^2\omega}{dx^2} + \frac{d^2\omega}{dy^2} + \frac{d^2\omega}{dz^2}\right)$$

$$-\left(\frac{2}{\varepsilon}\frac{dU}{dx} - \frac{U}{\varepsilon^2}\frac{d\varepsilon}{dx}\right)\frac{d\omega}{dx} - \left(\frac{2}{\varepsilon}\frac{dU}{dy} - \frac{U}{\varepsilon^2}\frac{d\varepsilon}{dy}\right)\frac{d\omega}{dy} - \left(\frac{2}{\varepsilon}\frac{dU}{dz} - \frac{U}{\varepsilon^2}\frac{d\varepsilon}{dz}\right)\frac{d\omega}{dz}$$

$$-\left(\frac{d}{dx}\cdot\frac{1}{\varepsilon}\frac{dU}{dx} + \frac{d}{dy}\cdot\frac{1}{\varepsilon}\frac{dU}{dy} + \frac{d}{dz}\cdot\frac{1}{\varepsilon}\frac{dU}{dz}\right)\omega. \qquad \text{(E)}$$

The terms without the sign of triple integration give the equation

$$\iint U'\left(\frac{d\xi'}{dx} + \frac{d\eta'}{dy} + \frac{d\zeta'}{dz}\right)(\delta\xi' dydz + \delta\eta' dxdz + \delta\zeta' dxdy) = 0.$$

Let α, β, γ be the angles made with the axes by the normal to the limiting surface, and let dS be the element of this surface. Then it is evident, from p. 216, that the foregoing equation may be written

$$\iint U'\left(\frac{d\xi'}{dx} + \frac{d\eta'}{dy} + \frac{d\zeta'}{dz}\right)(\cos\alpha\,\delta\xi' + \cos\beta\,\delta\eta' + \cos\gamma\,\delta\zeta')\,dS = 0. \qquad \text{(F)}$$

In applying this equation we may have any one of the following cases:

(1.) The external molecules may be absolutely fixed. In this case we have at the limiting surface

$$\delta\xi' = 0, \quad \delta\eta' = 0, \quad \delta\zeta' = 0,$$

and the equation (F) becomes identical.

(2.) The external molecules may be restricted to a given surface, admitting of motion along the surface, and excluding motion perpendicular to the surface. Let the equation of the limiting surface be

$$f(x, y, z) = 0.$$

Then since this equation is supposed to subsist between the co-ordinates of any one of the external molecules, however it may be displaced, we have

$$\frac{df}{dx}\xi + \frac{df}{dy}\eta + \frac{df}{dz}\zeta = 0;$$

and therefore, from p. 327,

$$\frac{df}{dx}\delta\xi + \frac{df}{dy}\delta\eta + \frac{df}{dz}\delta\zeta = 0,$$

or

$$\cos\alpha\delta\xi + \cos\beta\delta\eta + \cos\gamma\delta\zeta = 0.$$

The equation (F) is therefore again identically true.*

(3.) The external molecules may be completely free. In this case the limiting equation can only be satisfied by making

$$U'\left(\frac{d\xi'}{dx} + \frac{d\eta'}{dy} + \frac{d\zeta'}{dz}\right) = 0,$$

and therefore

$$V' = 0.$$

151. The most important case, however, is that in which there are two media in contact with each other, differing in the form of the function V. As an example of this case we shall consider the following question:

Let two elastic media (V_1, V_2) be in contact along an indefinite plane, which we shall take for that of xy. Let the values of V for these media be

$$\left.\begin{aligned} V_1 &= \tfrac{1}{2}\left(A_1\frac{d\xi^2}{dx^2} + B_1\frac{d\eta^2}{dy^2} + C_1\frac{d\zeta^2}{dz^2}\right), \\ V_2 &= \tfrac{1}{2}\left(A_2\frac{d\xi^2}{dx^2} + B_2\frac{d\eta^2}{dy^2} + C_2\frac{d\zeta^2}{dz^2}\right), \end{aligned}\right\} \quad (G)$$

A_1, B_1, C_1, A_2, B_2, C_2, being constants. Suppose now that the molecules of the first medium are disturbed by a plane wave represented by the equations

* Mécanique Analytique, tom. i. p. 207.

$$\xi_1 = L_1 \cos\frac{2\pi}{\lambda_1}(v_1 t - z),$$

$$\eta_1 = M_1 \cos\frac{2\pi}{\lambda_1}(v_1 t - z), \qquad \text{(H)}$$

$$\zeta_1 = N_1 \cos\frac{2\pi}{\lambda_1}(v_1 t - z).$$

Let it be required to determine the nature of the plane wave which is propagated in the second medium.

The action of the molecular forces being insensible at finite distances, it is evident that if we wish to find the motion of a molecule situated in the first medium at a finite distance from the plane of separation, we must make, in the general equation,

$$V = V_1.$$

For a molecule similarly situated in the second medium,

$$V = V_2.$$

For molecules situated in, and infinitely near to the plane of separation,

$$V = V_1 + V_2;$$

and in investigating the conditions to be observed at the limits of integration it is evident that we must make use of this last value. It is easy to see, indeed, that as each of the quantities, V_1, V_2, contains only displacements of its own molecule, the three questions alluded to, namely, that of the motion in the first medium, that of the motion in the second medium, and that of the transmission of the motion from the one medium to the other, will be fully solved by making

$$V = V_1 + V_2$$

in the general equation. This equation becomes, therefore,

$$\iiint \varepsilon_1 \left(\frac{d^2\xi_1}{dt^2}\delta\xi_1 + \frac{d^2\eta_1}{dt^2}\delta\eta_1 + \frac{d^2\zeta_1}{dt^2}\delta\zeta_1\right) dx dy dz$$

$$+ \iiint \varepsilon_2 \left(\frac{d^2\xi_2}{dt^2}\delta\xi_2 + \frac{d^2\eta_2}{dt^2}\delta\eta_2 + \frac{d^2\zeta_2}{dt^2}\delta\zeta_2\right) dx dy dz$$

$$= \iiint \delta V_1 dx dy dz + \iiint \delta V_2 dx dy dz. \qquad \text{(I)}$$

But

$$\delta V_1 = A_1 \frac{d\xi_1}{dx}\frac{d\delta\xi_1}{dx} + B_1 \frac{d\eta_1}{dy}\frac{d\delta\eta_1}{dy} + C_1 \frac{d\zeta_1}{dz}\frac{d\delta\zeta_1}{dz},$$

$$\delta V_2 = A_2 \frac{d\xi_2}{dx}\frac{d\delta\xi_2}{dx} + B_2 \frac{d\eta_2}{dy}\frac{d\delta\eta_2}{dy} + C_2 \frac{d\zeta_2}{dz}\frac{d\delta\zeta_2}{dz}.$$

Substituting these values in (I), integrating by parts, and equating to zero the coefficients of $\delta\xi_1$, $\delta\eta_1$, $\delta\zeta_1$, under the sign of triple integration, we find

$$-\varepsilon_1 \frac{d^2\xi_1}{dt^2} = A_1 \frac{d^2\xi_1}{dx^2}, \quad -\varepsilon_1 \frac{d^2\eta_1}{dt^2} = B_1 \frac{d^2\eta_1}{dy^2}, \quad -\varepsilon_1 \frac{d^2\zeta_1}{dt^2} = C_1 \frac{d^2\zeta_1}{dz^2}, \qquad \text{(K)}$$

the equations of motion in the first medium, and

$$-\varepsilon_2 \frac{d^2\xi_2}{dt^2} = A_2 \frac{d^2\xi_2}{dx^2}, \quad -\varepsilon_2 \frac{d^2\eta_2}{dt^2} = B_2 \frac{d^2\eta_2}{dy^2}, \quad -\varepsilon_2 \frac{d^2\zeta_2}{dt^2} = C_2 \frac{d^2\zeta_2}{dz^2}, \qquad \text{(L)}$$

the equations of motion in the second medium.

The terms without the sign of triple integration give the equation

$$\iint A_1 \frac{d\xi'_1}{dx}\delta\xi'_1 dydz + \iint B_1 \frac{d\eta'_1}{dy}\delta\eta'_1 dxdz + \iint C_1 \frac{d\zeta'_1}{dz}\delta\zeta'_1 dxdy$$

$$\text{(M)}$$

$$-\iint A_2 \frac{d\xi'_2}{dx}\delta\xi'_2\, dydz - \iint B_2 \frac{d\eta'_2}{dy}\delta\eta'_2 dxdz - \iint C_2 \frac{d\zeta'_2}{dz}\delta\zeta'_2 dxdy = 0,$$

the quantities ξ'_1, &c., being accented, to show that they refer to the limits, and a negative sign being interposed between the two classes of terms, inasmuch as the integrations are performed in opposite directions. Let a, β, γ be the angles which the normal to the bounding surface makes with the axes of co-ordinates. Then if the preceding equation be transformed as in p. 215, we shall have at the limit of integration,

$$\left(A_1 \frac{d\xi'_1}{dx}\delta\xi'_1 - A_2 \frac{d\xi'_2}{dx}\delta\xi_2\right)\cos a + \left(B_1 \frac{d\eta'_1}{dy}\delta\eta'_1 - B_2 \frac{d\eta'_2}{dy}\delta\eta'_2\right)\cos\beta$$

$$+\left(C_1 \frac{d\zeta'_1}{dz}\delta\zeta'_1 - C_2 \frac{d\zeta'_2}{dz}\delta\zeta'_2\right)\cos\gamma = 0.$$

In the present case, where the bounding surface is the plane of xy, we have

$$\cos a = 0, \quad \cos\beta = 0;$$

and the foregoing equation becomes

$$C_1 \frac{d\zeta'_1}{dz}\,\delta\zeta'_1 - C_2\,\frac{d\zeta'_2}{dz}\,\delta\zeta'_2 = 0.$$

But since the displacement of the same molecule must be the same to whichever medium it be referred, we must have

$$\delta\zeta'_1 = \delta\zeta'_2.$$

Hence we have, finally,

$$C_1\,\frac{d\zeta'_1}{dz} = C_2\,\frac{d\zeta'_2}{dz}. \qquad \text{(N)}$$

We have, besides, the three equations,

$$\xi'_1 = \xi'_2, \quad \eta'_1 = \eta'_2, \quad \zeta'_1 = \zeta'_2, \qquad \text{(O)}$$

which denote that the vibrating molecules at the bounding surface may be considered as belonging to either medium. There are, therefore, four conditions expressed by equations (N) and (O), to be fulfilled when the oscillation passes from the one medium to the other. When the nature of the oscillation in the first medium is given, these equations, in conjunction with the equations of propagation (L), serve to determine the undulation which is transmitted into the second.

In the present case, where the several quantities

$$A_1,\ B_1,\ C_1, \quad A_2,\ B_2,\ C_2, \quad \varepsilon_1,\ \varepsilon_2,$$

are constants, if we assume

$$-a_1^2\varepsilon_1 = A_1, \quad -b_1^2\varepsilon_1 = B_1, \quad -c_1^2\varepsilon_1 = C_1,$$
$$-a_2^2\varepsilon_2 = A_2, \quad -b_2^2\varepsilon_2 = B_2, \quad -c_2^2\varepsilon_2 = C_2,$$

the equations (K) and (L) will become

$$\frac{d^2\xi_1}{dt^2} = a_1^2\,\frac{d^2\xi_1}{dx^2}, \qquad \frac{d^2\xi_2}{dt^2} = a_2^2\,\frac{d^2\xi_2}{dx^2},$$
$$\frac{d^2\eta_1}{dt^2} = b_1^2\,\frac{d^2\eta_1}{dy^2}, \quad \text{(P)}, \qquad \frac{d^2\eta_2}{dt^2} = b_2^2\,\frac{d^2\eta_2}{dy^2}, \quad \text{(Q)},$$
$$\frac{d^2\zeta_1}{dt^2} = c_1^2\,\frac{d^2\zeta_1}{dz^2}, \qquad \frac{d^2\zeta_2}{dt^2} = c_2^2\,\frac{d^2\zeta_2}{dz^2}.$$

Now since the motion in the first medium is, by hypothesis, represented by the equations (H), if we substitute in (P) the values of ξ_1, η_1, ζ_1, derived from these equations, we shall have

$$L_1 = 0, \quad M_1 = 0, \quad v_1 = c_1.$$

To determine the motion in the second medium, we have, in the first place, the general solution of (Q), namely,

$$\xi_2 = f(x + a_2 t, y, z) + F(x - a_2 t, y, z),$$
$$\eta_2 = \phi(y + b_2 t, x, z) + \Phi(y - b_2 t, x, z),$$
$$\zeta_2 = \psi(z + c_2 t, x, y) + \Psi(z - c_2 t, x, y).$$

In order that these equations should represent plane wave motion, they must be reduced to the form

$$\xi_2 = f(my + nz + x + a_2 t) + F(m'y + n'z + x - a_2 t),$$
$$\eta_2 = \phi(lx + \&c.) + \Phi(l'x + \&c.),$$
$$\zeta_2 = \psi(px + \&c.) + \Psi(p'x + \&c.).$$

The conditions at the limits are

$$\xi'_1 = \xi'_2, \quad \eta'_1 = \eta'_2, \quad \zeta'_1 = \zeta'_2,$$

resulting from the equivalence of vibrations, and

$$C_1 \frac{d\zeta'_1}{dz} = C_2 \frac{d\zeta'_2}{dz}, \qquad \text{(R)}$$

derived from the quantities under the sign of double integration. The equations

$$\xi'_1 = \xi'_2, \quad \eta'_1 = \eta'_2,$$

combined with the equations

$$L_1 = 0, \quad M_1 = 0,$$

give generally, as is easily seen,

$$\xi_2 = 0, \quad \eta_2 = 0.$$

This reduces the general solution to

$$\xi_2 = 0, \quad \eta_2 = 0, \quad \zeta_2 = \psi(px + qy + z + c_2 t) + \Psi(p'x + q'y + z - c_2 t).$$

The equation

$$\zeta'_1 = \zeta'_2$$

becomes, therefore,

$$N_1 \cos \frac{2\pi}{\lambda_1} v_1 t = \psi(px + qy + c_2 t) + \Psi(p'x + q'y - c_2 t). \qquad \text{(S)}$$

Again, we have from equation (R), by the substitution of the values of ζ'_1, ζ'_2,

$$\frac{2\pi}{\lambda_1} N_1 C_1 \sin \frac{2\pi}{\lambda_1} v_1 t = C_2 \left(\frac{d\psi}{dz} + \frac{d\Psi}{dz}\right) = \frac{C_2}{c_2}\left(\frac{d\psi}{dt} - \frac{d\Psi}{dt}\right).$$

Integrating this equation with respect to t, we have

$$\Psi(p'x + q'y - c_2t) - \psi(px + qy + c_2t) = \frac{N_1 C_1 c_2}{C_2 v_1} \cos\frac{2\pi}{\lambda_1} v_1 t + \Omega, \quad \text{(T)}$$

Ω being an arbitrary function of x and y.

Solving equations (S) and (T) for ψ and Ψ, we find

$$\Psi(p'x + q'y - c_2t) = \tfrac{1}{2} N_1 \left(1 + \frac{C_1 c_2}{C_2 v_1}\right) \cos\frac{2\pi}{\lambda_1} v_1 t + \tfrac{1}{2}\Omega,$$

$$\psi(px + qy + c_2t) = \tfrac{1}{2} N_1 \left(1 - \frac{C_1 c_2}{C_2 v_1}\right) \cos\frac{2\pi}{\lambda_1} v_1 t - \tfrac{1}{2}\Omega.$$

From these equations it is easy to see that we must have

$$p = 0, \quad q = 0, \quad p' = 0, \quad q' = 0, \quad \Omega = 0;$$

and, therefore,

$$\Psi(-c_2t) = \tfrac{1}{2} N_1 \left(1 + \frac{C_1 c_2}{C_2 v_1}\right) \cos\frac{2\pi}{\lambda_1} v_1 t,$$

$$\psi(+c_2t) = \tfrac{1}{2} N_1 \left(1 - \frac{C_1 c_2}{C_2 v_1}\right) \cos\frac{2\pi}{\lambda_1} v_1 t.$$

Hence

$$\Psi(z - c_2t) = \tfrac{1}{2} N_1 \left(1 + \frac{C_1 c_2}{C_2 v_1}\right) \cos\frac{2\pi}{\lambda_1}\frac{v_1}{c_2}(z - c_2t),$$

$$\psi(z + c_2t) = \tfrac{1}{2} N_1 \left(1 - \frac{C_1 c_2}{C_2 v_1}\right) \cos\frac{2\pi}{\lambda_1}\frac{v_1}{c_2}(z + c_2t).$$

The wave motion, which is propagated into the second medium, is therefore represented by the equations

$$\xi_2 = 0, \quad \eta_2 = 0, \quad \text{(V)}$$

$$\zeta_2 = \tfrac{1}{2} N_1 \left\{ \left(1 + \frac{C_1 c_2}{C_2 v_1}\right) \cos\frac{2\pi}{\lambda_1}\frac{v_1}{c_2}(z - c_2t) + \left(1 - \frac{C_1 c_2}{C_2 v_1}\right) \cos\frac{2\pi}{\lambda_1}\frac{v_1}{c_2}(z + c_2t) \right\}.$$

CHAPTER X.

APPLICATION OF THE CALCULUS OF VARIATIONS TO THE INTEGRATION OF FUNCTIONS OF ONE OR MORE INDEPENDENT VARIABLES.

152. The general principle upon which depends the present application of the Calculus of Variations may be briefly stated as follows:

A differential expression of any kind is said to be *integrable* when it admits of being expressed as a function of quantities which refer solely to any assumed limits of integration. Thus, for example, the expression

$$Vdx$$

is integrable if it be possible to satisfy the equation

$$\int_{x_0}^{x_1} Vdx = F\left\{x_0, y_0, \left(\frac{dy}{dx}\right)_0 \ldots x_1, y_1, \left(\frac{dy}{dx}\right)_1 \ldots\right\},$$

without determining the form of the function y, or assigning any particular values to the limits x_0, x_1. It is evident, from the nature of the process of integration, that the function F must be of the form

$$f\left\{x_1, y_1, \left(\frac{dy}{dx}\right)_1 \ldots \left(\frac{d^{n-1}y}{dx^{n-1}}\right)_1\right\} - f\left\{x_0, y_0, \left(\frac{dy}{dx}\right)_0 \ldots \left(\frac{d^{n-1}y}{dx^{n-1}}\right)_0\right\}.$$

Now if this equation be admissible it is evident that the variation of the given integral will be of the form

$$\delta\int_{x_0}^{x_1} Vdx = A_1\delta y_1 + B_1\delta\left(\frac{dy}{dx}\right)_1 + \&c. + A_0\delta y_0 + B_0\delta\left(\frac{dy}{dx}\right)_0 + \&c.$$

If we compare this with the general expression of Art. 11,

$$\delta\int_{x_0}^{x_1} Vdx = N_1\delta y_1 + (P_1)_1\delta\left(\frac{dy}{dx}\right)_1 + \&c. + N_0\delta y_0 + \&c.$$

$$+\int_{x_0}^{x_1}\left(N - \frac{dP_1}{dx} + \frac{d^2P_2}{dx^2} - \&c.\right)\delta y dx,$$

it will immediately appear that the two expressions cannot be identical unless the term

$$\int_{x_0}^{x_1}\left(N-\frac{dP_1}{dx}+\&c.\right)\delta y dx$$

vanish without any determination of the function y. Hence it is evident that the form of V must be such as to render the equation

$$N-\frac{dP_1}{dx}+\&c.=0$$

identically true.

Conversely, if this equation be identically true, the given differential

$$Vdx$$

will be integrable. For in this case the variation of the integral

$$\int_{x_1}^{x_0} Vdx$$

will depend solely upon the variations of the limiting quantities,

$$x_1,\ y_1,\ \left(\frac{dy}{dx}\right)_1,\ \ldots\ x_0,\ y_0,\ \left(\frac{dy}{dx}\right)_0\ \ldots.$$

The given integral is therefore a function of these quantities only, and consequently the differential

$$Vdx$$

is integrable.

This reasoning is extended with equal facility to integrals of all orders. For we have seen that the variation of a definite integral consists of two classes of terms, namely: 1. A series of terms depending upon the variations of the *limiting* values of the indeterminate functions, *sc.* the primitive function and its differential coefficients which are contained in V. 2. A series of terms depending upon the variation in the *general* form of the primitive function itself, &c.

Now if the form of V be such as to cause these latter terms to vanish, it is plain that the value of the given integral will depend solely upon the limiting values of the several indeterminate quantities. The given differential expression is therefore *integrable*, that is to say, its integral admits of being reduced to another of a lower degree. Hence we have the following general rule:

Let V be a function containing any number (n) of independent variables, $x_1, x_2, \ldots . x_n$, one dependent variable, u, and its several differential coefficients with regard to x_1, x_2, &c. Let it be required to determine in what case the integral

$$\iint \ldots . \int V dx_1 dx_2 \ldots . dx_n$$

is reducible to another of the order $n-1$.

Determine, by the rules of the Calculus of Variations, the complete variation of the given integral, and reduce it until the quantity under the highest sign of integration contain but one variation, δu. Then if the form of the function V be such as to cause the coefficient of δu to vanish, the given differential admits of being once integrated.

We have thus seen that the rules of the Calculus of Variations give us the means of determining immediately the *criterion of integrability* of a differential function.

It is also evident that we shall be enabled, by the same rules, to determine in what case an integral such as, for example,

$$\iint V dx dy,$$

may be reduced to another,

$$\iint V' dx dy.$$

For if this reduction can be effected, it is plain that the differential

$$(V - V') dx dy$$

must be integrable.

We shall now proceed to consider successively the cases of single and double integrals. In the former of these the generality of the principle here laid down enables us to proceed at once to the most extended case, namely, to determine the conditions necessary, in order that a given differential function may be capable of being integrated any number of times successively. This will form the subject of the following Article.

Prop. I.

153. Let

$$V = f\left(x, y, \frac{dy}{dx}, \frac{d^2y}{dx^2}, \dots \frac{d^ny}{dx^n}\right),$$

and let it be required to determine the conditions necessary, in order that V may be integrable m times successively, m being less than n.

It is evident from the principles laid down in the preceding Article, that, if V be integrable m times successively, the variation of

$$\int^m V dx^m$$

will, when reduced as far as possible, consist entirely of terms free from the sign of integration. Now

$$\delta \int^m V dx^m = \int^m \delta V dx^m = \int^m \left(N\delta y + P_1 \frac{d\delta y}{dx} + \&c. + P_n \frac{d^n\delta y}{dx^n}\right) dx^m. \text{ (A)}$$

Consider any term of this series as

$$\int^m P_k \frac{d^k\delta y}{dx^k} dx^m.$$

We have, then, neglecting terms free from the sign of integration,

$$\int^m P_k \frac{d^k\delta y}{dx^k} dx^m = (-1)^k \left(\int^m \frac{d^k P_k}{dx^k} \delta y dx^m - k \int^{m-1} \frac{d^{k-1} P_k}{dx^{k-1}} \delta y dx^{m-1} + \frac{k.k-1}{1.2} \int^{m-2} \frac{d^{k-2} P_k}{dx^{k-2}} \delta y dx^{m-2} - \&c.\right)$$

Hence

$$\delta . \int^m V dx^m = \Sigma \int^m P_k \frac{d^k\delta y}{dx^k} dx^m$$

$$= \Sigma \left\{(-1)^k \left(\int^m \frac{d^k P_k}{dx^k} \delta y dx^m - k \int^{m-1} \frac{d^{k-1} P_k}{dx^{k-1}} \delta y dx^{m-1} + \frac{k.k-1}{1.2} \int^{m-2} \frac{d^{k-2} P_k}{dx^{k-2}} \delta y dx^{m-2} - \&c.\right)\right\}.$$

Now if this expression be free from all signs of integration, it is evident that the coefficients of δy under the several signs

$$\int^m, \int^{m-1}, \&c.$$

must vanish of themselves. Hence we have the equations

$$\Sigma(-1)^k \frac{d^k P_k}{dx^k} = 0, \quad \Sigma(-1)^{k-1} k \frac{d^{k-1}P_k}{dx^{k-1}} = 0,$$

$$\Sigma(-1)^{k-2} \frac{k \,.\, k-1}{1 \,.\, 2} \frac{d^{k-2}P_k}{dx^{k-2}} = 0, \text{ \&c.};$$

or, since $P_0 = N$,

$$N - \frac{dP_1}{dx} + \frac{d^2P_2}{dx^2} - \text{\&c.} + (-1)^n \frac{d^nP_n}{dx^n} = 0,$$

$$P_1 - 2\frac{dP_2}{dx} + 3\frac{d^2P_3}{dx^2} - \text{\&c.} + (-1)^{n-1} n \frac{d^{n-1}P_n}{dx^{n-1}} = 0,$$

$$P_2 - 3\frac{dP_3}{dx} + \text{\&c.} + (-1)^{n-2} \frac{n \,.\, n-1}{1 \,.\, 2} \frac{d^{n-2}P_n}{dx^{n-2}} = 0, \qquad \text{(B)}$$

$$\text{\&c.} \quad \text{\&c.} \quad \text{\&c.}$$

$$P_{m-1} - m\frac{dP_m}{dx} + \frac{m \,.\, m+1}{1 \,.\, 2} \frac{d^2P_{m+1}}{dx^2} - \text{\&c.}$$

$$+ (-1)^{n-m+1} \frac{m \,.\, m+1 \ldots . n}{1 \,.\, 2 \ldots n-m+1} \frac{d^{n-m+1}P_n}{dx^{n-m+1}} = 0.$$

If these m equations be satisfied independently of the form of the function y, the given differential function will be integrable m times successively.

The same method may be applied to the case in which V contains any number of dependent variables, y, z, u, &c. Each of these variables will introduce m equations similar to (B). Hence we infer generally that if V contain m' dependent variables, the number of equations which must be satisfied in order that it may be integrable m times successively will be mm'. If n, n', &c., be the orders of the function V in the several dependent variables which it contains, i. e. if

$$V = f\left(x, y, \frac{dy}{dx} \ldots \frac{d^ny}{dx^n}, z, \frac{dz}{dx} \ldots \frac{d^{n'}z}{dx^{n'}}, \text{\&c.}\right),$$

it is evident that m cannot be greater than the least of these quantities. If $m = 1$, that is to say, if it be required to find the condition to be satisfied in order that Vdx may be a perfect differential, all the equations (B) except the first disappear. The required criterion is therefore

$$N - \frac{dP_1}{dx} + \frac{d^2P_2}{dx^2} - \&c. + (-1)^n \frac{d^nP_n}{dx^n} = 0.$$

This agrees with the result obtained in the preceding Article.

PROP. II.

154. To find the form of the function V such that $\iint Vdxdy$ may be reduced to a single integral, where

$$V = f(x, y, z, p, q).$$

The method here adopted is precisely analogous to that already given for the case of a single integral. It consists in forming the expression of the function Ω, which remains under the sign of double integration in the variation of $\iint Vdxdy$, and then determining the form of V, so as to cause that function to disappear of itself.

Now it is readily shown, as in p. 250, that the differential coefficients of the first and second orders will disappear from Ω if V be a linear function of p, q, and in that case only. Hence we must have

$$V = \alpha p + \beta q + \gamma,$$

α, β, γ being functions of x, y, z. It is plain that in forming the expression for

$$\frac{dV_x}{dx}, \frac{dV_y}{dy}, N,$$

we may reject all terms involving differential coefficients, inasmuch as all such terms will disappear from the final result.

Forming the expression on this principle, we have

$$\left(\frac{dV_x}{dx}\right) = \frac{d\alpha}{dx}, \quad \left(\frac{dV_y}{dy}\right) = \frac{d\beta}{dy}, \quad N = \frac{d\gamma}{dz},$$

and, consequently,

$$\Omega = \frac{d\gamma}{dz} - \frac{d\alpha}{dx} - \frac{d\beta}{dy}.$$

We infer, therefore, finally, that

The double integral,

$$\iint f(x, y, z, p, q)\, dxdy,$$

may be reduced to a single integral, if

$$f(x, y, z, p, q) = pF_1(x, y, z) + qF_2(x, y, z) - F_3(x, y, z),$$

the functions F_1, F_2, F_3 being connected by the equation

$$\frac{dF_1}{dx} + \frac{dF_2}{dy} + \frac{dF_3}{dz} = 0. \qquad \text{(A)}$$

155. As an example of this, let

$$V = \mu(px + qy - z),$$

μ being a function of x, y, z. We have then

$$F_1 = \mu x, \quad F_2 = \mu y, \quad F_3 = \mu z.$$

Whence

$$\frac{dF_1}{dx} = \mu + x\frac{d\mu}{dx}, \quad \frac{dF_2}{dy} = \mu + y\frac{d\mu}{dy}, \quad \frac{dF_3}{dz} = \mu + z\frac{d\mu}{dz}.$$

Equation (A) becomes, therefore,

$$x\frac{d\mu}{dx} + y\frac{d\mu}{dy} + z\frac{d\mu}{dz} + 3\mu = 0.$$

The double integral,

$$\iint \mu(px + qy - z)\, dxdy,$$

will therefore be reducible to a single integral if μ be a *homogeneous* function of the order -3. Hence we may infer the following theorem:

Let a number of surfaces be described through the same closed curve, and let dS be the element of the superficial area of any one of these surfaces. Let also P be the perpendicular from the origin upon the tangent plane, and x, y, z the running co-ordinates. The value of the integral,

$$\iint \frac{P}{x^3}\phi\left(\frac{z}{x}, \frac{y}{x}\right) dS,$$

extended to the entire of the surface bounded by the closed curve, will be the same for all these surfaces. For since

$$PdS = -(px + qy - z)\, dxdy,$$

it is evident, from the foregoing discussion, that the value of this integral depends only upon the *limiting* values of x, y, z, i. e. upon the bounding curve, which is, by hypothesis, the same for all these surfaces.*

Prop. III.

156. To determine the conditions requisite in order that

$$\iint V dx dy$$

may be reducible to a single integral, where

$$V = f(x, y, z, p, q, r, s, t),$$

r, s, t being the differential coefficients of the second order.

Proceeding, as in the foregoing Proposition, we shall form the value of Ω, and then determine the form of V, so that this quantity may vanish of itself. Now we have already seen (pp. 249–252)

---

* This theorem may be differently proved as follows:

Let r, θ, ϕ be the polar co-ordinates of a point on the surface. Then since

$$\cos\theta = \frac{z}{r}, \quad \tan\phi = \frac{y}{x},$$

and, therefore,

$$\sin\theta d\theta = \frac{1}{r^2}\left(\frac{z}{r}(x+pz) - rp\right)dx + \frac{1}{r^2}\left(\frac{z}{r}(y+qz) - rq\right)dy,$$

$$d\phi = \frac{x}{y^2+x^2}dy - \frac{y}{x^2+y^2}dx;$$

we find, by the ordinary rule,

$$\sin\theta d\theta d\phi = \frac{px+qy-z}{r^3}dxdy.$$

Hence it is easy to see that the given integral may be put under the form

$$\iint \frac{r^3}{x^3}\psi\left(\frac{z}{x}, \frac{y}{x}\right)\sin\theta d\theta d\phi;$$

or, putting for x and y their values,

$$\iint \chi(\theta, \phi)\sin\theta d\theta d\phi,$$

an expression which, as it only contains the independent variables, is evidently reducible to a single integral. All methods of this nature are inferior in generality to that given in the text, by which we arrive at a rule for determining the integrability of *any* given function of x, y, z, p, q.

that the coefficients of the third and fourth orders will disappear from Ω when (and only when) V is of the form

$$A\,(rt - s^2) + Br + 2Cs + Dt + E, \tag{A}$$

where A, B, C, D, E are, in general, functions of p, q, x, y, z. This will give

$$\left.\begin{aligned} N &= (rt - s^2)\frac{dA}{dz} + r\frac{dB}{dz} + 2s\frac{dC}{dz} + t\frac{dD}{dz} + \frac{dE}{dz}, \\ V_x &= (rt - s^2)\frac{dA}{dp} + r\frac{dB}{dp} + 2s\frac{dC}{dp} + t\frac{dD}{dp} + \frac{dE}{dp}, \\ V_y &= (rt - s^2)\frac{dA}{dq} + r\frac{dB}{dq} + 2s\frac{dC}{dq} + t\frac{dD}{dq} + \frac{dE}{dq}, \end{aligned}\right\} \tag{B}$$

$$V_{x^2} = At + B, \quad V_{xy} = -2As + 2C, \quad V_{y^2} = Ar + D.$$

In forming the values of

$$\frac{dV_x}{dx}, \frac{dV_y}{dy}, \frac{d^2V_{x^2}}{dx^2}, \frac{d^2V_{xy}}{dxdy}, \frac{d^2V_{y^2}}{dy^2},$$

it is plain that we may reject all coefficients of an order higher than the second, inasmuch as we have before seen that all such terms will disappear from the final result.

Differentiating upon this principle, we have

$$\frac{dV_{x^2}}{dx} = t\left(\frac{dA}{dx} + p\frac{dA}{dz} + r\frac{dA}{dp} + s\frac{dA}{dq}\right) + \frac{dB}{dx} + p\frac{dB}{dz} + r\frac{dB}{dp} + s\frac{dB}{dq},$$

$$\frac{1}{2}\frac{dV_{xy}}{dy} = -s\left(\frac{dA}{dy} + q\frac{dA}{dz} + s\frac{dA}{dp} + t\frac{dA}{dq}\right) + \frac{dC}{dy} + q\frac{dC}{dz} + s\frac{dC}{dp} + t\frac{dC}{dq}.$$

Hence

$$\frac{dV_{x^2}}{dx} + \frac{1}{2}\frac{dV_{xy}}{dy} - V_x = \frac{dB}{dx} + p\frac{dB}{dz} + \frac{dC}{dy} + q\frac{dC}{dz} - \frac{dE}{dp}$$

$$+ t\left(\frac{dA}{dx} + p\frac{dA}{dz} + \frac{dC}{dq} - \frac{dD}{dp}\right) - s\left(\frac{dA}{dy} + q\frac{dA}{dz} - \frac{dB}{dq} + \frac{dC}{dp}\right).$$

Similarly,

$$\frac{dV_{y^2}}{dy} + \frac{1}{2}\frac{dV_{xy}}{dx} - V_y = \frac{dD}{dy} + q\frac{dD}{dz} + \frac{dC}{dx} + p\frac{dC}{dz} - \frac{dE}{dq}$$

$$+ r\left(\frac{dA}{dy} + q\frac{dA}{dz} + \frac{dC}{dp} - \frac{dB}{dq}\right) - \left(\frac{dA}{dx} + p\frac{dA}{dz} - \frac{dD}{dp} + \frac{dC}{dq}\right).$$

Hence

$$\Omega = \frac{d}{dx}\left(\frac{dV_{x^2}}{dx} + \tfrac{1}{2}\frac{dV_{xy}}{dy} - V_x\right) + \frac{d}{dy}\left(\frac{dV_{y^2}}{dy} + \tfrac{1}{2}\frac{dV_{xy}}{dx} - V_y\right) + N$$

$$= \Theta(rt - s^2) + Lr + 2Ms + Nt + \Pi; \qquad \text{(C)}$$

where

$$\Theta = 3\frac{dA}{dz} + \frac{d^2A}{dxdp} + \frac{d^2A}{dydq} + p\frac{d^2A}{dzdp} + q\frac{d^2A}{dzdq} + 2\frac{d^2C}{dpdq} - \frac{d^2B}{dq^2} - \frac{d^2D}{dp^2}.$$

$$L = \frac{d^2A}{dy^2} + 2q\frac{d^2A}{dydz} + q^2\frac{d^2A}{dz^2} + 2\frac{dB}{dz} + \frac{d^2B}{dxdp} - \frac{d^2B}{dydq} + p\frac{d^2B}{dzdp} - q\frac{d^2B}{dzdq} + 2\left(\frac{d^2C}{dydp} + q\frac{d^2C}{dzdp}\right) - \frac{d^2E}{dp^2}. \qquad \text{(D)}$$

$$M = -\left(\frac{d^2A}{dxdy} + p\frac{d^2A}{dydz} + q\frac{d^2A}{dxdz} + pq\frac{d^2A}{dz^2}\right) + 2\frac{dC}{dz} + \frac{d^2B}{dxdq} + p\frac{d^2B}{dxdq} + \frac{d^2D}{dydp} + q\frac{d^2D}{dzdp} - \frac{d^2E}{dpdq}.$$

$$N = \frac{d^2A}{dx^2} + 2p\frac{d^2A}{dxdz} + p^2\frac{d^2A}{dz^2} + 2\frac{dD}{dz} + \frac{d^2D}{dydq} - \frac{d^2D}{dxdp} + q\frac{d^2D}{dzdq} - p\frac{d^2D}{dzdp} + 2\left(\frac{d^2C}{dxdq} + p\frac{d^2C}{dzdq}\right) - \frac{d^2E}{dq^2}.$$

$$\Pi = \frac{d^2B}{dx^2} + 2p\frac{d^2B}{dxdz} + p^2\frac{d^2B}{dz^2} + \frac{d^2D}{dy^2} + 2q\frac{d^2D}{dydz} + q^2\frac{d^2D}{dz^2} + 2\left(\frac{d^2C}{dxdy} + q\frac{d^2C}{dxdz} + p\frac{d^2C}{dydz} + pq\frac{d^2C}{dz^2}\right) - \left(\frac{d^2E}{dxdp} + \frac{d^2E}{dydq} + p\frac{d^2E}{dzdp} + q\frac{d^2E}{dzdq}\right) + \frac{dE}{dz}.$$

If, therefore, the given function V be integrable, it must satisfy the equations

$$\Theta = 0, \quad L = 0, \quad M = 0, \quad N = 0, \quad \Pi = 0. \tag{E}$$

These equations, therefore, furnish the *criterion of integrability* of the given differential function

$$V dx dy.$$

We shall now consider some particular cases of the general problem, which will be found to lead to interesting results connected with the theory of surfaces.

(1.) Let

$$V = A(rt - s^2).$$

Here we have

$$B = 0, \quad C = 0, \quad D = 0, \quad E = 0,$$

and, therefore,

$$\Pi = 0.$$

The equations (E) are, therefore, reduced to

$$\left.\begin{aligned} 3\frac{dA}{dz} + \frac{d^2A}{dxdp} + \frac{d^2A}{dydq} + p\frac{d^2A}{dzdp} + q\frac{d^2A}{dzdq} &= 0, \\ \frac{d^2A}{dy^2} + 2q\frac{d^2A}{dydz} + q^2\frac{d^2A}{dz^2} &= 0, \\ \frac{d^2A}{dxdy} + p\frac{d^2A}{dydz} + q\frac{d^2A}{dxdz} + pq\frac{d^2A}{dz^2} &= 0, \\ \frac{d^2A}{dx^2} + 2p\frac{d^2A}{dxdz} + p^2\frac{d^2A}{dz^2} &= 0. \end{aligned}\right\} \tag{F}$$

Integrating the last three of these equations, which is easily effected by assuming

$$u = \frac{dA}{dx} + p\frac{dA}{dz}, \quad v = \frac{dA}{dy} + q\frac{dA}{dz},$$

we find

$$u = \phi(z - px - qy, p, q), \quad v = \psi(z - px - qy, p, q),$$

$$A = x\phi(z - px - qy, p, q) + y\psi(z - px - qy, p, q)$$
$$+ \chi(z - px - qy, p, q). \tag{G}$$

Substituting this value in the first of equations (F), and putting, for the sake of brevity,

$$z - px - qy = \omega,$$

we have

$$\frac{d\phi}{dp} + \frac{d\psi}{dq} = -\left(x\frac{d\phi}{dz} + y\frac{d\psi}{dz} + \frac{d\chi}{dz}\right) = -\left(x\frac{d\phi}{d\omega} + y\frac{d\psi}{d\omega} + \frac{d\chi}{d\omega}\right).$$

Integrating with regard to ω, we have

$$x\phi + y\psi + \chi = -\int\left(\frac{d\phi}{dp} + \frac{d\psi}{dq}\right)d\omega + F(p, q).$$

Substituting this value in (F) we have, putting $-\frac{d\Phi}{d\omega}$, $-\frac{d\Psi}{d\omega}$, for ϕ, ψ, and omitting, as is plainly allowable, $F(p, q)$,

$$A = \frac{d\Phi}{dp} + \frac{d\Psi}{dq}. \qquad \text{(H)}$$

If P be the perpendicular from the centre upon the tangent plane at any point of a surface it is known that

$$P = \frac{\omega}{\sqrt{(1 + p^2 + q^2)}}.$$

The foregoing value of A may therefore be written

$$A = \frac{d\Phi(P, p, q)}{dp} + \frac{d\Psi(P, p, q)}{dq}. \qquad \text{(I)}$$

This is the most general value of A, which renders

$$A(rt - s^2)\,dxdy$$

integrable, i. e. such that the value of the integral

$$\iint A\,(rt - s^2)\,dxdy$$

is dependent solely upon the nature of the limiting curve and the values which z and either of its differential coefficients have at the corresponding points of the surface. Hence we have the following theorem :

If a number of surfaces be described, touching along the same closed curve, the value of the integral

$$\iint A\,dpdq \qquad \text{(K)}$$

extended through that part of any one of the surfaces which is bounded by the curve of contact, will be the same for all the surfaces so described.

This theorem is evident, if we recollect that

$$dpdq = (rt - s^2)dxdy.$$

For it is plain, from the foregoing discussion, that the variation of the definite integral will contain only the variations of the *limiting* values of y, z, and p or q (p. 232). Since, then, the surfaces have a common curve of contact, this variation will disappear altogether. The value of the integral therefore remains constant in passing from one of these surfaces to another. The same conclusion holds if the surface be entirely closed, i. e. the value of the integral (K) will in this case be the same for all *closed* surfaces. For in this case the variations of the limits will vanish of themselves.*

Hence also,

If the tangent plane, at any point of a surface, have a closed curve of contact, the value of the integral (K), *extended to the entire of that part of the surface which is bounded by that curve, will be zero.*

For by the preceding theorem the value is the same for the surface as for the tangent plane.

(2.) Let

$$V = Br + 2Cs + Dt,$$

where B, C, D are functions of p and q.

Here we have

$$A = 0, \quad B = f(p, q), \quad C = F(p, q), \quad D = \phi(p, q).$$

---

* This will, perhaps, appear most readily if we suppose the definite integral to be transformed, by taking as co-ordinates the radius vector and the angles which determine its position, the origin being placed within the surface. It will then assume the form

$$\iint V' d\theta d\phi.$$

Now the equations of the limiting curves are, in general, of the form

$$\theta = f_1(\phi), \quad \theta = f_0(\phi);$$

and the single integrals which occur in the variation of

$$\iint V' d\theta d\phi$$

are supposed to be extended through the entire perimeter of each of these curves respectively. But in the case of a *closed* surface, these curves become points, and the single integrals consequently vanish.

Hence it is easy to see that the equations

$$L = 0, \quad M = 0, \quad N = 0, \quad \Pi = 0,$$

become identical, and that the equation

$$\Theta = 0$$

becomes

$$\frac{d^2B}{dq^2} - 2\frac{d^2C}{dpdq} + \frac{d^2D}{dp^2} = 0.$$

It may be readily shown that this equation is satisfied by making

$$B = \frac{1+q^2}{(1+p^2+q^2)^{\frac{3}{2}}}, \quad C = -\frac{pq}{(1+p^2+q^2)^{\frac{3}{2}}}, \quad D = \frac{1+p^2}{(1+p^2+q^2)^{\frac{3}{2}}}.$$

Hence, recollecting that

$$\frac{1}{R} + \frac{1}{R'} = -\frac{(1+q^2)r - 2pqs + (1+p^2)t}{(1+p^2+q^2)^{\frac{3}{2}}},$$

we infer that

If a number of surfaces be described as in (1), *the value of the definite integral,*

$$\iint \left(\frac{1}{R} + \frac{1}{R'}\right) dxdy,$$

*extended to the entire of that part of any one of the surfaces which is bounded by the curve of contact, will be the same for all these surfaces.**

---

* This theorem, which has been before referred to, is more easily proved by a particular method. For, if we assume

$$\xi = \frac{p}{\sqrt{(1+p^2+q^2)}}, \quad \eta = \frac{q}{\sqrt{(1+p^2+q^2)}},$$

we shall easily see that

$$\frac{1}{R} + \frac{1}{R'} = -\left(\frac{d\xi}{dx} + \frac{d\eta}{dy}\right).$$

Hence

$$\iint \left(\frac{1}{R} + \frac{1}{R'}\right) dxdy = \int \left(\xi \frac{dy}{dx} - \eta\right) dx = \int \frac{p\dfrac{dy}{dx} - q}{\sqrt{(1+p^2+q^2)}} dx.$$

It would not be difficult to multiply examples of the present application of the Calculus of Variations, but the length to which the present work has already extended forbids me from entering further into the subject. I shall, therefore, conclude this Chapter with the consideration of two questions, which seem to be of some importance.

Prop. IV.

157. To find whether it be possible to represent the superficial area of a surface by any formula such as

$$\Gamma + \iint F(P, \theta, \phi)\, d\theta d\phi, \tag{A}$$

where Γ is a quantity referring solely to the limits of integration, and P, θ, ϕ are the perpendicular from the origin upon the tangent plane, and the polar angles which determine its position.*

The method to be pursued in all such investigations as the present is as follows:

Let

$$\iint V dx dy$$

be an integral by which it is known that the required quantity may be represented, and let

$$\Gamma + \iint \Pi d\theta d\phi$$

be the expression by which it is proposed to be represented, Π being an indeterminate function of θ, ϕ, and some third quantity, which in the present case is taken to be the perpendicular on the tangent plane. Transform the double integral in the proposed expression (by changing the independent variables) into another of the form

$$\iint V' dx dy.$$

We have then

$$\iint V dx dy = \Gamma + \iint V' dx dy,$$

and, consequently,

$$\iint (V - V')\, dx dy = \Gamma.$$

If, therefore, it be possible to determine V' (or Π) such that

$$(V - V')\, dx dy$$

---

* It is known that in a plane curve the length of any portion may be represented by an expression of the form

$$\Gamma + \int P d\omega,$$

where P is the perpendicular on the tangent, and ω the angle it makes with any fixed line; Γ being a quantity which refers solely to the limits. The object of the present proposition is to determine whether there be an analogous formula for the quadrature of any portion of the surface.

may satisfy the criterion of integrability, the problem is possible; if otherwise, it is impossible.

In the present case, where the independent variables in the proposed expression are the angles which determine the position of the perpendicular upon the tangent plane, it is known that

$$\sin\theta d\theta d\phi = \frac{rt - s^2}{(1 + p^2 + q^2)^{\frac{3}{2}}} dxdy.$$

Hence it is evident that if the expression

$$F(P, \theta, \phi)\, d\theta d\phi$$

be transformed into one of the form

$$V' dxdy,$$

we shall have

$$V' = (rt - s^2) f(z - px - qy, p, q);$$

also,

$$V = \sqrt{(1 + p^2 + q^2)}.$$

If, therefore, the proposed representation be possible, the expression

$$\{\sqrt{(1 + p^2 + q^2)} - (rt - s^2) f(z - px - qy, p, q)\}\, dxdy$$

must be integrable. We have, then, in the equations of Prop. III.

$$A = -f(z - px - qy, p, q), \quad B = 0, \quad C = 0, \quad D = 0,$$
$$E = \sqrt{(1 + p^2 + q^2)}.$$

Substituting these values in the equations

$$L = 0, \quad M = 0, \quad N = 0,$$

it is easily seen that they become

$$\frac{d^2E}{dp^2} = 0, \quad \frac{d^2E}{dpdq} = 0, \quad \frac{d^2E}{dq^2} = 0,$$

which are manifestly impossible, inasmuch as E is a *given* function. It is, therefore, impossible to express the area of a surface by any such formula as (A).

The present example shows the utility of the Calculus of Variations in examining into the question of the *possibility* of effecting any proposed reduction of one multiple integral to another differing from it in form.*

PROP. V.

158. Let R, R' be the principal radii of curvature of a closed surface, P the perpendicular on the tangent plane, and $d\omega$ the element of the spherical surface described by a portion of this perpendicular whose length is equal to unity. Then

$$\iint (R+R')d\omega = 2\iint Pd\omega, \qquad \text{(A)}$$

the integrals being extended through the entire of the closed surface.

It is easy to show that

$$d\omega = \sin\theta d\theta d\phi = \frac{rt - s^2}{(1+p^2+q^2)^{\frac{3}{2}}} dxdy;$$

also,

$$R + R' = -\frac{(1+q^2)r - 2pqs + (1+p^2)t}{rt - s^2}\sqrt{(1+p^2+q^2)},$$

$$P = \frac{z - px - qy}{\sqrt{(1+p^2+q^2)}}.$$

If now we assume

$$\iint Vdxdy = \iint \{2P - (R+R')\}\, d\omega,$$

and substitute the foregoing values of R, R', P, $d\omega$, we shall have

$$V = 2\frac{z - px - qy}{(1+p^2+q^2)^2}(rt - s^2) + \frac{(1+q^2)r - 2pqs + (1+p^2)t}{1+p^2+q^2}. \qquad \text{(B)}$$

Comparing this with the general form (A), p. 343, we find

---

* I have investigated the general form which the function V should have, in order that the area of any portion of a surface may be capable of being represented by a formula such as

$$\Gamma + \iint Vd\theta d\phi,$$

but the result is altogether different from $F(\text{P}, \theta, \phi)$. The reader will find it considered in the note upon the present Article. It does not appear to me to be possible to establish, in the present question, any analogy between the case of the surface and that of the curve.

$$A = 2\,\frac{(z - px - qy)}{(1 + p^2 + q^2)^2}, \quad B = \frac{1 + q^2}{1 + p^2 + q^2},$$

$$C = -\frac{pq}{1 + p^2 + q^2}, \quad D = \frac{1 + p^2}{1 + p^2 + q^2}, \quad E = 0. \qquad \text{(C)}$$

Now since B, C, D are independent of x, y, z, and since, moreover, A satisfies the equations

$$\frac{dA}{dx} + p\,\frac{dA}{dz} = 0, \quad \frac{dA}{dy} + q\,\frac{dA}{dz} = 0, \qquad \text{(D)}$$

it is plain that the values (C) will render the quantities L, M, N, H, identically zero. It remains, therefore, to examine the result of these substitutions in Θ.

We have, from p. 344,

$$\Theta = 3\,\frac{dA}{dz} + \frac{d^2A}{dxdp} + \frac{d^2A}{dydq} + p\,\frac{d^2A}{dzdp} + q\,\frac{d^2A}{dzdq}$$

$$+ 2\,\frac{d^2C}{dpdq} - \frac{d^2B}{dq^2} - \frac{d^2D}{dp^2}.$$

Differentiating equations (D) with respect to p and q respectively, and adding them, we find

$$2\,\frac{dA}{dz} + \frac{d^2A}{dxdp} + \frac{d^2A}{dydq} + p\,\frac{d^2A}{dzdp} + q\,\frac{d^2A}{dzdq} = 0.$$

This reduces the value of Θ to

$$\Theta = \frac{dA}{dz} + 2\,\frac{d^2C}{dpdq} - \frac{d^2B}{dq^2} - \frac{d^2D}{dp^2}$$

$$= \frac{dA}{dz} + \frac{d}{dp}\left(\frac{dC}{dq} - \frac{dD}{dp}\right) + \frac{d}{dq}\left(\frac{dC}{dp} - \frac{dB}{dq}\right).$$

Now

$$\frac{dC}{dq} = \frac{-p}{1 + p^2 + q^2} + \frac{2pq^2}{(1 + p^2 + q^2)^2},$$

$$\frac{dD}{dp} = \frac{2pq^2}{(1 + p^2 + q^2)^2}.$$

Hence

$$\frac{dC}{dq} - \frac{dD}{dp} = \frac{-p}{1 + p^2 + q^2},$$

and, therefore,

$$\frac{d}{dp}\left(\frac{dC}{dq} - \frac{dD}{dp}\right) = -\frac{1}{1+p^2+q^2} + \frac{2p^2}{(1+p^2+q^2)^2}.$$

Similarly,

$$\frac{d}{dq}\left(\frac{dC}{dp} - \frac{dB}{dq}\right) = -\frac{1}{1+p^2+q^2} + \frac{2q^2}{(1+p^2+q^2)^2}.$$

Hence

$$\frac{d}{dp}\left(\frac{dC}{dq} - \frac{dD}{dp}\right) + \frac{d}{dq}\left(\frac{dC}{dp} - \frac{dB}{dq}\right) = -\frac{2}{(1+p^2+q^2)^2} = -\frac{dA}{dz}.$$

We have, therefore,

$$\Theta = 0.$$

Since, then, the integrals are supposed to be extended through the entire of the closed surface, it is plain that the complete variation of

$$\iint V dx dy, \text{ or } \iint \{2P - (R + R')\} d\omega,$$

will be identically zero. This integral will, therefore, have the same value for all closed surfaces. If, therefore, we can find its value for any *one* surface, we shall, by this theorem, find its general value. Let the surface be a sphere described round the origin as centre. We have then

$$R + R' = 2R = 2P,$$

and, consequently,

$$\iint \{2P - (R + R')\} d\omega = 0.$$

Hence, in general,

$$\iint (R + R') d\omega = 2\iint P d\omega.$$

If dS be the element of the surface, and if in the left-hand member of equation (A) we put for $d\omega$ its value

$$\frac{dS}{RR'},$$

we shall have the theorem

$$\iint\left(\frac{1}{R} + \frac{1}{R'}\right) dS = 2\iint P d\omega.$$

Other theorems of the same kind might be obtained in a similar way; but as the object of the present Chapter is rather to state a general method, than to examine particular problems, the foregoing illustrations will probably be considered sufficient.

NOTES.

NOTE A, Page 4.

THIS may be readily shown as follows:

$$(1.)\ F2\psi = F(\psi + \psi) = F\psi + F\psi = 2F\psi,$$
$$F3\psi = F(2\psi + \psi) = F2\psi + F\psi = 3F\psi;$$

and, in general,

$$Fi\psi = iF\psi,$$

i being an integer.

(2.) Let

$$i = \frac{m}{n}.$$

Then since

$$F.n\psi = nF\psi,$$

if we put $n\psi$ for ψ, we have

$$F.\frac{\psi}{n} = \frac{1}{n}F\psi;$$

and, therefore,

$$F.\frac{m}{n}\psi = F.m\frac{\psi}{n} = mF.\frac{\psi}{n} = \frac{m}{n}F\psi.$$

The theorem is, therefore, true for all rational values of i, and may easily be extended to irrational values by the method of exhaustions.

NOTE B, Page 27.

The term "second variation," as generally used, is ambiguous. 1. It may signify the variation of the variation. 2. It may denote the quadratic part of the series which is obtained by the substitution of $y + \delta y$ (or $y + i\psi$) for y. If this series be

$$A + Bi + Ci^2 + \&c.$$

the term Ci^2 is sometimes denominated the second variation. These significations will become identical if

$$\delta^2 y = 0. \qquad \text{(a)}$$

Now in the applications of the Calculus of Variations to the inves-igation of maxima and minima, we are only concerned with the latter ignification. I have, therefore, introduced the condition (a), in order o obviate any confusion which might arise from the double meaning.

I cannot but think that M. Delaunay has been misled by this am-iguity, when, in the theory of maxima and minima, he employs the econd variation in the first of the foregoing significations.* For in his signification of the term the second variation has not (as it seems o me) any connexion with the theory of maxima and minima.

Note C, Page 32.

The reader must not suppose that indeterminate functions admit of no maxima or minima except those which are discoverable by this rule. For, on referring to the reasoning employed in the Differential Cal-ulus (which is, in this point, identical with that which is used in the present case), it will at once appear that this reasoning is based upon he assumed possibility of taking the increment so small that any erm of the series may exceed the sum of all those which follow it. This is true only if the coefficients of all these terms be finite. If any f these coefficients be infinite, the condition is no longer necessarily possible. In fact in this case the development fails altogether. But we cannot conclude from thence that the function does not admit of a maximum or minimum. Thus in the curve ABA', Fig. 17, the ordi-nate BY is a maximum, although the value of the differential co-fficient,

$$\frac{dy}{dx},$$

s finite. This maximum is not therefore given by the ordinary rule, which fails for the above-mentioned reason, namely, that the second differential coefficient becomes infinite.

Similarly, in the theory of maxima and minima, as given by the Calculus of Variations, if the second variation (as defined in (2), Note B), become infinite, the reasoning upon which the rule depends will fail, and the given indeterminate function may admit of a maxi-mum or minimum which does not satisfy the condition

$$D\int V dx = 0.$$

Vid. note upon p. 163.

---

* Liouville, Journal de Math., tom. vi. p. 225. This error (if it be an error) does not affect M. Delaunay's conclusion.

NOTE D, Page 72.

It is evident, from the general discussion contained in Props. II., III., and IV., that, if either the conditions of the problem or the function which is to be made a maximum or minimum contain the *limiting* values of coefficients of an order higher than $n-1$, the Calculus of Variations does not appear to furnish a solution. It may naturally be asked, then, whether such problems admit of *any* solution? The answer to this is easy, if we recollect that, in the general discussion alluded to, *mixed* functions are tacitly excluded. It is assumed throughout that the function preserves the *same* form for all values of x, from x_0 to x_1 inclusive. The question discussed is not, therefore, whether, under the conditions of the problem, a given integral admit of *any* maximum or minimum value, but whether such a value can be given to it consistently with the supposition that the form of the function y remains unchanged. Problems of the class above alluded to do not admit of such a solution. But if mixed functions be admitted, the possibility of the solution will be restored.

This reasoning will, perhaps, be rendered more clear by a geometrical illustration. Let

$$V = f\left(x, y, \frac{dy}{dx}\right),$$

and let it be proposed to describe between two given points, A, B, a curve such that the differential coefficients,

$$\frac{dy}{dx}, \frac{d^2y}{dx^2}, \ldots \frac{d^my}{dx^m},$$

may have, at each of these points, given values, and that the definite integral,

$$\int V dx,$$

may be a maximum or minimum.

The solution of this problem is given by a mixed curve, consisting of (1) a finite portion satisfying the equation

$$N - \frac{dP_1}{dx} = 0,$$

and passing through the points A, B. (2.) An infinitesimal portion at each extremity, satisfying the other conditions of the problem.

Thus, for example, if it were required to determine a curve of minimum length passing through two given points, and touching two given lines, the solution would be a mixed curve, consisting of a finite right line connecting the given points, and terminated by two infinitesimal elements touching the given lines.

Note E, Page 82.

This theorem may readily be proved by the separation of the symbols of operation and quantity.

If we denote by Δ the symbol of differentiation as applied to K only, and by Δ' the same symbol as applied to u only, it is easily seen that the expression

$$\frac{d^m . K \dfrac{d^{m'} u}{dx^{m'}}}{dx^m} \pm \frac{d^{m'} . K \dfrac{d^m u}{dx^m}}{dx^{m'}} \qquad \text{(a)}$$

may be written

$$\{(\Delta + \Delta')^m \Delta'^{m'} \pm (\Delta + \Delta')^{m'} \Delta'^m\} Ku,$$

or

$$(\Delta + \Delta')^{m'} \Delta'^{m'} \{(\Delta + \Delta')^{m-m'} \pm \Delta'^{m-m'}\} Ku.$$

Now it is readily proved that either of the expressions

$$(\Delta + \Delta')^{2n} + \Delta'^{2n},$$
$$(\Delta + \Delta')^{2n+1} - \Delta'^{2n+1},$$

may be represented by a series of the form

$$\Delta^{2n} + A\Delta^{2n-2} (\Delta + \Delta') \Delta' + B\Delta^{2n-4} (\Delta + \Delta')^2 \Delta'^2 + \&c. + E (\Delta + \Delta')^n \Delta'^n, \qquad \text{(b)}$$

where A, B, and E are functions of n. For if we assume

$$X = (\Delta + \Delta') \Delta',$$

we shall have

$$\Delta' = \frac{-\Delta \pm \sqrt{(\Delta^2 + 4X)}}{2}, \quad \Delta + \Delta' = \frac{\Delta \pm \sqrt{(\Delta^2 + 4X)}}{2}.$$

Hence

$$(\Delta + \Delta')^{2n} + \Delta'^{2n} = \frac{1}{2^{2n}} [\{\sqrt{(\Delta^2 + 4X)} + \Delta\}^{2n} + \{\sqrt{(\Delta^2 + 4X)} - \Delta\}^{2n}]$$

$$= \frac{1}{2^{2n-1}} \left\{ \Delta^{2n} + \frac{2n . 2n - 1}{1.2} \Delta^{2n-2} (\Delta^2 + 4X) + \&c. + (\Delta^2 + 4X)^n \right\}.$$

If this expression be arranged according to the powers of X, and if we then replace X by its value $(\Delta + \Delta')\Delta'$, we shall evidently have a series of the form (b). The same method will obviously apply to the expression

$$(\Delta + \Delta')^{2n+1} - \Delta'^{2n+1}.$$

Substituting the value so found in

$$(\Delta + \Delta')^{m'} \Delta'^{m'} \{(\Delta + \Delta')^{m-m'} \pm \Delta'^{\,m-m'}\},$$

we have

$$\Delta^{2n}(\Delta + \Delta')^{m'} \Delta'^{m'} + A\Delta^{2n-2}(\Delta + \Delta')^{m'+1} \Delta'^{m'+1} + \&c.$$

where

$$n = \tfrac{1}{2}(m - m') \text{ or } = \tfrac{1}{2}(m - m' - 1),$$

according as $m - m'$ is even or odd.

Hence we have, finally,

$$\frac{d^m K \frac{d^{m'}u}{dx^{m'}}}{dx^m} \pm \frac{d^{m'} K \frac{d^m u}{dx^m}}{dx^{m'}} = \{\Delta^{2n}(\Delta + \Delta')^{m'} \Delta'^{\,m'} + \&c.\} Ku$$

$$= \frac{d^{m'} \frac{d^{2n}K}{dx^{2n}} \frac{d^{m'}u}{dx^{m'}}}{dx^{m'}} + \frac{d^{m'+1} A \frac{d^{2n-2}K}{dx^{2n-2}} \frac{d^{m'+1}u}{dx^{m'+1}}}{dx^{m'+1}}$$

$$= \frac{d^{m'} C_{m_\prime} \frac{d^{m'}u}{dx^{m'}}}{dx^{m'}} + \frac{d^{m'+1} C_{m_\prime+1} \frac{d^{m'+1}u}{dx^{m'+1}}}{dx^{m'+1}} + \&c.,$$

putting

$$C_{m_\prime} = \frac{d^{2n}K}{dx^{2n}}, \quad C_{m_\prime+1} = A \frac{d^{2n-2}K}{dx^{2n-2}} + \&c.$$

Note F, Page 84.

The following theorem is easily proved by the method of separation of symbols :

$$P \frac{d^m Q}{dx^m} = \frac{d^m PQ}{dx^m} - \frac{m}{1} \frac{d^{m-1} P'Q}{dx^{-1}} + \frac{m.m-1}{1.2} \frac{d^{m-2}P''Q}{dx^{m-2}} - \&c.$$

$$+ (-1)^m P^{(m)} Q; \tag{a}$$

putting, for the sake of brevity,

$$P' = \frac{dP}{dx}, \quad P'' = \frac{d^2P}{dx^2}, \dots P^{(m)} = \frac{d^mP}{dx^m}.$$

If now we multiply the equation

$$Au + \frac{d.A_1 \frac{du}{dx}}{dx} + \frac{d^2.A_2 \frac{d^2u}{dx^2}}{dx^2} + \&c. = 0 \tag{b}$$

by ut, and subtract it from the value of U, we shall have (putting for y its value, ut)

$$U = u \frac{d.A_1 \frac{d.ut}{dx}}{dx} + u \frac{d^2.A_2 \frac{d^2.ut}{dx^2}}{dx^2} + \&c.$$

$$- ut \frac{d.A_1 \frac{du}{dx}}{dx} - ut \frac{d^2.A_2 \frac{d^2u}{dx^2}}{dx^2} - \&c.$$

Or, as it may be more briefly expressed,

$$U = \Sigma \left(u \frac{d^m.A_m \frac{d^m ut}{dx^m}}{dx^m} - ut \frac{d^m.A_m \frac{d^m u}{dx^m}}{dx^m} \right), \qquad \text{(c)}$$

m having all values from 1 to n. We shall now proceed to show that each term in the sum Σ may be represented by a series of the form

$$\frac{d.b_1 \frac{dt}{dx}}{dx} + \frac{d^2.b_2 \frac{d^2t}{dx^2}}{dx^2} + \&c.$$

Applying the theorem (a), and putting

$$\frac{d^m.ut}{dx^m} = (ut)^m, \quad \frac{du}{dx} = u', \ \&c.,$$

we have

$$u \frac{d^m.A_m(ut)^m}{dx^m} = \frac{d^m.uA_m(ut)^m}{dx^m} - \frac{m}{1} \frac{d^{m-1}.u'A_m(ut)^m}{dx^{m-1}}$$

$$+ \frac{m.m-1}{1.2} \frac{d^{m-2}.u''A_m(ut)^m}{dx^{m-2}} - \&c. \qquad \text{(d)}$$

We have also

$$(ut)^m = ut^{(m)} + mu't^{(m-1)} + \frac{m.m-1}{1.2} u''t^{(m-2)} + \&c.$$

If this value of $(ut)^m$ be substituted in (d), the general term of the resulting series may be represented by

$$M \frac{d^p.u^{(m-p)}A_m u^{(m-q)} t^{(q)}}{dx^p},$$

where it is easily seen that

$$M = \pm \frac{m.m-1 \ldots p+1}{1.2.3 \ldots m-p} \cdot \frac{m.m-1 \ldots q+1}{1.2.3 \ldots m-q},$$

the upper or lower sign being taken according as $m-p$ is even or odd.

Now if p be not equal to q, it is plain, from the form of this expression, that there must be another term,

$$\pm M \frac{d^q . u^{(m-q)} A_m u^{(m-p)} t^{(p)}}{dx^q},$$

the upper or lower sign being taken according as $p - q$ is even or odd. Hence if we assume

$$K = MA_m u^{(m-p)} u^{(m-q)},$$

it will at once appear that, with the exception of the terms in which $p = q$ (which have already the required form), all the terms of the series (c) may be arranged in groups of the form

$$\left(\frac{d^p . K \frac{d^q t}{dx^q}}{dx^p} \pm \frac{d^q . K \frac{d^p t}{dx^p}}{dx^q} \right).$$

But we have before seen (Note E) that a group of this form may always be expressed by the series

$$\frac{d^q . C_q \frac{d^q t}{dx^q}}{dx^q} + \frac{d^{q+1} . C_{q+1} \frac{d^{q+1} t}{dx^{q+1}}}{dx^{q+1}} + \&c.$$

We have, therefore,

$$\Sigma u \frac{d^m . A_m (ut)^m}{dx^m} = Ct + \frac{d . C_1 \frac{dt}{dx}}{dx} + \frac{d^2 . C_2 \frac{d^2 t}{dx^2}}{dx^2} + \&c. + \frac{d^n . C_n \frac{d^n t}{dx^n}}{dx^n}. \qquad \text{(e)}$$

Now it is evident that the coefficient of t in

$$\Sigma . u \frac{d^m . A_m (ut)^m}{dx^m}$$

will be

$$\Sigma . u \frac{d^m . A_m \frac{d^m u}{dx^m}}{dx^m}$$

Hence

$$\Sigma . ut \frac{d^m . A_m \frac{d^m u}{dx^m}}{dx^m} = Ct.$$

Substituting this value in (e), and putting b_1, b_2, &c. for C_1, C_2, &c., we have

$$U = \Sigma \left(u \frac{d^m . A_m \frac{d^m . ut}{dx^m}}{dx^m} - ut \frac{d^m . A_m \frac{d^m u}{dx^m}}{dx^m} \right)$$

$$= \frac{d . b_1 \frac{dt}{dx}}{dx} + \frac{d^2 . b_2 \frac{d^2 t}{dx^2}}{dx^2} + \&c. + \frac{d^n . b_n \frac{d^n t}{dx^n}}{dx^n},$$

and therefore, finally,

$$\int U dx = b_1 \frac{dt}{dx} + \frac{d . b_2 \frac{d^2 t}{dx^2}}{dx} + \&c. + \frac{d^{n-1} . b_n \frac{d^n t}{dx^n}}{dx^{n-1}}. \qquad \text{(f)}$$

This demonstration is taken, with some modifications, from M. Delaunay's Memoir.

Note G, Page 127.

M. Delaunay, in reasoning upon this problem, concludes that if the order of the equation

$$L = 0$$

be higher than that of the function V, the conditions

$$\lambda_1 = 0, \quad \lambda_0 = 0, \quad \left(\frac{d\lambda}{dx}\right)_1 = 0, \quad \left(\frac{d\lambda}{dx}\right)_0 = 0, \quad \&c.$$

are necessary.*

But his reasoning upon this point does not appear to me conclusive. A careful examination will, I think, show the reader that, admitting the truth of the *infinitesimal* conditions given by M. Delaunay, these conditions may be satisfied by the equations

$$(\beta_{n'})_1 = 0, \quad (\beta_{n'})_0 = 0, \ \&c.,$$

as well as by the equations

$$\lambda_1 = 0, \quad \lambda_0 = 0, \ \&c.$$

I have given in Chapter IV. some examples which appear to be altogether inconsistent with the truth of M. Delaunay's result. Thus, in treating the problem of the shortest line by the method of Lagrange, the function V is of the order 0, and the equation

$$L = 0$$

of the order 1. Yet the conditions

$$\lambda_1 = 0, \quad \lambda_0 = 0,$$

are altogether inadmissible.

---

* Journal de l'Ecole Polytechnique, tom. xvii. p. 85.

NOTE H, Page 136.

The remark in the text is to be understood as applying only to the method of treating the equation

$$D.\int V dx = 0.$$

The two cases differ essentially in the mode of applying the theory of Jacobi to the investigation of the conditions which relate to the *second* variation. This part of the subject has been considered by M. Delaunay. But, in that part of his memoir which regards relative maxima and minima, the reasoning does not appear to me quite satisfactory, and the conclusion is far less perfect than in the case of absolute maxima and minima.

I have already pointed out (Note B) a misconception which he appears to me to have formed as to the meaning of the term "second variation." There is, however, no difficulty in modifying the reasoning so as to remove this error, and that without affecting the results arrived at. It is the imperfection of the results themselves which has induced me to omit this part of the theory. The reader who is curious on the subject may consult M. Delaunay's Memoir, which has been already frequently referred to.

NOTE I, Page 152.

This theorem, as well as the corresponding theorems of pp. 184 and 278, apply properly to *integral* functions only. There is no difficulty, however, in extending them to the case of fractional functions. Thus if

$$\mu = \frac{\mu'}{\mu''},$$

where μ' is an integral function of the degree m', and μ'' an integral function of the degree m'', we should have

$$\frac{1}{\mu}\left(\cos\alpha\,\frac{d\mu}{dx} + \cos\beta\,\frac{d\mu}{dy}\right) = \frac{1}{\mu'}\left(\cos\alpha\,\frac{d\mu'}{dx} + \cos\beta\,\frac{d\mu'}{dy}\right)$$
$$- \frac{1}{\mu''}\left(\cos\alpha\,\frac{d\mu''}{dx} + \cos\beta\,\frac{d\mu''}{dy}\right).$$

Hence if the construction given in the text be made separately for the two curves,

$$\mu' = c', \quad \mu'' = c'',$$

and if the corresponding portions of the normal be denoted by

$$PN',\ PN'',$$

we shall find by proceeding as in the text,

$$-\frac{1}{\rho} = \frac{m'}{PN'} - \frac{m''}{PN''}.$$

A similar extension is readily obtained for the theorems of pp. 184 and 278.

Note K, Page 163.

An elegant construction has been given for this curve by M. Delaunay, viz.:

If an ellipse or hyperbola, whose transverse axis is a, be supposed to roll upon the axis of x, its focus will generate the required curve.

Let the polar equation of the rolling curve, the generating point being the pole, be

$$r = f(\omega), \qquad \text{(a)}$$

and let the equation of the curve generated be

$$y = \phi(x), \text{ or } dy = pdx. \qquad \text{(b)}$$

Then it is easy to prove the following equations,

$$\frac{dr}{rd\omega} = p, \quad \frac{r^2 d\omega}{\sqrt{(dr^2 + r^2 d\omega^2)}} = y. \qquad \text{(c)}$$

These formulæ enable us to pass from the rolling curve to the curve generated, or *vice versâ*. For if the rolling curve be given, the equation of the curve generated will be found by eliminating r and ω between the equations (a) and (c). If the equation of the generated curve be given, that of the rolling curve will be found by eliminating x and y between the equations (b) and (c). By this method we are sometimes furnished with a means of constructing a curve given by the differential equation

$$y = f(p).$$

Thus, in the present case, where the equation of the generated curve is

$$ay = (y^2 + c)\sqrt{(1 + p^2)},$$

if we put for y and p their values given by equations (c), we shall have for the differential equation of the rolling curve,

$$d\omega = \frac{\sqrt{(c)}dr}{r\sqrt{(ar - r^2 - c)}}.$$

Integrating this equation, we find

$$\frac{1}{r} = \frac{a}{2c} - \sqrt{\left(\frac{a^2}{4c^2} - \frac{1}{c}\right)} \cos \omega,$$

the equation of a focal conic section. Comparing this with the ordinary equation

$$\frac{1}{r} = \frac{1 - e \cos \omega}{A(1 - e^2)},$$

we find

$$A = \tfrac{1}{2} a, \quad e = \sqrt{\left(1 - \frac{4c}{a^2}\right)}.$$

Hence the proposition is evident. If $c = 0$, the conic section will become a finite right line, and the generated curve a circle, as in p. 164.

There is a remarkable peculiarity connected with this problem, which has been already alluded to in the note upon p. 32, namely, that the equation furnished by the Calculus of Variations does not include all the cases which may arise. For we have seen, in p. 164, that if the curve intersect the axis of revolution, we must have $c = 0$, and that in this case the curve will become a circle. The same thing will appear from the foregoing construction. For it is plain that a curve, generated by the focus of a conic section which rolls upon a right line, can never (unless the conic section become a right line) intersect the axis of revolution. Now, suppose the original problem to have been given as follows:

To construct upon a given base AB, Fig. 18, a curve such that the superficial area of the surface generated by its revolution round AB may be given, and that its solid contents may be a maximum.

This problem evidently admits of a solution. But this solution is not given by the sphere, inasmuch as its superficial area is a determinate function of AB, and cannot therefore be made equal to any other given quantity. The solution of this problem is, therefore, not contained in the equation

$$\frac{1}{\rho} + \frac{1}{n} = \frac{2}{a}.$$

It is easy to see that in this case the development of the new form of V, according to powers of i, fails. For the second variation is, in general,

$$\int \left(\frac{d^2 V}{dy^2} \delta y^2 + 2 \frac{d^2 V}{dy dp} \delta y \delta p + \frac{d^2 V}{dp^2} \delta p^2\right) dx.$$

But since

$$V = y^2 - ay\sqrt{(1+p^2)},$$

we have

$$\frac{d^2V}{dy^2} = 2, \quad \frac{d^2V}{dydp} = -\frac{ap}{\sqrt{(1+p^2)}}, \quad \frac{d^2V}{dp^2} = -\frac{ay}{(1+p^2)^{\frac{3}{2}}}.$$

If for $(1+p^2)^{\frac{3}{2}}$ we put its value derived from the equation

$$ay = (y^2 + c)\sqrt{(1+p^2)},$$

we shall find

$$\frac{d^2V}{dp^2} = -\frac{(y^2+c)^3}{a^2y^2}.$$

Now if c do not vanish, the supposition $y = 0$ renders this quantity infinite. The method therefore fails altogether.

Note L, Page 171.

There is another case, which has been omitted in the text, that, namely, in which the maximum or minimum curve is required to touch the limiting curve at each extremity.

None of the variations,

$$\delta x_1, \delta y_1, \left(\frac{d\delta x}{ds}\right)_1, \left(\frac{d\delta y}{ds}\right)_1, \delta x_0, \delta y_0, \left(\frac{d\delta x}{ds}\right)_0, \left(\frac{d\delta y}{ds}\right)_0,$$

vanishing in this case, it is easy to see that the terms outside the sign of integration will give the equations

$$\left.\begin{aligned} a\delta x_1 + b\delta y_1 + \mu'_1\rho_1^3\left(\frac{d^2x}{ds^2}\frac{d\delta x}{ds} + \frac{d^2y}{ds^2}\frac{d\delta y}{ds}\right)_1 &= 0, \\ a\delta x_0 + b\delta y_0 + \mu'_0\rho_0^3\left(\frac{d^2x}{ds^2}\frac{d\delta x}{ds} + \frac{d^2y}{ds^2}\frac{d\delta y}{ds}\right)_0 &= 0. \end{aligned}\right\} \quad (a)$$

Let the equations of the limiting curves be

$$y = f_1(x), \quad y = f_0(x). \quad (b)$$

Then since the required curve touches both these curves, we have

$$\left(\frac{dy}{ds}\right)_1 = \left(\frac{df_1}{dx}\right)_1\left(\frac{dx}{ds}\right)_1, \quad \frac{dy}{ds_0} = \left(\frac{df_0}{dx}\right)_0\left(\frac{dx}{ds}\right)_0. \quad (c)$$

And since equations (b) and (c) are supposed to hold however the curve be varied, we shall have from equations (b),

$$\delta y_1 = \left(\frac{df_1}{dx}\right)_1 \delta x_1, \quad \delta y_0 = \left(\frac{df_0}{dx}\right)_0 \delta x_0, \tag{d}$$

and from equations (c),

$$\left(\frac{d\delta y}{ds}\right)_1 = \left(\frac{df_1}{dx}\right)_1 \left(\frac{d\delta x}{ds}\right)_1 + \left(\frac{d^2 f_1}{dx^2}\right)_1 \left(\frac{dx}{ds}\right)_1 \delta x_1$$

$$\left(\frac{d\delta y}{ds}\right)_0 = \left(\frac{df_0}{dx}\right)_0 \left(\frac{d\delta x}{ds}\right)_0 + \left(\frac{d^2 f_0}{dx^2}\right)_0 \left(\frac{dx}{ds}\right)_0 \delta x_0; \tag{e}$$

we have also

$$\left(\frac{dx}{ds}\right)_1 \left(\frac{d\delta x}{ds}\right)_1 + \left(\frac{dy}{ds}\right)_1 \left(\frac{d\delta y}{ds}\right)_1 = 0,$$

$$\left(\frac{dx}{ds}\right)_0 \left(\frac{d\delta x}{ds}\right)_0 + \left(\frac{dy}{ds}\right)_0 \left(\frac{d\delta y}{ds}\right)_0 = 0. \tag{f}$$

Solving equations (e) and (f) for

$$\left(\frac{d\delta x}{ds}\right)_1, \left(\frac{d\delta y}{ds}\right)_1, \left(\frac{d\delta x}{ds}\right)_0, \left(\frac{d\delta y}{ds}\right)_0,$$

we find

$$\left(\frac{d\delta x}{ds}\right)_1 = -\frac{\left(\frac{dx}{ds}\right)_1 \left(\frac{dy}{ds}\right)_1 \left(\frac{d^2 f_1}{dx^2}\right)_1}{\left(\frac{dx}{ds}\right)_1 + \left(\frac{df_1}{dx}\right)_1 \left(\frac{dy}{ds}\right)_1} \delta x_1 = -\left(\frac{dx^2}{ds^2}\right)_1 \left(\frac{dy}{ds}\right)_1 \left(\frac{d^2 f_1}{dx^2}\right)_1 \delta x_1$$

$$\left(\frac{d\delta y}{ds}\right)_1 = \frac{\left(\frac{dx^2}{ds^2}\right)_1 \left(\frac{d^2 f_1}{dx^2}\right)_1}{\left(\frac{dx}{ds}\right)_1 + \left(\frac{df_1}{dx}\right)_1 \left(\frac{dy}{ds}\right)_1} \delta x_1 = \left(\frac{dx^3}{ds^3}\right)_1 \left(\frac{d^2 f_1}{dx^2}\right)_1 \delta x_1 \tag{g}$$

$$\left(\frac{d\delta x}{ds}\right)_0 = \&c., \quad \left(\frac{d\delta y}{ds}\right)_0 = \&c.$$

Eliminating, by means of the equations (d) and (g), the several variations

$$\delta y_1, \left(\frac{d\delta y}{ds}\right)_1, \left(\frac{d\delta x}{ds}\right)_1, \delta y_0, \left(\frac{d\delta y}{ds}\right)_0, \left(\frac{d\delta x}{ds}\right)_0,$$

from the equations (a), we find

$$a + b\left(\frac{df_1}{dx}\right)_1 - \mu' \rho_1^2 \left(\frac{d^2 f_1}{dx^2}\right)_1 \left(\frac{dx^2}{ds^2}\right)_1 = 0,$$

$$a + b\left(\frac{df_0}{dx}\right)_0 - \mu' \rho_0^2 \left(\frac{d^2 f_0}{dx^2}\right)_0 \left(\frac{dx^2}{ds^2}\right)_0 = 0;$$

or, on account of equation (b), p. 168,

$$\left. \begin{aligned} a + b\left(\frac{df_1}{dx}\right)_1 - (ay_1 - bx_1 + c)\left(\frac{d^2f_1}{dx^2}\right)_1\left(\frac{dx^2}{ds^2}\right)_1 = 0, \\ a + b\left(\frac{df_0}{dx}\right)_0 - (ay_0 - bx_0 + c)\left(\frac{d^2f_0}{dx^2}\right)_0\left(\frac{dx^2}{ds^2}\right)_0 = 0. \end{aligned} \right\} \quad (h)$$

To interpret these equations let AB, Fig. 19, be a line whose equation is

$$ay - bx + c = 0.$$

Let P_1 be the point x_1y_1, P_1N_1 the normal to the limiting curve, r_1 its radius of curvature at the point P_1, and P_1p_1 a perpendicular on the line AB. We have then, since the curves touch at P_1,

$$\sqrt{(a^2 + b^2)} \cos N_1P_1p_1 = a\left(\frac{dx}{ds}\right)_1 + b\left(\frac{dy}{ds}\right)_1,$$

$$\sqrt{(a^2 + b^2)}\, P_1p_1 = ay_1 - bx_1 + c,$$

$$\frac{1}{r_1} = -\left(\frac{d^2f_1}{dx^2}\right)_1\left(\frac{dx^3}{ds^3}\right)_1.$$

Making these substitutions in the first of equations (h), we have

$$r_1 = -P_1p_1 \sec N_1P_1p_1 = -P_1N_1,$$

and, similarly,

$$r_0 = -P_0p_0 \sec N_0P_0p_0 = -P_0N_0.$$

Note M, Page 195.

Although the equations (C) are not integrable generally, the following remarkable property of the curve in question may be deduced from them:

Let two planes be drawn making angles with the co-ordinate planes whose cosines are

$$\frac{a}{\sqrt{(a^2 + b^2 + c^2)}}, \quad \frac{b}{\sqrt{(a^2 + b^2 + c^2)}}, \quad \frac{c}{\sqrt{(a^2 + b^2 + c^2)}}, \qquad (1)$$

$$\frac{f''}{\sqrt{(f^2 + f'^2 + f''^2)}}, \quad \frac{f'}{\sqrt{(f^2 + f'^2 + f''^2)}}, \quad \frac{f}{\sqrt{(f^2 + f'^2 + f''^2)}}, \qquad (2)$$

respectively. Let O (Fig. 20) be the origin, PP' any arc of the curve, and pp', $\pi\pi'$ its projections upon the planes (1) and (2) respectively. The sector Opp' is proportional to the difference between the perpendiculars $P\pi$, $P'\pi'$.

The truth of this theorem will at once appear from equation (G), p. 199, which may be written

$$c(ydx - xdy) + b(xdz - zdx) + a(zdy - ydz) = f''dx + f'dy + fdz.$$

For if we denote the element of Opp' by dA, it is evident that

$$dA = \left\{\frac{c(ydx - xdy) + b(xdz - zdx) + a(zdy - ydz)}{\sqrt{(a^2 + b^2 + c^2)}}\right\}.$$

Also,

$$d.P\pi = \frac{f''dx + f'dy + fdz}{\sqrt{(f^2 + f'^2 + f''^2)}}.$$

Making these substitutions in equation (G), we have

$$\sqrt{(a^2 + b^2 + c^2)}\,dA = \sqrt{(f^2 + f'^2 + f''^2)}\,d.P\pi.$$

Integrating, and putting

$$\sqrt{(f^2 + f'^2 + f''^2)} = k\sqrt{(a^2 + b^2 + c^2)},$$

we have

$$A = Opp' = K(P\pi - P'\pi').$$

If the planes (1) (2) be at right angles, it is easily seen that the projection of the curve on (1) will be a right line, and, therefore, that the curve will be plane. This agrees with the conclusion stated in p. 198.

The curve may also be represented by two differential equations of the first order. For if we take the plane (1) for the plane of yz, it is easily seen that equation (G), p. 199, may be written

$$ydz - zdy = ldx + mdy + ndz,$$

or

$$(y - n)dz - (z + m)dy = ldx;$$

or by transforming the co-ordinates y, z, and taking, as before, s for the independent variable,

$$y\frac{dz}{ds} - z\frac{dy}{ds} = l\frac{dx}{ds}. \qquad \text{(a)}$$

Differentiating this, we have

$$y\frac{d^2z}{ds^2} - z\frac{d^2y}{ds^2} = l\frac{d^2x}{ds^2}.$$

We have also the identical condition

$$\frac{dx}{ds}\frac{d^2x}{ds^2} + \frac{dy}{ds}\frac{d^2y}{ds^2} + \frac{dz}{ds}\frac{d^2z}{ds^2} = 0.$$

Eliminating, by means of these equations,

$$\frac{d^2y}{ds^2},\ \frac{d^2z}{ds^2},$$

from the equation

$$\left(\frac{d^2x}{ds^2}\right)^2+\left(\frac{d^2y}{ds^2}\right)^2+\left(\frac{d^2z}{ds^2}\right)^2=\frac{1}{m^2},$$

and reducing, we have

$$\frac{d^2x}{ds^2}=\frac{1}{m}\,\frac{y\dfrac{dy}{ds}+z\dfrac{dz}{ds}}{\sqrt{(l^2+y^2+z^2)}}.$$

Integrating, and adding an arbitrary constant, we have

$$\frac{dx}{ds}=\frac{1}{m}\{C+\sqrt{(l^2+y^2+z^2)}\}. \tag{b}$$

The curve is therefore represented by the system of equations (a) and (b).

Note N, Page 242.

The discussion given in the text is, of course, incomplete. In order to prove generally that the solution given by the Calculus of Variations is perfectly definite, it would be necessary to show that a function of two independent variables is completely determined by a partial differential equation, combined with a number of particular conditions equal to the order of the equation, and referring only to particular systems of values of the independent variables. Of this important proposition I have not succeeded in obtaining a complete demonstration, independent of the *nature* of the particular conditions. The following proof, however, although applicable only to a particular class of conditions, may be considered of some importance:

Let

$$L=0$$

be a partial differential equation of the nth order. Suppose also that the values of z and of one of its differential coefficients of each order, as far as the order $n-1$ inclusive, corresponding to a given system of values of x and y, be also given. The general value of z, in terms of x and y, will in this case be perfectly determinate.

Let the given system of values of x and y be represented by the equation

$$f(x, y)=0. \tag{a}$$

Let also the particular conditions be

$$z = X, \quad \frac{dz}{dx} = X_1, \ \&c., \quad \frac{d^{n-1}z}{dx^{n-1}} = X_{n-1}, \tag{b}$$

Now since these equations are true for all values of x and y which satisfy the equation (a), it is plain that we must have from the equation $z = X$,

$$\frac{dz}{dx} + \frac{dz}{dy}\frac{dy}{dx} = \frac{dX}{dx},$$

$$\frac{d^2z}{dx^2} + 2\frac{dy}{dx}\frac{d^2z}{dxdy} + \frac{dy^2}{dx^2}\frac{d^2z}{dy^2} + \frac{d^2y}{dx^2}\frac{dz}{dy} = \frac{d^2X}{dx^2},$$

&c.,

the values of $\frac{dy}{dx}$, $\frac{d^2y}{dx^2}$, &c., being derived from (a). If these differentiations be continued up to the nth order inclusive, it is evident that the equation $z = X$, combined with its several differentials, will give $n+1$ relations among the quantities which enter into L. Similarly we shall have from the equation

$$\frac{dz}{dx} = X_1$$

n relations among these quantities.

Treating in the same way the equations

$$\frac{d^2z}{dx^2} = X_2, \ \&c., \quad \frac{d^{n-1}z}{dx^{n-1}} = X_{n-1},$$

we have among the several differential coefficients:

From the equation $L = 0$ . . . . 1 equation.

. . . . $\frac{d^{n-1}z}{dx^{n-1}} = X_{n-1}$ . . . 2 equations.

. . . .

. . . . $z = X$ . . . . $n+1$.

The total number of equations which subsist among these quantities is, therefore,

$$\frac{(n+1)\,(n+2)}{1\,.\,2}.$$

This is evidently also the number of the quantities themselves. Hence we infer that if the values of z and one differential coefficient of each order, as far as $n-1$ inclusive, corresponding to the system

$$f(x, y) = 0,$$

be given, the values of all the differential coefficients, as far as the order n inclusive, corresponding to the same system, will be given also.

Again, since the equation

$$L = 0$$

is supposed to hold for all values of x and y, it is plain that we must have

$$\frac{dL}{dx} = 0, \quad \frac{dL}{dy} = 0.$$

The introduction of differential coefficients of the order $n+1$ will therefore furnish us with *two* new equations derived from L. And as each of the n equations (b) admits of *one* more differentiation, it is plain that we shall obtain from them n new equations. We shall thus have in all $n+2$ new equations. This is precisely the number of the new quantities introduced, namely, the $n+2$ differential coefficients of the order $n+1$. These coefficients are therefore determined. And the same reasoning may evidently be extended to differential coefficients of all orders. Hence we infer generally as follows:

If z be a function of x and y which satisfies the partial differential equation (of the nth order)

$$L = 0,$$

and if, moreover, the values of z and of one differential coefficient of each order, as far as $n-1$ inclusive, corresponding to a given system of values of x and y, be given, the values of all the differential coefficients of all orders corresponding to the same system will be also given.

The function z is therefore perfectly determinate. Hence it is plain that the n conditions (b) are necessary and sufficient to determine the arbitrary quantities which enter into the solution of the equation

$$L = 0.$$

As these conditions are independent of each other, it appears natural to conclude that the determinate character of the problem results solely from their *number*. It would be desirable, however, to have a more general discussion of the question.

Note O, Page 252.

This reasoning may be extended to the general case, in which V contains differential coefficients of any order. The method of investigation is precisely similar to that which has been given in the text, and the result may be generally stated as follows:

If V be a function of x, y, z, and of the differentials of this latter quantity, as far as the nth order inclusively, the equation

$$\Omega = 0$$

will, in general, be a partial differential equation of the $2n^{th}$ order. But this equation will be reduced to one of the order $2n - 2$, by the disappearance of the two highest order of terms, when (and only when) V is a function of

$$\frac{d^n z}{dx^n}, \frac{d^n z}{dx^{n-1}dy} \ldots \frac{d^n z}{dy^n},$$

of a *degree* not higher than the second, and in which the coefficients of the terms of the second degree are connected by this condition, that the sum of the coefficients of all such terms as

$$\left(\frac{d^n z}{dx^{n-1}.\,dy}\right)^2, \frac{d^n z}{dx^{n-2}.\,dy^2} \cdot \frac{d^n z}{dx^n},$$

in which the indices of dx and dy respectively are equal, shall be equal to zero. Thus, for example, if A be the coefficient of

$$\left(\frac{d^n z}{dx^{n-4}.\,dy^4}\right)^2,$$

since there are three other terms equivalent to this in the sense just defined, namely,

$$\frac{d^n z}{dx^{n-7}.\,dy^7} \cdot \frac{d^n z}{dx^{n-1}.\,dy}, \quad \frac{d^n z}{dx^{n-6}.\,dy^6} \cdot \frac{d^n z}{dx^{n-2}.\,dy^2}, \quad \frac{d^n z}{dx^{n-5}.\,dy^5} \cdot \frac{d^n z}{dx^{n-3}.\,dy^3},$$

if B, C, D be the coefficients of these terms respectively, the above-mentioned condition requires that

$$A + B + C + D = 0.$$

Note P, Page 286.

The reader will observe that one important case has been omitted, that, namely, in which it is required to determine, among *all* closed surfaces of equal superficial area, that one whose solid content is a maximum. This is known to be a sphere. But this solution has not yet been obtained from the equations furnished by the Calculus of Variations. The terms which refer to the limits of integration will in this case disappear altogether (vid. p. 347), and the only equation furnished by the Calculus of Variations will be

$$\frac{1}{R} + \frac{1}{R'} = \frac{1}{a}.$$

It would seem, therefore, that the sphere ought to be the *only* closed surface included in this equation. But this has never been proved.*

NOTE Q, Page 351.

It may be worth while to investigate generally the form which V should have in order that the area of any portion of a surface may be capable of being represented by

$$\Gamma + \iint V d\theta d\phi.$$

This is readily effected by the method given in the text. For if we assume

$$A(1+p^2+q^2)^{\frac{3}{2}} = V,$$

it will be evident, from the general principle laid down in pp. 349–50, that the expression

$$A(rt-s^2) - \sqrt{(1+p^2+q^2)}$$

must satisfy the criterion of integrability. We have then, in the general formulæ of p. 344,

$$B=0,\ C=0,\ D=0,\ E=-\sqrt{(1+p^2+q^2)}.$$

The equation

$$\Pi = 0$$

becomes therefore identical; and the equations

$$L=0,\ M=0,\ N=0,$$

give

$$\frac{d^2A}{dy^2} + 2q\frac{d^2A}{dydz} + q^2\frac{d^2A}{dz^2} = -\frac{1+q^2}{(1+p^2+q^2)^{\frac{3}{2}}},$$

$$\frac{d^2A}{dxdy} + p\frac{d^2A}{dydz} + q\frac{d^2A}{dxdz} + pq\frac{d^2A}{dz^2} = -\frac{pq}{(1+p^2+q^2)^{\frac{3}{2}}},$$

$$\frac{d^2A}{dx^2} + 2p\frac{d^2A}{dxdz} + p^2\frac{d^2A}{dz^2} = -\frac{1+p^2}{(1+p^2+q^2)^{\frac{3}{2}}}.$$

These equations, which are easily integrated by the method of p. 345, give

$$A = -\tfrac{1}{2}\frac{\{x^2+y^2+(px+qy)^2\}}{(1+p^2+q^2)^{\frac{3}{2}}} + x\phi(z-px-qy, p, q)$$
$$+ y\psi(z-px-qy, p, q) + \chi(z-px-qy, p, q). \qquad \text{(a)}$$

---

* Delaunay, p. 111.

It remains, then, to consider the equation

$$\Theta = 0;$$

or, as it may be written,

$$\frac{dA}{dz} + \frac{d}{dp}\left(\frac{dA}{dx} + p\,\frac{dA}{dz}\right) + \frac{d}{dq}\left(\frac{dA}{dy} + q\,\frac{dA}{dz}\right) = 0. \qquad \text{(b)}$$

Now equation (a) gives

$$\frac{dA}{dx} + p\,\frac{dA}{dz} = -\frac{x + (px + qy)\,p}{(1 + p^2 + q^2)^{\frac{3}{2}}} + \phi,$$

$$\frac{dA}{dy} + q\,\frac{dA}{dz} = -\frac{y + (px + qy)\,q}{(1 + p^2 + q^2)^{\frac{3}{2}}} + \psi.$$

Hence

$$\frac{d}{dp}\left(\frac{dA}{dx} + p\,\frac{dA}{dz}\right) = -\frac{2px + qy}{(1 + p^2 + q^2)^{\frac{3}{2}}} + 3p\,\frac{x + (px + qy)p}{(1 + p^2 + q^2)^{\frac{5}{2}}} + \frac{d\phi}{dp},$$

$$\frac{d}{dq}\left(\frac{dA}{dy} + q\,\frac{dA}{dz}\right) = -\frac{2qy + px}{(1 + p^2 + q^2)^{\frac{3}{2}}} + 3q\,\frac{y + (px + qy)q}{(1 + p^2 + q^2)^{\frac{5}{2}}} + \frac{d\psi}{dq}.$$

Adding these expressions, we find

$$\frac{d}{dp}\left(\frac{dA}{dx} + p\,\frac{dA}{dz}\right) + \frac{d}{dq}\left(\frac{dA}{dy} + q\,\frac{dA}{dz}\right) = \frac{d\phi}{dp} + \frac{d\psi}{dq};$$

also,

$$\frac{dA}{dz} = \frac{dA}{dw}\ (\text{p. } 346) = x\,\frac{d\phi}{dw} + y\,\frac{d\psi}{dw} + \frac{d\chi}{dw}.$$

Substituting these values in equation (b), and integrating with regard to w, we have

$$x\phi + y\psi + \chi = -\frac{d}{dp}\int\phi dw - \frac{d}{dq}\int\psi dw + F(p, q).$$

Omitting $F(p, q)$, which may evidently be considered to be included in ϕ or ψ, and assuming, as before,

$$\Phi = -\int\phi dw, \quad \Psi = -\int\psi dw,$$

we have

$$A = -\tfrac{1}{2}\,\frac{x^2 + y^2 + (px + qy)^2}{(1 + p^2 + q^2)^{\frac{3}{2}}} + \frac{d\Phi}{dp} + \frac{d\Psi}{dq},$$

and, therefore,

$$V = -\tfrac{1}{2}\{x^2 + y^2 + (px + qy)^2\} + (1 + p^2 + q^2)^{\frac{3}{2}}\left(\frac{d\Phi}{dp} + \frac{d\Psi}{dq}\right).$$

If we put

$$\Phi = 0, \quad \Psi = 0,$$

shall have the area of any portion of the surface represented by

$$\Gamma - \tfrac{1}{2}\iint\{x^2 + y^2 + (px + qy)^2\}\, d\omega.$$

ence we may deduce the following theorem:

If a tangent plane be drawn at any point of a closed surface, and if denote by T the distance between the point of contact and the point which it cuts the line from which θ is measured, the total superficial ea will be given by the expression

$$S = -\tfrac{1}{2}\iint T^2 d\omega.$$

e truth of this is obvious, if we recollect that

$$T^2 = x^2 + y^2 + (px + qy)^2.$$

r it is plain, from the foregoing investigation, that the value of the tegral

$$\iint(dS + \tfrac{1}{2}T^2 d\omega)$$

the same for all closed surfaces. Let the surface be a sphere whose uation is

$$x^2 + y^2 + z^2 = 1.$$

eserving the usual significations of θ and ϕ, we have

$$T = \tan\theta,$$

d, therefore,

$$\iint T^2 d\omega = \int_0^{\pi}\int_0^{2\pi} \tan^2\theta \sin\theta\, d\theta\, d\phi = 2\pi\int_0^{\pi} \tan^2\theta \sin\theta\, d\theta$$

$$= 2\pi\int_0^{\pi}\left(\frac{\sin\theta}{\cos^2\theta} - \sin\theta\right) d\theta = -8\pi.$$

ence, in the case of a sphere, we have

$$\iint(dS + \tfrac{1}{2}T^2 d\omega) = 0,$$

d therefore, in general,

$$S = -\tfrac{1}{2}\iint T^2 d\omega.$$

It may be interesting to verify this theorem for the case of a surface of revolution.

Let AB (Fig. 21) be the axis of revolution, APB the generating rve, and PE a tangent at any point. Suppose the angle θ to be ckoned from the line AB, then if PN be the normal, and PY the dinate, to the curve, we shall have

$$PNA = \theta, \quad T = PE = -PY\sec\theta.$$

Let the equation of the curve be

$$dy = pdx\,;$$

then

$$\cos\theta = \frac{p}{\sqrt{(1+p^2)}}, \quad T = -y\sec\theta = -y\,\frac{\sqrt{(1+p^2)}}{p}.$$

Now since T is evidently independent of ϕ, we have

$$\tfrac{1}{2}\iint T^2 d\omega = \tfrac{1}{2}\int_0^\pi\!\int_0^{2\pi} T^2 \sin\theta d\theta d\phi = \pi\int_0^\pi T^2 \sin\theta d\theta.$$

Substituting for θ and T their values, we have

$$\tfrac{1}{2}\iint T^2 d\omega = -\pi\int \frac{y^2 dp}{p^2\sqrt{(1+p^2)}}\,;$$

or (integrating by parts)

$$\tfrac{1}{2}\iint T^2 d\omega = \pi y^2 \sqrt{\left(1+\frac{1}{p^2}\right)} - 2\pi\int y\sqrt{\left(1+\frac{1}{p^2}\right)}dy$$

$$= \pi y^2 \sqrt{\left(1+\frac{1}{p^2}\right)} - 2\pi\int y\sqrt{(1+p^2)}\,dx.$$

If the limiting points be A, B, it is plain that the term free from the sign of integration vanishes, and consequently that

$$-\tfrac{1}{2}\iint T^2 d\omega = 2\pi\int y\sqrt{(1+p^2)}\,dx.$$

But the second member of this equation manifestly represents the superficial area of the surface generated by the revolution of APB. Hence the proposition is evident.

THE END.

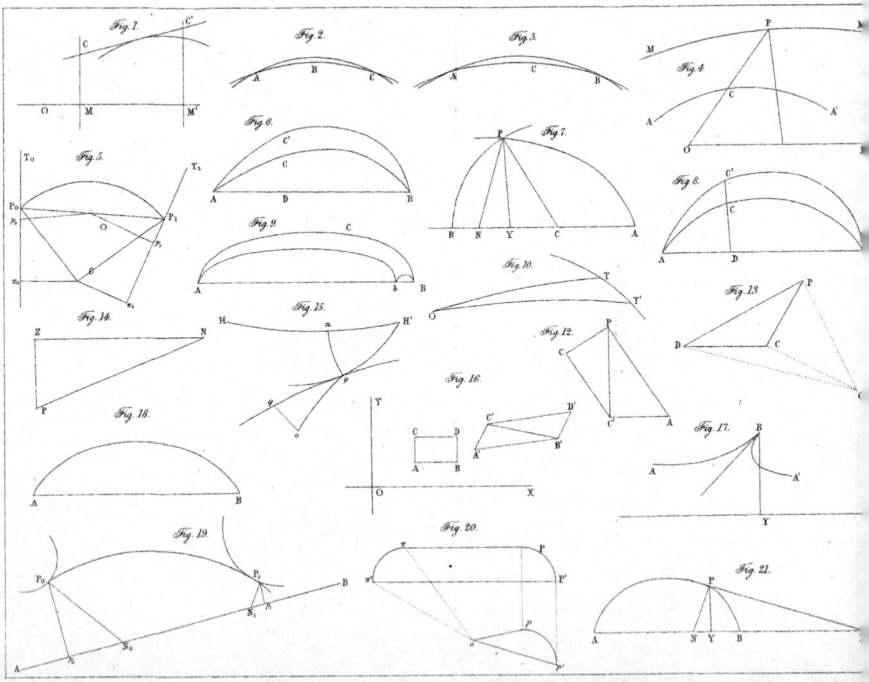
Fig. 1.
Fig. 2.
Fig. 3.
Fig. 4.
Fig. 5.
Fig. 6.
Fig. 7.
Fig. 8.
Fig. 9.
Fig. 10.
Fig. 12.
Fig. 13.
Fig. 14.
Fig. 15.
Fig. 16.
Fig. 17.
Fig. 18.
Fig. 19.
Fig. 20.
Fig. 21.

Printed in Dunstable, United Kingdom

91220106R00228